Mathematical and Physical Data, Equations, and Rules of Thumb

Mathematical and Physical Data, Equations, and Rules of Thumb

Stan Gibilisco

McGraw-Hill

New York Chicago San Francisco Lisbon London Madrid
Mexico City Milan New Delhi San Juan Seoul
Singapore Sydney Toronto

Cataloging-in-Publication Data is on file with the Library of Congress

McGraw-Hill

A Division of The ***McGraw-Hill*** *Companies*

1 2 3 4 5 6 7 8 9 0 DOC/DOC 0 7 6 5 4 3 2 1

ISBN 0-07-136148-0

The sponsoring editor for this book was Scott Grillo and the production supervisor was Pamela A. Pelton. It was set in Century Schoolbook by Pro-Image Corporation.

Printed and bound by R. R. Donnelley & Sons Company.

McGraw-Hill books are available at special quantity discounts to use as premiums and sales promotions, or for use in corporate training programs. For more information, please write to the Director of Special Sales, Professional Publishing, McGraw-Hill, Two Penn Plaza, New York, NY 10121-2298. Or contact your local bookstore.

This book is printed on recycled, acid-free paper containing a minimum of 50% recycled, de-inked fiber.

To Tony, and Samuel, and Tim
from Uncle Stan

Contents

Preface

This is a comprehensive sourcebook of definitions, formulas, units, constants, symbols, conversion factors, and miscellaneous data for use by engineers, technicians, hobbyists, and students. Some information is provided in the fields of mathematics, physics, and chemistry. Lists of symbols are included.

Every effort has been made to arrange the material in a logical manner, and to portray the information in concise but fairly rigorous terms. Special attention has been given to the index. It was composed with the goal of making it as easy as possible for you to locate specific definitions, formulas, and data.

Feedback concerning this edition is welcome, and suggestions for future editions are encouraged.

Stan Gibilisco

Acknowledgments

I extend thanks to Dr. Emma Previato, Professor of Mathematics and Statistics at Boston University, for her help in proofreading the pure mathematics sections.

Illustrations were generated with *CorelDRAW*. Some clip art is courtesy of Corel Corporation, 1600 Carling Avenue, Ottawa, Ontario, Canada K1Z 8R7.

Mathematical and Physical Data, Equations, and Rules of Thumb

Chapter

1

Algebra, Functions, Graphs, and Vectors

This chapter contains data pertaining to sets, functions, arithmetic, real-number algebra, complex-number algebra, coordinate systems, graphs, and vector algebra.

Sets

A *set* is a collection or group of definable unique *elements* or *members*. Set elements commonly include:

- Points on a line
- Instants in time
- Coordinates in a plane
- Coordinates in space
- Coordinates on a display
- Curves on a graph or display
- Physical objects
- Chemical elements
- Digital logic states
- Locations in memory or storage
- Data bits, bytes, or characters
- Subscribers to a network

If an element a is contained in a set $\boldsymbol{A}$, then this fact is written as:

$$a \in \boldsymbol{A}$$

Set intersection

The *intersection* of two sets $\boldsymbol{A}$ and $\boldsymbol{B}$, written $\boldsymbol{A} \cap \boldsymbol{B}$, is the set $\boldsymbol{C}$ such that the following statement is true for every element x:

$$x \in \boldsymbol{C} \leftrightarrow x \in \boldsymbol{A} \text{ and } x \in \boldsymbol{B}$$

Set union

The *union* of two sets $\boldsymbol{A}$ and $\boldsymbol{B}$, written $\boldsymbol{A} \cup \boldsymbol{B}$, is the set $\boldsymbol{C}$ such that the following statement is true for every element x:

$$x \in \boldsymbol{C} \leftrightarrow x \in \boldsymbol{A} \text{ or } x \in \boldsymbol{B}$$

Subsets

A set $\boldsymbol{A}$ is a *subset* of a set $\boldsymbol{B}$, written $\boldsymbol{A} \subseteq \boldsymbol{B}$, if and only if the following holds true:

$$x \in \boldsymbol{A} \rightarrow x \in \boldsymbol{B}$$

Proper subsets

A set $\boldsymbol{A}$ is a *proper subset* of a set $\boldsymbol{B}$, written $\boldsymbol{A} \subset \boldsymbol{B}$, if and only if both the following hold true:

$$x \in \boldsymbol{A} \rightarrow x \in \boldsymbol{B}$$

$$\boldsymbol{A} \neq \boldsymbol{B}$$

Disjoint sets

Two sets $\boldsymbol{A}$ and $\boldsymbol{B}$ are *disjoint* if and only if all three of the following conditions are met:

$$\boldsymbol{A} \neq \varnothing$$

$$\boldsymbol{B} \neq \varnothing$$

$$\boldsymbol{A} \cap \boldsymbol{B} = \varnothing$$

where $\varnothing$ denotes the *empty set,* also called the *null set.*

Coincident sets

Two non-empty sets $\boldsymbol{A}$ and $\boldsymbol{B}$ are *coincident* if and only if, for all elements x:

$$x \in \boldsymbol{A} \leftrightarrow x \in \boldsymbol{B}$$

Cardinality

The *cardinality* of a set is defined as the number of elements in the set. The null set has cardinality zero. The set of people in a city, stars in a galaxy, or atoms in the observable universe has finite cardinality.

Most commonly used number sets have infinite cardinality. Some number sets have cardinality that is denumerable; such a set can be completely defined in terms of a sequence, even though there might be infinitely many elements in the set. Some infinite number sets have non-denumerable cardinality; such a set cannot be completely defined in terms of a sequence.

One-one function

Let $\boldsymbol{A}$ and $\boldsymbol{B}$ be two non-empty sets. Suppose that for every member of $\boldsymbol{A}$, a function f assigns some member of $\boldsymbol{B}$. Let a_1 and a_2 be members of $\boldsymbol{A}$. Let b_1 and b_2 be members of $\boldsymbol{B}$, such that f assigns $f(a_1) = b_1$ and $f(a_2) = b_2$. Then f is a *one-one function* if and only if:

$$a_1 \neq a_2 \rightarrow b_1 \neq b_2$$

Onto function

A function f from set $\boldsymbol{A}$ to set $\boldsymbol{B}$ is an *onto function* if and only if:

$$b \in \boldsymbol{B} \rightarrow f(a) = b \text{ for some } a \in \boldsymbol{A}$$

One-to-one correspondence

A function f from set $\boldsymbol{A}$ to set $\boldsymbol{B}$ is a *one-to-one correspondence,* also known as a *bijection,* if and only if f is both one-one and onto.

Domain and range

Let f be a function from set $\boldsymbol{A}$ to set $\boldsymbol{B}$. Let $\boldsymbol{A}'$ be the set of all elements a in $\boldsymbol{A}$ for which there is a corresponding element b in $\boldsymbol{B}$. Then $\boldsymbol{A}'$ is called the *domain* of f.

Let f be a function from set $\boldsymbol{A}$ to set $\boldsymbol{B}$. Let $\boldsymbol{B}'$ be the set of all elements b in $\boldsymbol{B}$ for which there is a corresponding element a in $\boldsymbol{A}$. Then $\boldsymbol{B}'$ is called the *range* of f.

Continuity

A function f is *continuous* if and only if, for every point a in the domain $\boldsymbol{A}'$ and for every point $b = f(a)$ in the range $\boldsymbol{B}'$, $f(x)$ approaches b as x approaches a. If this requirement is not met for every point a in $\boldsymbol{A}'$, then the function f is *discontinuous,* and each point or value a_{d} in $\boldsymbol{A}'$ for which the requirement is not met is called a *discontinuity*.

Denumerable Number Sets

Numbers are abstract expressions of physical or mathematical quantity, extent, or magnitude. Mathematicians define numbers in terms of set cardinality. *Numerals* are the written symbols that are mutually agreed upon to represent numbers.

Natural numbers

The *natural numbers,* also called the *whole numbers* or *counting numbers,* are built up from a starting point of zero. Zero is defined as the null set $\varnothing$. On this basis:

$$0 = \varnothing$$

$$1 = \{\varnothing\}$$

$$2 = \{0, 1\} = \{\varnothing,\{\varnothing\}\}$$

$$3 = \{0, 1, 2\} = \{\varnothing,\{\varnothing\},\{\varnothing,\{\varnothing\}\}\}$$

$$\downarrow$$

Etc.

The set of natural numbers is denoted $\boldsymbol{N}$, and is commonly expressed as:

$$\boldsymbol{N} = \{0, 1, 2, 3, ..., n, ...\}$$

In some instances, zero is not included, so the set of natural numbers is defined as:

$$\boldsymbol{N} = \{1, 2, 3, 4, ..., n, ...\}$$

Natural numbers can be expressed as points along a geometric ray or half-line, where quantity is directly proportional to displacement (Fig. 1.1).

Decimal numbers

The *decimal number system* is also called *modulo 10, base 10,* or *radix 10*. Digits are representable by the set {0, 1, 2, 3, 4, 5, 6, 7, 8, 9}. The digit immediately to the left of the radix point is multiplied by 10^0, or 1. The next digit to the left is multiplied by 10^1, or 10. The power of 10 increases as you move further to the left. The first digit to the right of the radix point is multiplied by a factor of 10^{-1}, or 1/10. The next digit to the right is multiplied by 10^{-2}, or 1/100. This continues as you go further to the right. Once the process of multiplying each digit is completed, the resulting values are added. This is what is represented when you write a decimal number. For example,

$$2704.53816 = 2 \times 10^3 + 7 \times 10^2 + 0 \times 10^1 + 4 \times 10^0$$
$$+ 5 \times 10^{-1} + 3 \times 10^{-2} + 8 \times 10^{-3} + 1 \times 10^{-4} + 6 \times 10^{-5}$$

Figure 1.1 The natural numbers can be depicted as points on a ray.

Binary numbers

The *binary number system* is a method of expressing numbers using only the digits 0 and 1. It is sometimes called *base 2, radix 2,* or *modulo 2*. The digit immediately to the left of the radix point is the "ones" digit. The next digit to the left is a "twos" digit; after that comes the "fours" digit. Moving further to the left, the digits represent 8, 16, 32, 64, etc., doubling every time. To the right of the radix point, the value of each digit is cut in half again and again, that is, 1/2, 1/4, 1/8, 1/16, 1/32, 1/64, etc.

Consider an example using the decimal number 94:

$$94 = (4 \times 10^0) + (9 \times 10^1)$$

In the binary number system the breakdown is:

$$1011110 = 0 \times 2^0 + 1 \times 2^1 + 1 \times 2^2$$

$$+ 1 \times 2^3 + 1 \times 2^4 + 0 \times 2^5 + 1 \times 2^6$$

When you work with a computer or calculator, you give it a decimal number that is converted into binary form. The computer or calculator does its operations with zeros and ones. When the process is complete, the machine converts the result back into decimal form for display.

Octal and hexadecimal numbers

Another numbering scheme is the *octal number system,* which has eight symbols, or 2^3. Every digit is an element of the set {0, 1, 2, 3, 4, 5, 6, 7}. Counting thus proceeds from 7 directly to 10, from 77 directly to 100, from 777 directly to 1000, etc.

Yet another scheme, commonly used in computer practice, is the *hexadecimal number system,* so named because it has 16 symbols, or 2^4. These digits are the usual 0 through 9 plus six more, represented by A through F, the first six letters of the alphabet. The digit set is {0, 1, 2, 3, 4, 5, 6, 7, 8, 9, A, B, C, D, E, F}.

Integers

The set of natural numbers can be duplicated and inverted to form an identical, mirror-image set:

$$-\boldsymbol{N} = \{0, -1, -2, -3, \ldots, -n, \ldots\}$$

The union of this set with the set of natural numbers produces the set of integers, commonly denoted $\boldsymbol{Z}$:

$$\boldsymbol{Z} = \boldsymbol{N} \cup -\boldsymbol{N}$$

$$= \{\ldots, -n, \ldots, -2, -1, 0, 1, 2, \ldots, n, \ldots\}$$

Integers can be expressed as points along a line, where quantity is directly proportional to displacement (Fig. 1.2). In the illustration, integers correspond to points where hash marks cross the line. The set of natural numbers is a proper subset of the set of integers:

$$\boldsymbol{N} \subset \boldsymbol{Z}$$

For any number a, if $a \in \boldsymbol{N}$, then $a \in \boldsymbol{Z}$. This is formally written:

$$\forall a\colon a \in \boldsymbol{N} \rightarrow a \in \boldsymbol{Z}$$

The converse of this is not true. There are elements of $\boldsymbol{Z}$ (namely, the negative integers) that are not elements of $\boldsymbol{N}$.

Operations with integers

Several arithmetic operations are defined for pairs of integers. The basic operations include *addition, subtraction, multiplication, division,* and *exponentiation.*

Addition is symbolized by a cross or plus sign (+). The result of this operation is a *sum*. On the number line of Fig. 1.2, sums are depicted by moving to the right. For example, to illustrate the fact that $-2 + 5 = 3$, start at the point corresponding to -2, then move to the right 5 units, ending up at the point corresponding to 3. In general, to illustrate $a + b = c$, start at the point corresponding to a, then move to the right b units, ending up at the point corresponding to c.

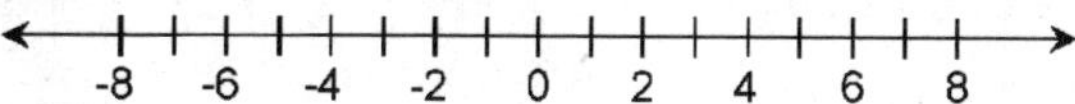

Figure 1.2 The integers can be depicted as points on a horizontal line.

Subtraction is symbolized by a dash (−). The result of this operation is a *difference*. On the number line of Fig. 1.2, differences are depicted by moving to the left. For example, to illustrate the fact that $3 - 5 = -2$, start at the point corresponding to 3, then move to the left 5 units, ending up at the point corresponding to −2. In general, to illustrate $a - b = c$, start at the point corresponding to a, then move to the left b units, ending up at the point corresponding to c.

Multiplication is symbolized by a tilted cross (×), a small dot (·), or sometimes in the case of variables, by listing the numbers one after the other (for example, ab). Occasionally an asterisk (*) is used. The result of this operation is a *product*. On the number line of Fig. 1.2, products are depicted by moving away from the zero point, or *origin,* either toward the left or toward the right depending on the signs of the numbers involved. To illustrate $a \times b = c$, start at the origin, then move away from the origin a units b times. If a and b are both positive or both negative, move toward the right; if a and b have opposite sign, move toward the left. The finishing point corresponds to c.

The preceding three operations are closed over the set of integers. This means that if a and b are integers, then $a + b$, $a - b$, and $a \times b$ are integers.

Division, also called the *ratio operation,* is symbolized by a forward slash (/) or a dash with dots above and below (÷). Occasionally it is symbolized by a colon (:). The result of this operation is a *quotient* or *ratio*. On the number line of Fig. 1.2, quotients are depicted by moving in toward the zero point, or *origin,* either toward the left or toward the right depending on the signs of the numbers involved. To illustrate $a/b = c$, it is easiest to envision the product $b \times c = a$ performed "backwards." But division, unlike addition, subtraction, or multiplication, is not closed over the set of integers. If a and b are integers, then a/b might be an integer, but this is not necessarily the case. The ratio operation gives rise to a more inclusive, but still denumerable, set of numbers. The quotient a/b is not defined if $b = 0$.

Exponentiation, also called *raising to a power,* is symbolized by a superscript numeral. The result of this operation is known as a *power*. If a is an integer and b is a positive integer, then a^b is the result of multiplying a by itself b times.

Rational numbers

A *rational number* (the term derives from the word *ratio*) is a quotient of two integers, where the denominator is positive. The standard form for a rational number r is:

$$r = a/b$$

Any such quotient is a rational number. The set of all possible such quotients encompasses the entire set of rational numbers, denoted $\boldsymbol{Q}$. Thus,

$$\boldsymbol{Q} = \{x|x = a/b\}$$

where $a \in \boldsymbol{Z}$, $b \in \boldsymbol{Z}$, and $b > 0$. The set of integers is a proper subset of the set of rational numbers. Thus natural numbers, integers, and rational numbers have the following relationship:

$$\boldsymbol{N} \subset \boldsymbol{Z} \subset \boldsymbol{Q}$$

Decimal expansions

Rational numbers can be denoted in decimal form as an integer followed by a period (radix point) followed by a sequence of digits. (See **Decimal numbers** above for more details concerning this notation.) The digits following the radix point always exist in either of two forms:

- A finite string of digits beyond which all digits are zero
- An infinite string of digits that repeat in cycles

Examples of the first type of rational number, known as *terminating decimals,* are:

$$3/4 = 0.750000 \ldots$$

$$-9/8 = -\ 1.1250000 \ldots$$

Examples of the second type of rational number, known as *nonterminating, repeating decimals,* are:

$$1/3 = 0.33333 \ldots$$

$$-123/999 = -0.123123123 \ldots$$

Non-denumerable Number Sets

The elements of non-denumerable number sets cannot be listed. In fact, it is impossible to even define the elements of such a set by writing down a list or sequence.

Irrational numbers

An *irrational number* is a number that cannot be expressed as the ratio of two integers. Examples of irrational numbers include:

- The length of the diagonal of a square that is one unit on each edge
- The circumference-to-diameter ratio of a circle

All irrational numbers share the property of being inexpressible in decimal form. When an attempt is made to express such a number in this form, the result is a *nonterminating, nonrepeating* decimal. No matter how many digits are specified to the right of the radix point, the expression is only an approximation of the actual value of the number. The set of irrational numbers can be denoted $\boldsymbol{S}$. This set is entirely disjoint from the set of rational numbers:

$$\boldsymbol{S} \cap \boldsymbol{Q} = \varnothing$$

Real numbers

The set of real numbers, denoted $\boldsymbol{R}$, is the union of the sets of rational and irrational numbers:

$$\boldsymbol{R} = \boldsymbol{Q} \cup \boldsymbol{S}$$

For practical purposes, $\boldsymbol{R}$ can be depicted as the set of points on a continuous geometric line, as shown in Fig. 1.2. In theoretical mathematics, the assertion that the points on a geometric line correspond one-to-one with the real numbers is known as the *Continuum Hypothesis*. The real numbers are related to rational numbers, integers, and natural numbers as follows:

$$\boldsymbol{N} \subset \boldsymbol{Z} \subset \boldsymbol{Q} \subset \boldsymbol{R}$$

The operations addition, subtraction, multiplication, division, and exponentiation can be defined over the set of real numbers. If # represents any one of these operations and x and y are elements of $\boldsymbol{R}$ with $y \neq 0$, then:

$$x \# y \in \boldsymbol{R}$$

The symbol $\aleph_0$ (aleph-null or aleph-nought) denotes the cardinality of the sets of natural numbers, integers, and rational numbers. The cardinality of the real numbers is denoted $\aleph_1$ (aleph-one). These "numbers" are called *infinite cardinals* or *transfinite cardinals*. Around the year 1900, the German mathematician Georg Cantor proved that these two "numbers" are not the same:

$$\aleph_1 > \aleph_0$$

This reflects the fact that the elements of $\boldsymbol{N}$ can be paired off one-to-one with the elements of $\boldsymbol{Z}$ or $\boldsymbol{Q}$, but not with the elements of $\boldsymbol{S}$ or $\boldsymbol{R}$. Any attempt to pair off the elements of $\boldsymbol{N}$ and $\boldsymbol{S}$ or $\boldsymbol{N}$ and $\boldsymbol{R}$ results in some elements of $\boldsymbol{S}$ or $\boldsymbol{R}$ being left over without corresponding elements in $\boldsymbol{N}$.

Imaginary numbers

The set of real numbers, and the operations defined above for the integers, give rise to some expressions that do not behave as real numbers. The best known example is the number i such that $i \times i = -1$. No real number satisfies this equation. This entity i is known as the *unit imaginary number*. Sometimes it is denoted j. If i is used to represent the unit imaginary number common in mathematics, then the real number x is written before i. Examples: $3i$, $-5i$, $2.787i$. If j is used to represent the unit imaginary number common in engineering, then x is written after j if $x \geq 0$, and x is written after $-j$ if $x < 0$. Examples: $j3$, $-j5$, $j2.787$.

The set $\boldsymbol{J}$ of all real-number multiples of i or j is the set of *imaginary numbers*:

$$\boldsymbol{J} = \{k | k = jx\} = \{k | k = xi\}$$

For practical purposes, the set $\boldsymbol{J}$ can be depicted along a number line corresponding one-to-one with the real number line. However, by convention, the imaginary number line is oriented vertically (Fig. 1.3). The sets of imaginary and real numbers have one element in common. That element is zero:

$$0i = j0 = 0$$

$$\boldsymbol{J} \cap \boldsymbol{R} = \{0\}$$

Complex numbers

A *complex number* consists of the sum of two separate components, a real number and an imaginary number. The general form for a complex number c is:

$$c = a + bi = a + jb$$

The set of complex numbers is denoted $\boldsymbol{C}$. Individual complex numbers can be depicted as points on a coordinate plane as shown in Fig. 1.4. According to the Continuum Hypothesis, the points on the so-called *complex-number plane* exist in a one-to-one correspondence with the elements of $\boldsymbol{C}$.

The set of imaginary numbers, $\boldsymbol{J}$, is a proper subset of $\boldsymbol{C}$. The set of real numbers, $\boldsymbol{R}$, is also a proper subset of $\boldsymbol{C}$. Formally:

$$\boldsymbol{J} \subset \boldsymbol{C}$$

$$\boldsymbol{N} \subset \boldsymbol{Z} \subset \boldsymbol{Q} \subset \boldsymbol{R} \subset \boldsymbol{C}$$

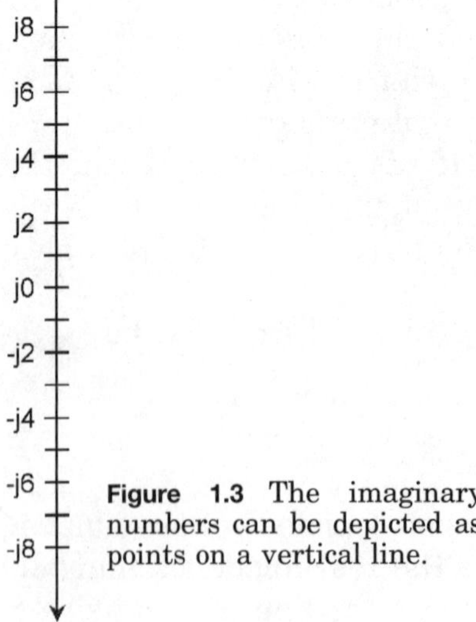

Figure 1.3 The imaginary numbers can be depicted as points on a vertical line.

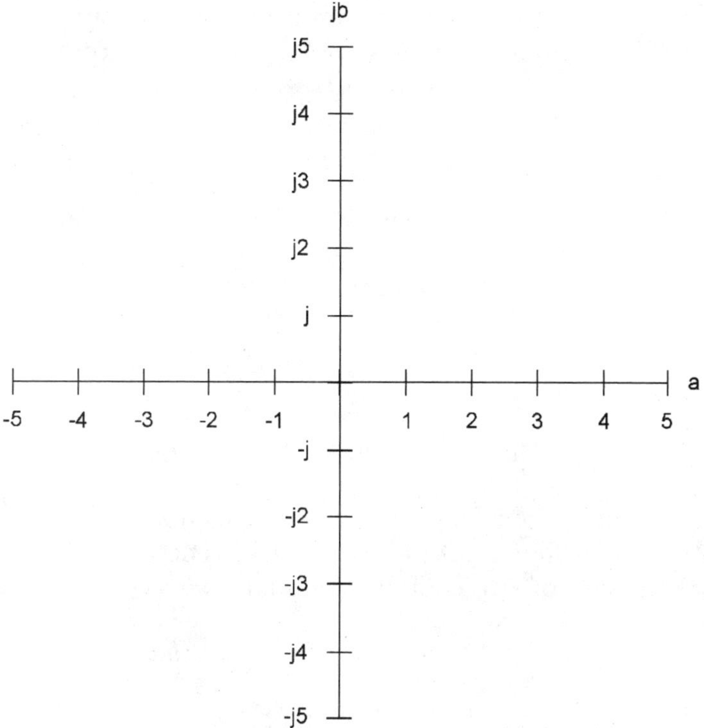

Figure 1.4 The complex numbers can be depicted as points on a plane.

Equality of complex numbers

Let x_1 and x_2 be complex numbers such that:

$$x_1 = a_1 + jb_1$$

$$x_2 = a_2 + jb_2$$

Then the two complex numbers are said to be equal if and only if their real and imaginary components are both equal:

$$x_1 = x_2 \leftrightarrow a_1 = a_2 \ \& \ b_1 = b_2$$

Operations with complex numbers

The operations of addition, subtraction, multiplication, division, and exponentiation are defined for the set of complex numbers as follows.

Complex addition: The real and imaginary parts are summed independently. The general formula for the sum of two complex numbers is:

$$(a + jb) + (c + jd) = (a + c) + j(b + d)$$

Complex subtraction: The second complex number is multiplied by -1, and then the resulting two numbers are summed. The general formula for the difference of two complex numbers is:

$$(a + jb) - (c + jd) = (a + jb) + (-1(c + jd))$$

$$= (a - c) + j(b - d)$$

Complex multiplication: The product of two complex numbers consists of a sum of four individual products. The general formula for the product of two complex numbers is:

$$(a + jb)(c + jd) = ac + jad + jbc + j^2bd$$

$$= (ac - bd) + j(ad + bc)$$

Complex division: This formula can be derived from the formula for complex multiplication. The general formula for the quotient of two complex numbers is:

$$(a + jb)/(c + jd)$$

$$= (ac + bd)/(c^2 + d^2) + j(bc - ad)/(c^2 + d^2)$$

The above formula assumes that the denominator is not zero. For complex division to be defined, the following must hold:

$$c + jd \neq 0 + j0$$

Complex exponentiation to a positive integer: This is symbolized by a superscript numeral. The result of this operation is known as a *power*. If $a + jb$ is an integer and c is a positive integer, then $(a + jb)^c$ is the result of multiplying $(a + jb)$ by itself c times.

Complex conjugates

Let x_1 and x_2 be complex numbers such that:

$$x_1 = a + jb$$

$$x_2 = a - jb$$

Then x_1 and x_2 are said to be *complex conjugates,* and the following equations hold true:

$$x_1 + x_2 = 2a$$

$$x_1 x_2 = a^2 + b^2$$

Complex vectors

Complex numbers can be represented as vectors in rectangular coordinates. This gives each complex number a unique *magnitude* and *direction*. The magnitude is the distance of the point $a + jb$ from the origin $0 + j0$. The direction is the angle of the vector, measured counterclockwise from the $+a$ axis. This is shown in Fig. 1.5.

The *absolute value* or *magnitude* of a complex number $a + jb$, written $|a + jb|$, is the length of its vector in the complex plane, measured from the origin (0,0) to the point (a,b). In the case of a pure real number $a + j0$:

$$|a + j0| = a \text{ if } a \geq 0$$

$$|a + j0| = -a \text{ if } a < 0$$

In the case of a pure imaginary number $0 + jb$:

$$|0 + jb| = b \text{ if } b \geq 0$$

$$|0 + jb| = -b \text{ if } b < 0$$

If a complex number is neither pure real nor pure imaginary, the absolute value is the length of the vector as shown in Fig. 1.6. The general formula is:

$$|a + jb| = (a^2 + b^2)^{1/2}$$

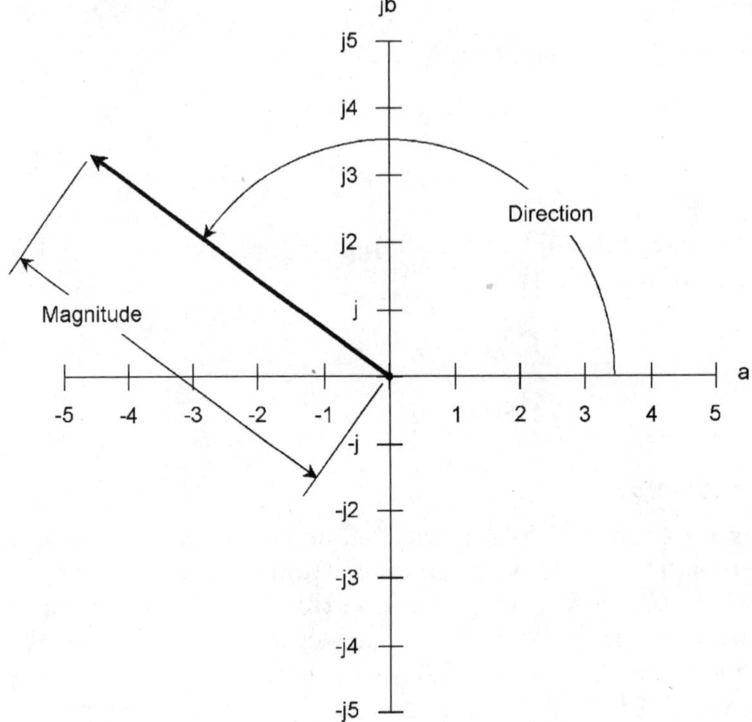

Figure 1.5 Magnitude and direction of a vector in the complex plane.

Polar form of complex numbers

Consider the polar plane defined in terms of radius r and angle θ counterclockwise from the $+a$ axis as shown in Fig. 1.7. The expression for a Cartesian vector (a,b), representing the complex number $a + jb$ in polar coordinates (r,θ) is obtained by these conversions:

$$r = (a^2 + b^2)^{1/2}$$

$$\theta = \tan^{-1} (b/a)$$

The expression for a polar vector (r,θ) in Cartesian coordinates (a,b) is obtained by these conversions:

$$a = r \cos \theta$$

$$b = r \sin \theta$$

Therefore the following equation holds:

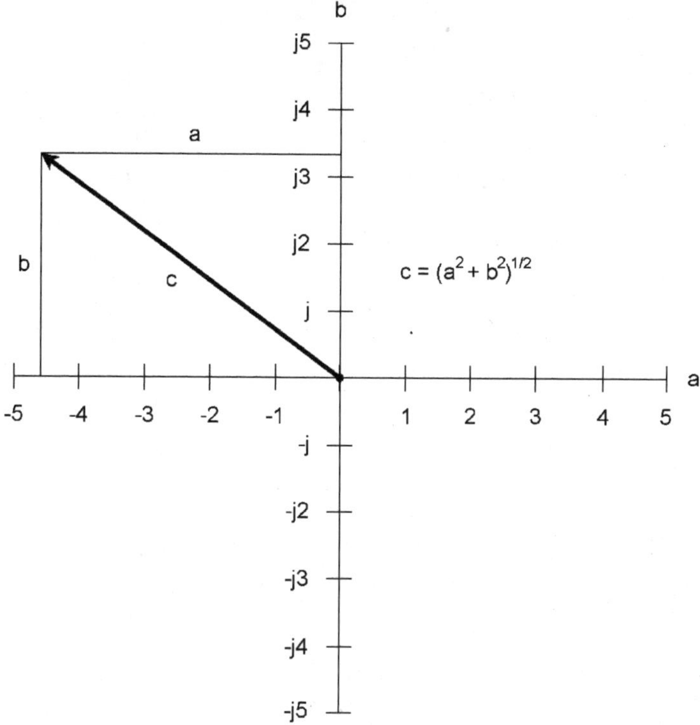

Figure 1.6 Calculation of absolute value (vector length) of a complex number.

$$a + jb = r \cos \theta + j(r \sin \theta)$$
$$= r(\cos \theta + j \sin \theta)$$

The value of r, corresponding to the magnitude of the vector, is called the *modulus*. The angle θ, corresponding to the direction of the vector, is called the *amplitude*.

Product of complex numbers in polar form

Let x_1 and x_2 be complex numbers in polar form such that:

$$x_1 = r_1(\cos \theta_1 + j \sin \theta_1)$$

$$x_2 = r_2(\cos \theta_2 + j \sin \theta_2)$$

Then the product of the complex numbers in polar form is given by the following formula:

$$x_1 x_2 = r_1 r_2(\cos (\theta_1 + \theta_2) + j \sin (\theta_1 + \theta_2))$$

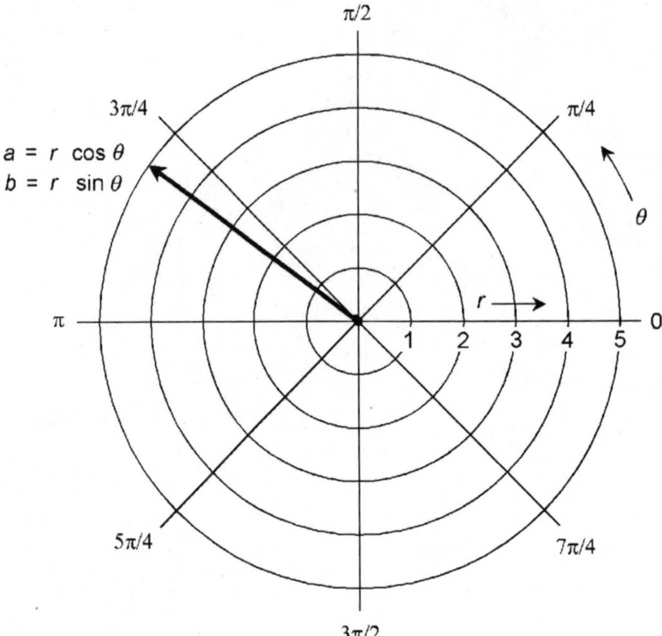

Figure 1.7 Polar form of a complex number.

Quotient of complex numbers in polar form

Let x_1 and x_2 be complex numbers in polar form such that:

$$x_1 = r_1(\cos\,\theta_1 + j\,\sin\,\theta_1)$$

$$x_2 = r_2(\cos\,\theta_2 + j\,\sin\,\theta_2)$$

Then the quotient of the complex numbers in polar form is given by the following formula:

$$x_1/x_2 = (r_1/r_2)(\cos\,(\theta_1 - \theta_2) + j\,\sin\,(\theta_1 - \theta_2))$$

De Moivre's Theorem

Let x be a complex number in polar form:

$$x = r(\cos\,\theta + j\,\sin\,\theta)$$

Then x raised to any real-number power p is given by the following formula:

$$x^p = r^p\,(\cos\,p\theta + j\,\sin\,p\theta)$$

Properties of Operations

Several properties, also called laws, are recognized as valid for the operations of addition, subtraction, multiplication, and division for all real, imaginary, and complex numbers.

Additive identity

When 0 is added to any real number a, the sum is always equal to a. When $0 + j0$ is added to any complex number $a + jb$, the sum is always equal to $a + jb$. The numbers 0 and $0 + j0$ are *additive identity elements*:

$$a + 0 = a$$

$$(a + jb) + (0 + j0) = a + jb$$

Multiplicative identity

When any real number a is multiplied by 1, the product is always equal to a. When any complex number $a + jb$ is multiplied by $1 + j0$, the product is always equal to $a + jb$. The numbers 1 and $1 + j0$ are *multiplicative identity elements*:

$$a \times 1 = a$$

$$(a + jb) \times (1 + j0) = a + jb$$

Additive inverse

For every real number a, there exists a unique real number $-a$ such that the sum of the two is equal to 0. For every complex number $a + jb$, there exists a unique complex number $-a - jb$ such that the sum of the two is equal to $0 + j0$. Formally:

$$a + (-a) = 0$$

$$(a + jb) + (-a - jb) = 0 + j0$$

Multiplicative inverse

For every nonzero real number a, there exists a unique real number $1/a$ such that the product of the two is equal to 1. For every complex number $a + jb$ except $0 + j0$, there exists a unique complex number $a/(a^2 + b^2) - jb/(a^2 + b^2)$ such that the product of the two is equal to $1 + j0$. Formally:

$$a \times (1/a) = 1$$

$$(a + jb) \times (a/(a^2 + b^2) - jb/(a^2 + b^2)) = 1 + j0$$

Commutativity of addition

When summing any two real or complex numbers, it does not matter in which order the sum is performed. For all real numbers a and b, and for all complex numbers $a + jb$ and $c + jd$, the following equations hold:

$$a + b = b + a$$

$$(a + jb) + (c + jd) = (c + jd) + (a + jb)$$

Commutativity of multiplication

When multiplying any two real or complex numbers, it does not matter in which order the product is performed. For all real numbers a and b, and for all complex numbers $a + jb$ and $c + jd$, the following equations hold:

$$ab = ba$$

$$(a + jb)(c + jd) = (c + jd)(a + jb)$$

Associativity of addition

When adding any three real or complex numbers, it does not matter how the addends are grouped. For all real numbers a_1, a_2, and a_3, and for all complex numbers $a_1 + jb_1$, $a_2 + jb_2$, and $a_3 + jb_3$, the following equations hold:

$$(a_1 + a_2) + a_3 = a_1 + (a_2 + a_3)$$

$$((a_1 + jb_1) + (a_2 + jb_2)) + (a_3 + jb_3)$$
$$= (a_1 + jb_1) + ((a_2 + jb_2) + (a_3 + jb_3))$$

Associativity of multiplication

When multiplying any three real or complex numbers, it does not matter how the multiplicands are grouped. For all real numbers a_1, a_2, and a_3, and for all complex numbers $a_1 + jb_1$, $a_2 + jb_2$, and $a_3 + jb_3$, the following equations hold:

$$(a_1a_2)a_3 = a_1(a_2a_3)$$

$$((a_1 + jb_1)(a_2 + jb_2))(a_3 + jb_3) = (a_1 + jb_1)((a_2 + jb_2)(a_3 + jb_3))$$

Distributivity of multiplication over addition

For all real numbers a_1, a_2, and a_3, and for all complex numbers $a_1 + jb_1$, $a_2 + jb_2$, and $a_3 + jb_3$, the following equations hold:

$$a_1(a_2 + a_3) = a_1a_2 + a_1a_3$$

$$(a_1 + jb_1)((a_2 + jb_2) + (a_3 + jb_3))$$
$$= (a_1 + jb_1)(a_2 + jb_2) + (a_1 + jb_1)(a_3 + jb_3)$$

Miscellaneous Principles

The following rules and definitions apply to arithmetic operations for real and complex numbers, with the constraint that no denominator be equal to zero, and no denominator contain any variable that can attain a value that renders the denominator equal to zero.

Zero numerator

For all nonzero real numbers a and all complex numbers $a + jb$ such that $a + jb \neq 0 + j0$:

$$0/a = 0$$

$$0/(a + jb) = 0 + j0$$

Zero denominator

For all real numbers a and all complex numbers $a + jb$:

$$a/0 \text{ is undefined}$$

$$a/(0 + j0) \text{ is undefined}$$

$$(a + jb)/0 \text{ is undefined}$$

$$(a + jb)/(0 + j0) \text{ is undefined}$$

Multiplication by zero

For all real numbers a and all complex numbers $a + jb$:

$$a \times 0 = 0$$

$$(a + jb) \times 0 = 0 + j0$$

Zeroth power

For all real numbers a and all complex numbers $a + jb$:

$$a^0 = 1$$

$$(a + jb)^0 = 1 + j0$$

Positive integer roots

If x is a real or complex number and x is multiplied by itself n times to obtain another real or complex number y, then x is defined as an nth root of y:

$$x^n = y$$

$$x = y^{1/n}$$

Factorial

If n is a natural number and $n \geq 1$, the value of $n!$ (n factorial) is the product of all natural numbers less than or equal to n:

$$n! = 1 \times 2 \times 3 \times 4 \times \ldots \times n$$

Arithmetic mean

Let $a_1, a_2, a_3, ..., a_n$ be real numbers. Then the *arithmetic mean,* denoted m_A, also known as the *average,* of $a_1, a_2, a_3, ...,$ and a_n is given by the following formula:

$$m_A = (a_1 + a_2 + a_3 + ... + a_n)/n$$

Geometric mean

Let $a_1, a_2, a_3, ..., a_n$ be positive reals. Then the *geometric mean,* denoted m_G, of $a_1, a_2, a_3, ...,$ and a_n is given by the following formula:

$$m_G = (a_1 a_2 a_3 \ldots a_n)^{1/n}$$

Arithmetic interpolation

Let $y = f(x)$ represent a function in which the value of a quantity y depends on the value of an independent variable x. Let y_1 and y_2 represent two values of the function such that:

$$y_1 = f(x_1)$$

$$y_2 = f(x_2)$$

Then the value y_a of the function at a point x_m midway between x_1 and x_2 can be estimated as follows via *arithmetic interpolation*:

$$y_a = (y_1 + y_2) / 2 = (f(x_1) + f(x_2))/2$$

Geometric interpolation

Let $y = f(x)$ represent a function in which the value of a quantity y depends on the value of an independent variable x. Let y_1 and y_2 represent two values of the function such that:

$$y_1 = f(x_1)$$

$$y_2 = f(x_2)$$

Then the value y_g of the function at a point x_m midway between x_1 and x_2 can be estimated as follows via *geometric interpolation*:

$$y_g = (y_1 y_2)^{1/2} = (f(x_1) \times f(x_2))^{1/2}$$

Product of signs

When numbers with plus and minus signs are multiplied, the following rules apply:

$$(+)(+) = (+)$$

$$(+)(-) = (-)$$

$$(-)(+) = (-)$$

$$(-)(-) = (+)$$

$$(-)^n = (+) \text{ if } n \text{ is even}$$

$$(-)^n = (-) \text{ if } n \text{ is odd}$$

Quotient of signs

When numbers with plus and minus signs are divided, the following rules apply:

$$(+)/(+) = (+)$$

$$(+)/(-) = (-)$$

$$(-)/(+) = (-)$$

$$(-)/(-) = (+)$$

Power of signs

When numbers with signs are raised to a positive integer power n, the following rules apply:

$$(+)^n = (+)$$

$$(-)^n = (-) \text{ if } n \text{ is odd}$$

$$(-)^n = (+) \text{ if } n \text{ is even}$$

Unit imaginary quadrature

The following equations hold for the unit imaginary number j:

$$j^2 = -1$$

$$j^3 = -j$$

$$j^4 = 1$$

$$j^5 = j$$

$$\vdots$$

$$j^n = j^{(n-4)}$$

$$\vdots$$

Reciprocal of reciprocal

For all nonzero real numbers a and all complex numbers $a + jb$, such that $a + jb \neq 0 + j0$:

$$1/(1/a) = a$$

$$1/(1/(a + jb)) = a + jb$$

Product of sums

For all real or complex numbers w, x, y, and z:

$$(w + x)(y + z) = wy + wz + xy + xz$$

Distributivity of division over addition

For all real or complex numbers x, y, and z, where $x \neq 0 + j0$:

$$(xy + xz)/x = xy/x + xz/x = x + y$$

Cross multiplication

For all real or complex numbers w, x, y, and z, where $x \neq 0 + j0$ and $z \neq 0 + j0$, the following statement is logically valid:

$$w/x = y/z \leftrightarrow wz = xy$$

Reciprocal of product

For all real or complex numbers x and y, where $x \neq 0 + j0$ and $y \neq 0 + j0$:

$$1/(xy) = (1/x)(1/y)$$

Reciprocal of quotient

For all real or complex numbers x and y, where $x \neq 0 + j0$ and $y \neq 0 + j0$:

$$1/(x/y) = y/x$$

Product of quotients

For all real or complex numbers w, x, y, and z, where $x \neq 0 + j0$ and $z \neq 0 + j0$:

$$(w/x)(y/z) = (wy)/(xz)$$

Quotient of products

For all real or complex numbers w, x, y, and z, where $yz \neq 0 + j0$:

$$(wx)/(yz) = (w/y)(x/z) = (w/z)(x/y)$$

Quotient of quotients

For all real or complex numbers w, x, y, and z, where $x \neq 0 + j0$, $y \neq 0 + j0$, and $z \neq 0 + j0$:

$$(w/x)/(y/z) = (w/x)(z/y) = (w/y)(z/x) = (wz)/(xy)$$

Sum of quotients (common denominator)

For all real or complex numbers x, y, and z, where $z \neq 0 + j0$:

$$x/z + y/z = (x + y)/z$$

Sum of quotients (general)

For all real or complex numbers w, x, y, and z, where $x \neq 0 + j0$ and $z \neq 0 + j0$:

$$w/x + y/z = (wz + xy)/(xz)$$

Prime numbers

Let p be a natural number. Suppose $ab = p$, where a and b are natural numbers. Further suppose that the following statement is true for all a and b:

$$a = 1 \;\&\; b = p$$

or

$$a = p \;\&\; b = 1$$

Then p is defined as a *prime number*. In other words, p is prime if and only if its only two factors are 1 and itself.

Prime factors

Let n be a natural number. Then there is a unique, increasing set of prime numbers $\{p_1, p_2, p_3, \ldots p_m\}$ such that the following equation, also known as the *Fundamental Theorem of Arithmetic*, holds true:

$$p_1 \times p_2 \times p_3 \times \ldots \times p_m = n$$

Rational-number powers

Let x be a real or complex number. Let y be a rational number such that $y = a/b$, where a and b are integers and $b \neq 0$. Then the following formula holds:

$$x^y = x^{a/b} = (x^a)^{1/b} = (x^{1/b})^a$$

Negative powers

Let x be a complex number where $x \neq 0 + j0$. Let y be a rational number. Then the following formula holds:

$$x^{-y} = (1/x)^y = 1/x^y$$

Sum of powers

Let x be a complex number. Let y and z be rational numbers. Then the following formula holds:

$$x^{(y+z)} = x^y x^z$$

Difference of powers

Let x be a complex number where $x \neq 0 + j0$. Let y and z be rational numbers. Then the following formula holds:

$$x^{(y-z)} = x^y / x^z$$

Product of powers

Let x be a complex number. Let y and z be rational numbers. Then the following formula holds:

$$x^{yz} = (x^y)^z = (x^z)^y$$

Quotient of powers

Let x be a complex number. Let y and z be rational numbers, with the constraint that $z \neq 0$. Then the following formula holds:

$$x^{y/z} = (x^y)^{1/z} = (x^{1/z})^y$$

Powers of sum

Let x and y be complex numbers. Then the following formulas hold:

$$(x + y)^2 = x^2 + 2xy + y^2$$

$$(x + y)^3 = x^3 + 3x^2y + 3xy^2 + y^3$$

$$(x + y)^4 = x^4 + 4x^3y + 6x^2y^2 + 4xy^3 + y^4$$

Powers of difference

Let x and y be complex numbers. Then the following formulas hold:

$$(x - y)^2 = x^2 - 2xy + y^2$$

$$(x - y)^3 = x^3 - 3x^2y + 3xy^2 - y^3$$

$$(x - y)^4 = x^4 - 4x^3y + 6x^2y^2 - 4xy^3 + y^4$$

Binomial formula

Let x and y be complex numbers, and let n be a natural number with $n \geq 1$. Then the value of $(x + y)^n$ is equal to the sum of the following expressions:

$$x^n$$

$$+$$

$$nx^{(n-1)}y$$

$$+$$

$$(n(n - 1)/2!)\ x^{(n-2)}y^2$$

$$+$$

$$(n(n - 1)(n - 2)/3!)\ x^{(n-3)}y^3$$

$$+$$

$$\vdots$$

$$+$$

$$y^n$$

Precedence of operations

When various operations appear in an expression and that expression is to be simplified, perform the operations in the following sequence:

- Simplify all expressions within parentheses from the inside out.
- Perform all exponential operations, proceeding from left to right.
- Perform all products and quotients, proceeding from left to right.
- Perform all sums and differences, proceeding from left to right.

The following are examples of this process, in which the order of the numerals and operations is the same in each case, but the groupings differ.

$$((2 + 3)(-3 - 1)^2)^2$$

$$(5 \times (-4)^2)^2$$

$$(5 \times 16)^2$$

$$80^2$$

$$6400$$

$$((2 + 3 \times (-3) - 1)^2)^2$$

$$((2 + (-9) - 1)^2)^2$$

$$(-8^2)^2$$

$$64^2$$

$$4096$$

Inequalities

The following general rules and definitions apply to inequalities, with the constraint that denominators of quotients must be nonzero because division by zero is undefined.

Transitivity

Inequalities are transitive when they all have the same sense. For all real numbers x, y, and z, the following statements are logically valid:

$$(x < y) \;\&\; (y < z) \rightarrow x < z$$

$$(x \leq y) \;\&\; (y \leq z) \rightarrow x \leq z$$

$$(x > y) \;\&\; (y > z) \rightarrow x > z$$

$$(x \geq y) \;\&\; (y \geq z) \rightarrow x \geq z$$

Corollaries to transitivity

For all real numbers x, y, and z, the following statements are logically valid:

$$(x < y) \& (y \leq z) \rightarrow x < z$$

$$(x \leq y) \& (y < z) \rightarrow x < z$$

$$(x > y) \& (y \geq z) \rightarrow x > z$$

$$(x \geq y) \& (y > z) \rightarrow x > z$$

Additive inverses

For all real numbers x and y, the following statements are logically valid:

$$x < y \rightarrow -x > -y$$

$$x \leq y \rightarrow -x \geq -y$$

$$x > y \rightarrow -x < -y$$

$$x \geq y \rightarrow -x < -y$$

Reciprocals of positive reals

For all real numbers x and y where $x > 0$ and $y > 0$, the following statements are logically valid:

$$x > y \rightarrow 1/x < 1/y$$

$$x \geq y \rightarrow 1/x \leq 1/y$$

$$x < y \rightarrow 1/x > 1/y$$

$$x \leq y \rightarrow 1/x \geq 1/y$$

Reciprocals of negative reals

For all real numbers x and y where $x < 0$ and $y < 0$, the following statements are logically valid:

$$x > y \rightarrow 1/x < 1/y$$

$$x \geq y \rightarrow 1/x \leq 1/y$$

$$x < y \rightarrow 1/x > 1/y$$

$$x \leq y \rightarrow 1/x \geq 1/y$$

Powers of positive reals

For all positive real numbers x, and for all natural numbers n where $n \geq 1$, the following statements are logically valid:

$$0 < x < 1 \rightarrow 0 < x^{(n+1)} < x^n$$

$$x = 1 \rightarrow x^{(n+1)} = x^n = 1$$

$$x > 1 \rightarrow x^{(n+1)} > x^n > 1$$

Even powers of negative reals

For all negative real numbers x, and for all even natural numbers n, the following statements are logically valid:

$$x < -1 \rightarrow x^{(n+2)} > x^n > 1$$

$$x = -1 \rightarrow x^{(n+2)} = x^n = 1$$

$$-1 < x < 0 \rightarrow 0 < x^{(n+2)} < x^n$$

Odd powers of negative reals

For all negative real numbers x, and for all odd natural numbers n, the following statements are logically valid:

$$x < -1 \rightarrow x^{(n+2)} < x^n < -1$$

$$x = -1 \rightarrow x^{(n+2)} = x^n = -1$$

$$-1 < x < 0 \rightarrow 0 > x^{(n+2)} > x^n$$

Addition property

For all real numbers x, y, and z, the following statements are logically valid:

$$x < y \rightarrow x + z < y + z$$

$$x \leq y \rightarrow x + z \leq y + z$$

$$x > y \rightarrow x + z > y + z$$

$$x \geq y \rightarrow x + z \geq y + z$$

Multiplication property

For all real numbers x, y, and z, the following statements are logically valid:

$$x < y \ \& \ z > 0 \rightarrow xz < yz$$

$$x \leq y \ \& \ z > 0 \rightarrow xz \leq yz$$

$$x > y \ \& \ z > 0 \rightarrow xz > yz$$

$$x \geq y \ \& \ z > 0 \rightarrow xz \geq yz$$

$$x < y \ \& \ z < 0 \rightarrow xz > yz$$

$$x \leq y \ \& \ z < 0 \rightarrow xz \geq yz$$

$$x > y \ \& \ z < 0 \rightarrow xz < yz$$

$$x \geq y \ \& \ z < 0 \rightarrow xz \leq yz$$

Complex-number magnitudes

Inequalities for complex numbers are defined according to their real-number relative absolute values (magnitudes). Let $a_1 + jb_1$ and $a_2 + jb_2$ be complex numbers. Then the following statements are logically valid:

$$|a_1 + jb_1| < |a_2 + jb_2| \leftrightarrow {a_1}^2 + {b_1}^2 < {a_2}^2 + {b_2}^2$$

$$|a_1 + jb_1| \leq |a_2 + jb_2| \leftrightarrow {a_1}^2 + {b_1}^2 \leq {a_2}^2 + {b_2}^2$$

$$|a_1 + jb_1| > |a_2 + jb_2| \leftrightarrow {a_1}^2 + {b_1}^2 > {a_2}^2 + {b_2}^2$$

$$|a_1 + jb_1| \geq |a_2 + jb_2| \leftrightarrow {a_1}^2 + {b_1}^2 \geq {a_2}^2 + {b_2}^2$$

Simple Equations

The objective of solving a single-variable equation is to get it into a form where the expression on the left-hand side of the equality symbol is exactly equal to the variable being sought (for example, x), and a defined expression *not containing that variable* is on the right.

Elementary rules

There are several ways in which an equation in one variable can be manipulated to obtain a solution, assuming a solution exists. Any and all of the aforementioned principles can be applied toward this result. In addition, the following rules can be applied in any order, and any number of times.

Addition of a quantity to each side: Any defined constant, variable, or expression can be added to both sides of an equation, and the result is equivalent to the original equation.

Subtraction of a quantity from each side: Any defined constant, variable, or expression can be subtracted from both sides of an equation, and the result is equivalent to the original equation.

Multiplication of each side by a quantity: Both sides of an equation can be multiplied by a defined constant, variable, or expression, and the result is equivalent to the original equation.

Division of each side by a quantity: Both sides of an equation can be divided by a nonzero constant, by a variable that cannot attain a value of zero, or by an expression that cannot attain a value of zero over the range of its variable(s), and the result is equivalent to the original equation.

Basic equation in one variable

Consider an equation of the following form:

$$ax + b = cx + d$$

where a, b, c, and d are complex numbers, and $a \neq c$. This equation is solved as follows:

$$ax + b = cx + d$$

$$ax = cx + d - b$$

$$ax - cx = d - b$$

$$(a - c)x = d - b$$

$$x = (d - b)/(a - c)$$

Factored equations in one variable

Consider an equation of the following factored form:

$$(x - a_1)(x - a_2)(x - a_3) \cdots (x - a_n) = 0$$

where $a_1, a_2, a_3, \ldots, a_n$ are complex numbers. Then the solutions of this equation are:

$$\begin{aligned} x_1 &= a_1 \\ x_2 &= a_2 \\ x_3 &= a_3 \\ &\vdots \\ x_n &= a_n \end{aligned}$$

Quadratic formula

Consider a quadratic equation in standard form:

$$ax^2 + bx + c = 0$$

where a, b, and c are complex numbers, with $a \neq 0 + j0$. The solutions of this equation can be found according to the following formula:

$$x = (-b \pm (b^2 - 4ac)^{1/2})/2a$$

Discriminant

Consider a quadratic equation in standard form:

$$ax^2 + bx + c = 0$$

where a, b, and c are real numbers. Define the discriminant, d, as follows:

$$d = b^2 - 4ac$$

Let the solutions to the quadratic equation be denoted as follows:

$$x_1 = a_1 + jb_1$$

$$x_2 = a_2 + jb_2$$

Then the following statements hold true:

$$d > 0 \rightarrow b_1 = 0 \; \& \; b_2 = 0 \; \& \; a_1 \neq a_2$$

$$d = 0 \rightarrow b_1 = 0 \; \& \; b_2 = 0 \; \& \; a_1 = a_2$$

$$d < 0 \rightarrow a_1 = a_2 \; \& \; b_1 = -b_2$$

These three principles are often stated as follows:

- If d > 0, then there are two distinct real-number solutions.
- If d = 0, then there is a single real-number solution.
- If d < 0, then there are two complex-conjugate solutions.

Simultaneous Equations

A linear equation in n variables takes the following form:

$$a_1x_1 + a_2x_2 + a_3x_3 + \cdots + a_nx_n + a_0 = 0$$

where x_1 through x_n represent the variables, and a_0 through a_n represent constants, usually real numbers.

Existence of solutions

Suppose there exists a set of m linear equations in n variables. If $m < n$, there exists no unique solution to the set of equations. If $m = n$ or $m > n$, there might exist a unique solution, but not necessarily. When solving sets of linear equations, it is first necessary to see if the number of equations is greater than or equal to the number of variables. If this is the case, any of the following methods can be used in an attempt to find a solution. If there exists no unique solution, this fact will become apparent as the steps are carried out.

2×2 substitution method

Consider the following set of two linear equations in two variables:

$$a_1x + b_1y + c_1 = 0$$

$$a_2x + b_2y + c_2 = 0$$

where a_1, a_2, b_1, b_2, c_1, and c_2 are real-number constants, and the variables are represented by x and y. The substitution method of solving these equations consists in performing either of the following sequences of steps. If $a_1 \neq 0$, use Sequence A. If $a_1 = 0$, use Sequence B. (If both $a_1 = 0$ and $a_2 = 0$, the set of equations is in fact a pair of equations in terms of a single variable, and the following steps are irrelevant.)

Sequence A: First, solve the first equation for x in terms of y:

$$a_1x + b_1y + c_1 = 0$$

$$a_1x = -b_1y - c_1$$

$$x = (-b_1y - c_1)/a_1$$

Next, substitute the above-derived solution for x in place of x in the second equation, obtaining:

$$a_2(-b_1y - c_1)/a_1 + b_2y + c_2 = 0$$

Solve this single-variable equation for y, using the previously outlined rules for solving single-variable equations. Assuming a solution exists, it can be substituted for y in either of the original equations, deriving a single-variable equation in terms of x. Solve for x, using the previously outlined rules for solving single-variable equations.

Sequence B: Because $a_1 = 0$, the first equation has only one variable, and is in the following form:

$$b_1y + c_1 = 0$$

Solve this equation for y:

$$b_1y = -c_1$$

$$y = -c_1/b_1$$

This can be substituted for y in the second equation, obtaining:

$$a_2x + b_2(-c_1/b_1) + c_2 = 0$$

$$a_2x - b_2(c_1/b_1) + c_2 = 0$$

$$a_2x = b_2(c_1/b_1) - c_2$$

$$x = (b_2(c_1/b_1) - c_2)/a_2$$

2×2 addition method

Consider the following set of two linear equations in two variables:

$$a_1x + b_1y + c_1 = 0$$

$$a_2x + b_2y + c_2 = 0$$

where a_1, a_2, b_1, b_2, c_1, and c_2 are real-number constants, and the variables are represented by x and y. The addition method of solving these equations consists in performing two separate and independent steps:

- Multiply one or both equations through by constant values to cancel out the coefficients of x, and then solve for y.
- Multiply one or both equations through by constant values to cancel out the coefficients of y, and then solve for x.

The scheme for solving for y begins by multiplying the first equation through by $-a_2$, and the second equation through by a_1, and then adding the two resulting equations:

$$\begin{array}{r} -a_2a_1x - a_2b_1y - a_2c_1 = 0 \\ a_1a_2x + a_1b_2y + a_1c_2 = 0 \\ \hline (a_1b_2 - a_2b_1)y + a_1c_2 - a_2c_1 = 0 \end{array}$$

Next, add a_2c_1 to each side, obtaining:

$$(a_1b_2 - a_2b_1)y + a_1c_2 = a_2c_1$$

Next, subtract a_1c_2 from each side, obtaining:

$$(a_1b_2 - a_2b_1)y = a_2c_1 - a_1c_2$$

Finally, divide through by $a_1b_2 - a_2b_1$, obtaining:

$$y = (a_2c_1 - a_1c_2)/(a_1b_2 - a_2b_1)$$

For this to be valid, the denominator must be nonzero; that is, $a_1b_2 \neq a_2b_1$. (If it turns out that $a_1b_2 = a_2b_1$, then there are not two distinct solutions to the set of equations.)

The process of solving for x is similar. Consider again the original set of simultaneous linear equations:

$$a_1x + b_1y + c_1 = 0$$

$$a_2x + b_2y + c_2 = 0$$

Multiply the first equation through by $-b_2$, and the second equation through by b_1, and then add the two resulting equations:

$$\begin{array}{r} -a_1b_2x - b_1b_2y - b_2c_1 = 0 \\ a_2b_1x + b_1b_2y + b_1c_2 = 0 \\ \hline (a_2b_1 - a_1b_2)x + b_1c_2 - b_2c_1 = 0 \end{array}$$

Next, add b_2c_1 to each side, obtaining:

$$(a_2b_1 - a_1b_2)x + b_1c_2 = b_2c_1$$

Next, subtract b_1c_2 from each side, obtaining:

$$(a_2b_1 - a_1b_2)x = b_2c_1 - b_1c_2$$

Finally, divide through by $b_2c_1 - b_1c_2$, obtaining:

$$x = (b_2c_1 - b_1c_2)/(a_2b_1 - a_1b_2)$$

For this to be valid, the denominator must be nonzero; that is, $a_1b_2 \neq a_2b_1$. (If it turns out that $a_1b_2 = a_2b_1$, then there are not two distinct solutions to the set of equations.)

Solving $n \times n$ sets of linear equations

In general, *matrices* are used for solving sets of equations larger than 2×2, because the above mentioned methods become too complex. *Linear algebra,* also known as *matrix algebra,* uses rules similar to those of the addition method described above. Please consult college-level texts on linear algebra for details.

Solving 2×2 general equations

When one or both of the equations in a 2×2 set are nonlinear, the substitution method generally works best. Two examples follow.

Example A. Consider the following two equations:

$$y = x^2 + 2x + 1$$

$$y = -x + 1$$

The first equation is quadratic, and the second equation is linear. Either equation can be directly substituted into the other to solve for x. Substituting the second equation into the first yields this result.

$$-x + 1 = x^2 + 2x + 1$$

This equation can be put into standard quadratic form as follows:

$$-x + 1 = x^2 + 2x + 1$$

$$-x = x^2 + 2x$$

$$0 = x^2 + 3x$$

$$x^2 + 3x + 0 = 0$$

Using the quadratic formula, let $a = 1$, $b = 3$, and $c = 0$:

$$x = (-3 \pm (3^2 - 4 \times 1 \times 0)^{1/2})/(2 \times 1)$$

$$x = (-3 \pm (9 - 0)^{1/2})/2$$

$$x = (-3 \pm 3)/2$$

$$x_1 = -3 \text{ and } x_2 = 0$$

These values can be substituted into the original linear equation to obtain the y-values:

$$y_1 = 3 + 1 \text{ and } y_2 = 0 + 1$$

$$y_1 = 4 \text{ and } y_2 = 1$$

The solutions are therefore:

$$(x_1,y_1) = (-3,4)$$

$$(x_2,y_2) = (0,1)$$

Example B. Consider the following two equations:

$$y = -2x^2 + 4x - 5$$

$$y = -2x - 5$$

The first equation is quadratic, and the second equation is linear. Either equation can be directly substituted into the other to solve for x. Substituting the second equation into the first yields this result.

$$-2x - 5 = -2x^2 + 4x - 5$$

This equation can be put into standard quadratic form as follows:

$$-2x - 5 = -2x^2 + 4x - 5$$

$$-2x = -2x^2 + 4x$$

$$0 = -2x^2 + 6x$$

$$-2x^2 + 6x + 0 = 0$$

Using the quadratic formula, let $a = -2$, $b = 6$, and $c = 0$:

$$x = (-6 \pm (6^2 - 4 \times -2 \times 0)^{1/2})/(2 \times -2)$$

$$x = (-6 \pm (36 - 0)^{1/2})/-4$$

$$x = (-6 \pm 6)/-4$$

$$x_1 = 3 \text{ and } x_2 = 0$$

These values can be substituted into the original linear equation to obtain the y-values:

$$y_1 = -2 \times 3 - 5 \text{ and } y_2 = -2 \times 0 - 5$$

$$y_1 = -11 \text{ and } y_2 = -5$$

The solutions are therefore:

$$(x_1,y_1) = (3,-11)$$

$$(x_2,y_2) = (0,-5)$$

The Cartesian Plane

The most common two-dimensional coordinate system is the *Cartesian plane* (Fig. 1.8), also called *rectangular coordinates* or the *xy-plane*. The independent variable is plotted along the x axis or *abscissa*; the dependent variable is plotted along the y axis or *ordinate*. The scales of the abscissa and ordinate are normally linear, although the divisions need not represent the same increments. Variations of this scheme include the *semilog graph,* in which one scale is linear and the other scale is logarithmic, and the *log-log graph,* in which both scales are logarithmic.

Slope-intercept form of linear equation

A linear equation in two variables can be rearranged from standard form to a conveniently graphable form as follows:

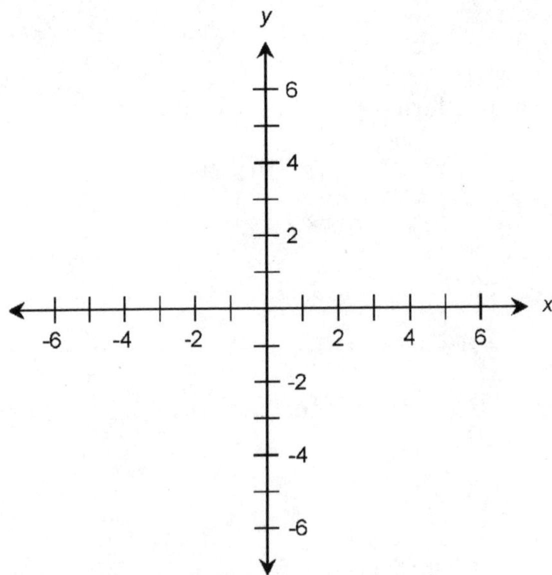

Figure 1.8 The Cartesian coordinate plane.

$$ax + by + c = 0$$

$$ax + by = -c$$

$$by = -ax - c$$

$$y = (-a/b)x - (c/b)$$

where a, b, and c are real-number constants, and $b \neq 0$. Such an equation appears as a straight line when graphed on the Cartesian plane. Let Δx represent a small change in the value of x on such a graph; let Δy represent the change in the value of y that results from this change in x. The ratio $\Delta y/\Delta x$ is defined as the slope of the line, and is commonly symbolized m. Let k represent the y-value of the point where the line crosses the ordinate. Then the following equations hold:

$$m = -a/b$$

$$k = -c/b$$

Thus, the linear equation can be rewritten in *slope-intercept form* as:

$$y = mx + k$$

To plot a graph of a linear equation in Cartesian coordinates, proceed as follows:

- Convert the equation to slope-intercept form.
- Plot the point $y = k$ and $x = 0$.
- Move to the right by n units on the graph.
- Move upward by mn units (or downward by $-mn$ units).
- Plot the resulting point $y = mn + k$.
- Connect the two points with a straight line.

Figures 1.9 and 1.10 illustrate the following linear equations as graphed in slope-intercept form:

$$y = 5x - 3$$

$$y = -x + 2$$

Note that a positive slope indicates that the graph "ramps up-

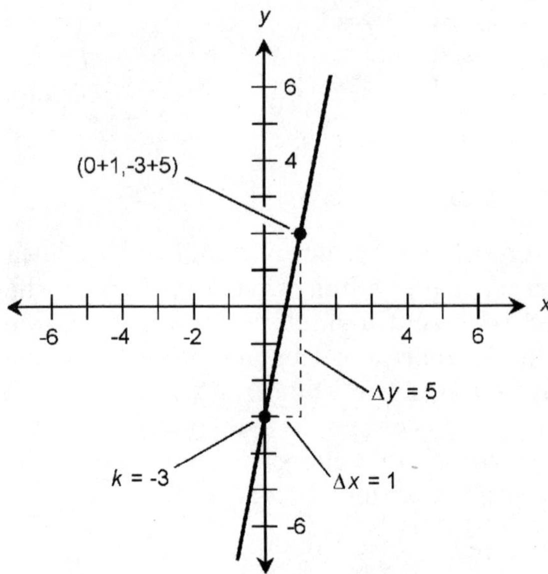

Figure 1.9 Slope-intercept plot of the equation $y = 5x - 3$.

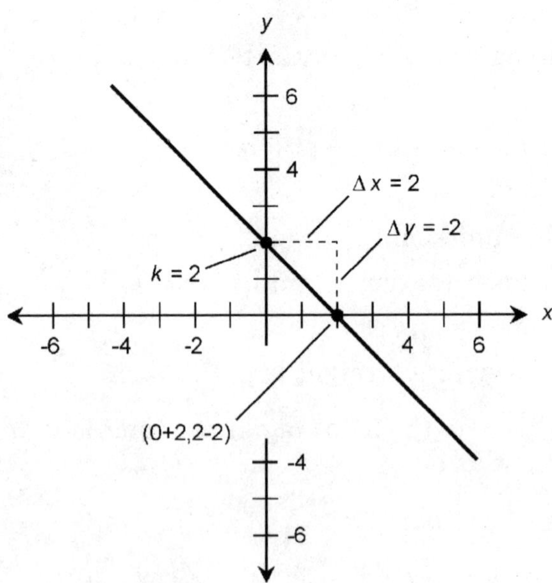

Figure 1.10 Slope-intercept plot of the equation $y = -x + 2$.

ward" and a negative slope indicates that the graph "ramps downward" as the point moves toward the right. A zero slope indicates a horizontal line. The slope of a vertical line is undefined.

Point-slope form of linear equation

It is not always convenient to plot a graph of a line based on the y-intercept point because the relevant part of the graph might lie at a great distance from that point. In this situation, the *point-slope form* of a linear equation can be used. This form is based on the slope m of the line and the coordinates of a known point (x_0,y_0):

$$y - y_0 = m(x - x_0)$$

To plot a graph of a linear equation using the point-slope method, proceed as follows:

- Convert the equation to point-slope form
- Determine a point (x_0,y_0) by "plugging in" values
- Plot (x_0,y_0) on the plane
- Move to the right by n units on the graph
- Move upward by mn units (or downward my $-mn$ units)
- Plot the resulting point (x_1,y_1)
- Connect the points (x_0,y_0) and (x_1,y_1) with a straight line

Figures 1.11 and 1.12 illustrate the following linear equations as graphed in point-slope form for regions in the immediate vicinity of points far removed from the origin:

$$y - 104 = 3(x - 72)$$

$$y + 55 = -2(x + 85)$$

Finding linear equation based on graph

Suppose the rectangular coordinates of two points P and Q are known; suppose further that these two points lie along a straight line (but not a vertical line). Let the coordinates of the points be:

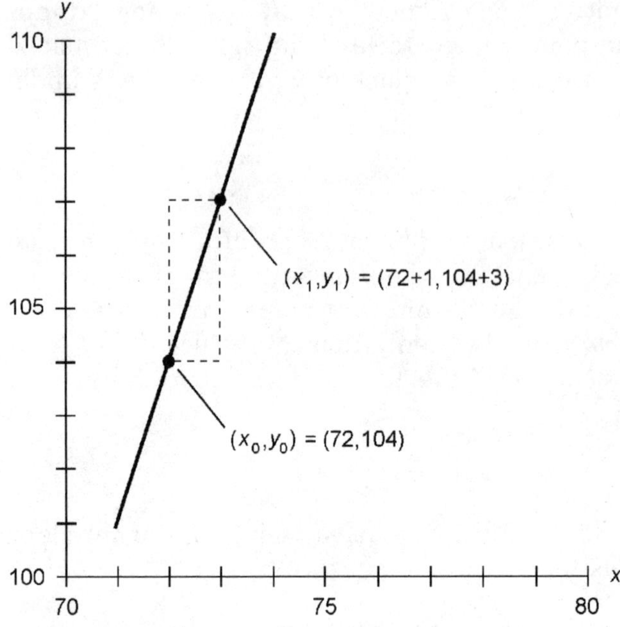

Figure 1.11 Point-slope plot of the equation $y - 104 = 3(x - 72)$.

$$P = (x_p,y_p)$$

$$Q = (x_q,y_q)$$

Then the slope m of the line is given by the either of the following formulas:

$$m = (y_q - y_p)/(x_q - x_p)$$

$$m = (y_p - y_q)/(x_p - x_q)$$

The point-slope equation of the line can be determined based on the known coordinates of P or Q. Therefore, either of the following formulas represent the line:

$$y - y_p = m(x - x_p)$$

$$y - y_q = m(x - x_q)$$

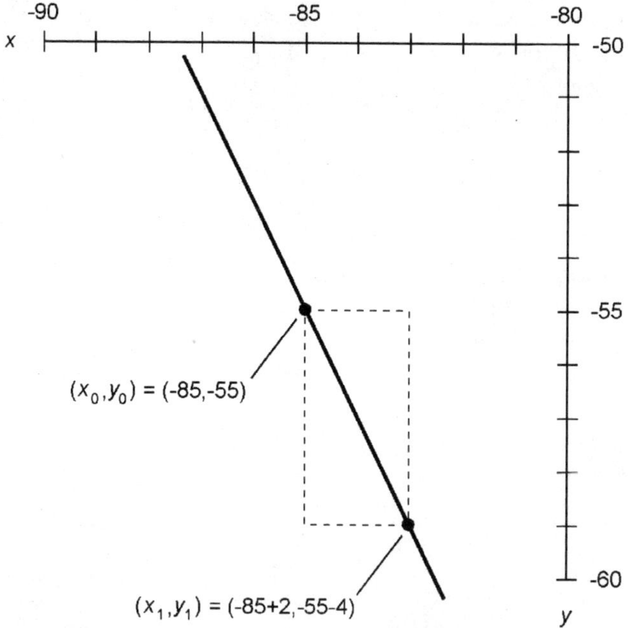

Figure 1.12 Point-slope plot of the equation $y + 55 = -2(x + 85)$.

Equation of parabola

The Cartesian-coordinate graph of a quadratic equation takes the form of a *parabola*. Suppose the following equation is given:

$$y = ax^2 + bx + c$$

where $a \neq 0$. (If $a = 0$, then the equation is linear, not quadratic.) To plot a graph of the above equation, first determine the coordinates of the point (x_0,y_0) where:

$$x_0 = -b/(2a)$$

$$y_0 = c - b^2/(4a)$$

This point represents the *base point* of the parabola; that is, the point at which the curvature is sharpest, and at which the slope of a line tangent to the curve is zero. Once this point is known, find four more points by "plugging in" values of x somewhat greater than and less than x_0 and determining the corresponding y-values. These x-values, call them x_{-2}, x_{-1}, x_1, and x_2, should be equally spaced on either side of x_0, such that:

$$x_{-2} < x_{-1} < x_0 < x_1 < x_2$$

$$x_{-1} - x_{-2} = x_0 - x_{-1} = x_1 - x_0 = x_2 - x_1$$

This will yield five points that lie along the parabola, and that are symmetrical relative to the axis of the curve. The graph can then be inferred, provided that the points are judiciously chosen. Some trial and error might be required. If $a > 0$, the parabola will open upward. If $a < 0$, the parabola will open downward.

Example A. Consider the following formula:

$$y = x^2 + 2x + 1$$

The base point is:

$$x_0 = -2/2 = -1$$

$$y_0 = 1 - 4/4 = 1 - 1 = 0$$

$$\therefore$$

$$(x_0,y_0) = (-1,0)$$

This point is plotted first. Next, consider the following points:

$$x_{-2} = x_0 - 2 = -3$$

$$y_{-2} = (-3)^2 + 2(-3) + 1 = 9 - 6 + 1 = 4$$

$$\therefore$$

$$(x_{-2},y_{-2}) = (-3,4)$$

$$x_{-1} = x_0 - 1 = -2$$

$$y_{-1} = (-2)^2 + 2(-2) + 1 = 4 - 4 + 1 = 1$$

$$\therefore$$

$$(x_{-1},y_{-1}) = (-2,1)$$

$$x_1 = x_0 + 1 = 0$$

$$y_1 = (0)^2 + 2(0) + 1 = 0 + 0 + 1 = 1$$

$$\therefore$$

$$(x_1,y_1) = (0,1)$$

$$x_2 = x_0 + 2 = 1$$

$$y_2 = (1)^2 + 2(1) + 1 = 1 + 2 + 1 = 4$$

$$\therefore$$

$$(x_2,y_2) = (1,4)$$

The five known points are plotted as shown in Fig. 1.13. From these, the curve can be inferred.

Example B. Consider the following formula:

$$y = -2x^2 + 4x - 5$$

The base point is:

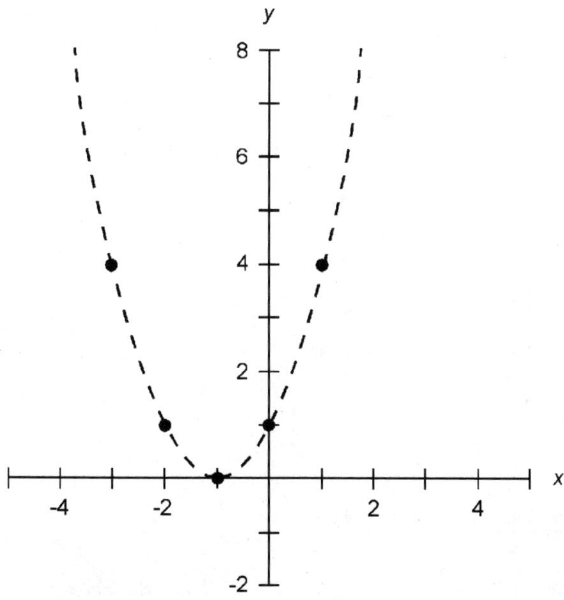

Figure 1.13 Plot of the parabola $y = x^2 + 2x + 1$.

$$x_0 = -4/-4 = 1$$

$$y_0 = -5 - 16/-8 = -5 + 2 = -3$$

$$\therefore$$

$$(x_0,y_0) = (1,-3)$$

This point is plotted first. Next, consider the following points:

$$x_{-2} = x_0 - 2 = -1$$

$$y_{-2} = -2(-1)^2 + 4(-1) - 5 = -2 - 4 - 5 = -11$$

$$\therefore$$

$$(x_{-2},y_{-2}) = (-1,-11)$$

$$x_{-1} = x_0 - 1 = 0$$

$$y_{-2} = -2(0)^2 + 4(0) - 5 = -5$$

$$\therefore$$

$$(x_{-1},y_{-1}) = (0,-5)$$

$$x_1 = x_0 + 1 = 2$$

$$y_{-2} = -2(2)^2 + 4(2) + 5 = -8 + 8 - 5 = -5$$

$$\therefore$$

$$(x_1,y_1) = (2,-5)$$

$$x_2 = x_0 + 2 = 3$$

$$y_{-2} = -2(3)^2 + 4(3) + 5 = -18 + 12 - 5 = -11$$

$$\therefore$$

$$(x_2,y_2) = (3,-11)$$

The five known points are plotted as shown in Fig. 1.14. From these, the curve can be inferred.

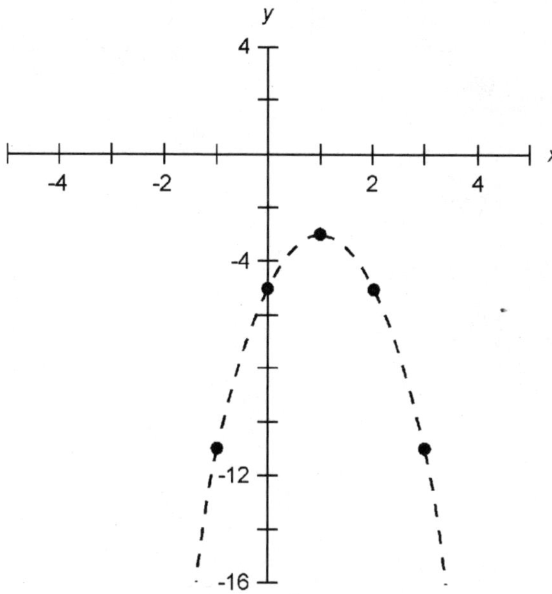

Figure 1.14 Plot of $y = -2x^2 + 4x - 5$.

Equation of circle

The general form for the equation of a *circle* in the *xy*-plane is given by the following formula:

$$(x - x_0)^2 + (y - y_0)^2 = r^2$$

where (x_0, y_0) represents the coordinates of the center of the circle, and r represents the *radius*. This is illustrated in Fig. 1.15. In the special case where the circle is centered at the origin, the formula becomes:

$$x^2 + y^2 = r^2$$

Such a circle intersects the x axis at the points $(r,0)$ and $(-r,0)$; it intersects the y axis at the points $(0,r)$ and $(0,-r)$. An even more specific case is the so-called *unit circle*:

$$x^2 + y^2 = 1$$

This curve intersects the x axis at the points $(1,0)$ and $(-1,0)$; it also intersects the y axis at the points $(0,1)$ and $(0,-1)$. The

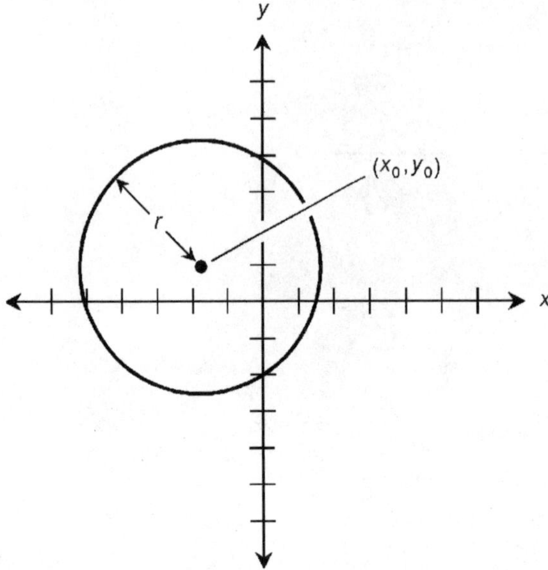

Figure 1.15 Plot of $(x - x_0)^2 + (y - y_0)^2 = r^2$.

unit circle is the basis for the definitions of the *circular trigonometric functions*.

Equation of ellipse

The general form for the equation of an *ellipse* in the *xy*-plane is given by the following formula:

$$(x - x_0)^2/a^2 + (y - y_0)^2/b^2 = 1$$

where (x_0,y_0) represents the coordinates of the center of the ellipse, a represents the distance from (x_0,y_0) to the curve as measured parallel to the x axis, and b represents the distance from (x_0,y_0) to the curve as measured parallel to the y axis. This is illustrated in Fig. 1.16. The values $2a$ and $2b$ represent the lengths of the *axes* of the ellipse; the greater value is the length of the *major axis,* and the lesser value is the length of the *minor axis*. In the special case where the ellipse is centered at the origin, the formula becomes:

$$x^2/a^2 + y^2/b^2 = 1$$

Such an ellipse intersects the x axis at the points $(a,0)$ and $(-a,0)$; it intersects the y axis at the points $(0,b)$ and $(0,-b)$.

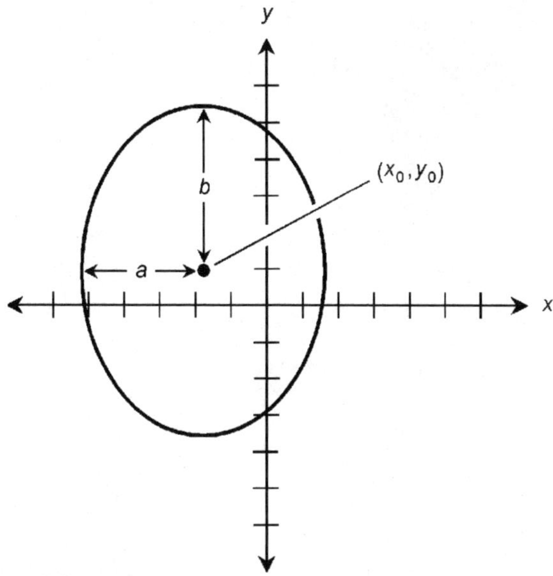

Figure 1.16 Plot of the ellipse $(x - x_0)^2/a^2 + (y - y_0)^2/b^2 = 1$.

Equation of hyperbola

The general form for the equation of a *hyperbola* in the *xy*-plane is given by the following formula:

$$(x - x_0)^2/a^2 - (y - y_0)^2/b^2 = 1$$

where (x_0,y_0) represents the coordinates of the center of the hyperbola. Let D represent a rectangle whose center is at (x_0,y_0), whose vertical edges are tangent to the hyperbola, and whose vertices (corners) lie on the *asymptotes* of the hyperbola (Fig. 1.17). Then a represents the distance from (x_0,y_0) to D as measured parallel to the x axis, and b represents the distance from (x_0,y_0) to D as measured parallel to the y axis. The values $2a$ and $2b$ represent the lengths of the *axes* of the hyperbola; the greater value is the length of the *major axis,* and the lesser value is the length of the *minor axis*. In the special case where the hyperbola is centered at the origin, the formula becomes:

$$x^2/a^2 - y^2/b^2 = 1$$

An even more specific case is the so-called *unit hyperbola,* the basis for the definitions of the *hyperbolic trigonometric functions*:

$$x^2 - y^2 = 1$$

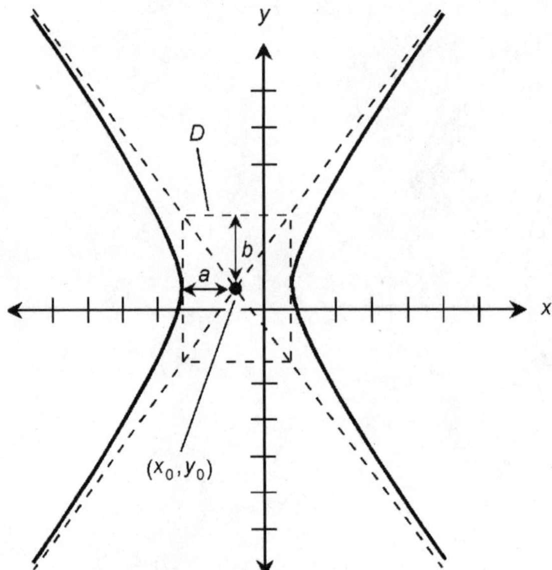

Figure 1.17 Plot of the hyperbola $(x - x_0)^2/a^2 - (y - y_0)^2/b^2 = 1$.

Graphic solution of pairs of equations

The solutions of pairs of equations can be depicted by graphs. Solutions appear as intersection points between the graphs of the equations in question.

Example A. Refer to Example A from "Solving 2×2 general equations" above:

$$y = x^2 + 2x + 1$$

$$y = -x + 1$$

These equations are graphed in Fig. 1.18. The line crosses the parabola at two points, indicating that there are two solutions of this set of simultaneous equations. The coordinates of the points, corresponding to the solutions, are:

$$(x_1,y_1) = (-3,4)$$

$$(x_2,y_2) = (0,1)$$

Example B. Refer to Example B from "Solving 2×2 general equations" above:

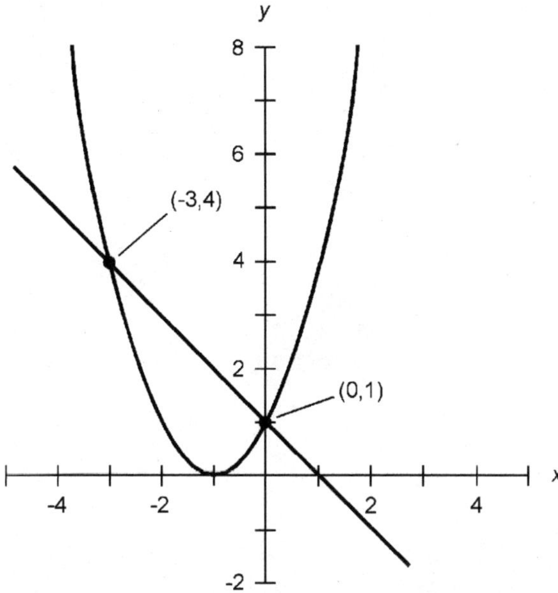

Figure 1.18 Graphic depiction of solutions of $y = x^2 + 2x + 1$ and $y = -x + 1$.

$$y = -2x^2 + 4x - 5$$

$$y = -2x - 5$$

These equations are graphed in Fig. 1.19. The line crosses the parabola at two points, indicating that there are two solutions of this set of simultaneous equations. The coordinates of the points, corresponding to the solutions, are:

$$(x_1,y_1) = (3,-11)$$

$$(x_2,y_2) = (0,-5)$$

The Polar Plane

The *polar coordinate plane* is an alternative way of expressing the positions of points, and of graphing equations and relations, in two dimensions. The independent variable is plotted as the distance or radius r from the origin, and the dependent variable is plotted as an angle θ relative to a reference axis. Figure 1.20 shows the polar system generally used in mathematics, physical science, and engineering; θ is in radians and is plotted counter-

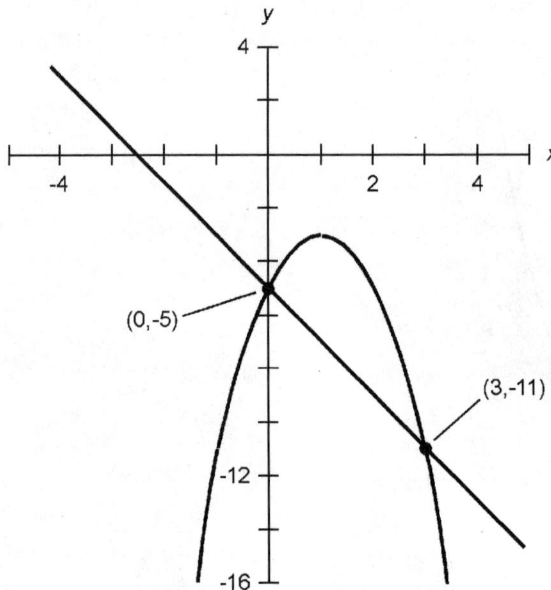

Figure 1.19 Graphic depiction of solutions of $y = -2x^2 + 4x - 5$ and $y = -2x - 5$.

clockwise from the ray extending to the right. In the sections that follow, this scheme is used. Figure 1.21 shows the polar system employed in wireless communications, broadcasting, location, and navigation; θ is in degrees and is plotted clockwise from the ray extending upwards (corresponding to geographic north).

Cartesian vs. polar coordinates

Let (x_0, y_0) represent the coordinates of a point in the Cartesian plane. The coordinates (r_0, θ_0) of the same point in the polar plane are given by:

$$r_0 = (x_0^{\ 2} + y_0^{\ 2})^{1/2}$$

$$\theta_0 = \tan^{-1} (y_0/x_0)$$

Polar coordinates are converted to Cartesian coordinates by the following formulas:

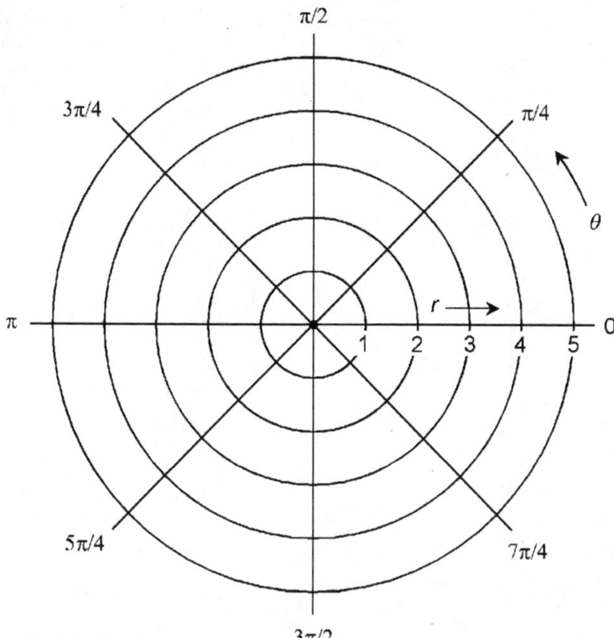

Figure 1.20 The polar plane for mathematics and physical sciences.

$$x_0 = r_0 \cos \theta_0$$

$$y_0 = r_0 \sin \theta_0$$

Equation of circle

The equation of a *circle* centered at the origin in the polar plane is given by the following formula:

$$r = a$$

where a is a real number and $a > 0$. This is illustrated in Fig. 1.22.

The general form for the equation of a circle passing through the origin, and centered at the point (r_0, θ_0) in the polar plane is given by:

$$r = 2r_0 \cos (\theta - \theta_0)$$

This is illustrated in Fig. 1.23.

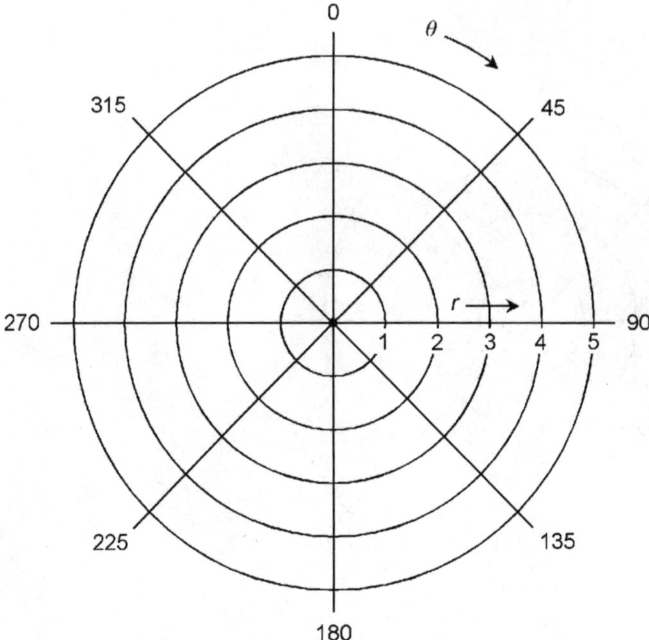

Figure 1.21 The polar plane for wireless engineering, broadcast engineering, navigation, and location.

Equation of ellipse centered at origin

The equation of an *ellipse* centered at the origin in the polar plane is given by the following formula:

$$r = ab/(a^2 \sin^2 \theta + b^2 \cos^2 \theta)^{1/2}$$

where a represents the distance from the origin to the curve as measured along the "horizontal" ray $\theta = 0$, and b represents the distance from the origin to the curve as measured along the "vertical" ray $\theta = \pi/2$. This is illustrated in Fig. 1.24. The values $2a$ and $2b$ represent the lengths of the *axes* of the ellipse; the greater value is the length of the *major axis*, and the lesser value is the length of the *minor axis*.

Equation of hyperbola centered at origin

The equation of a *hyperbola* centered at the origin in the polar plane is given by the following formula:

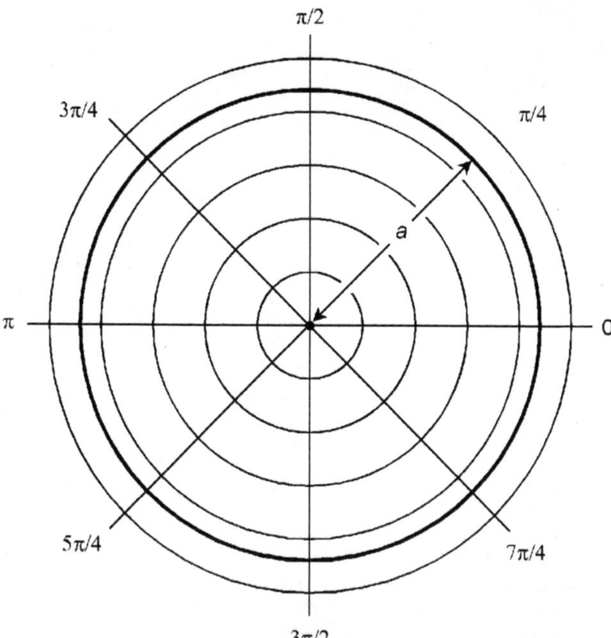

Figure 1.22 Polar graph of circle centered at the origin.

$$r = ab/(a^2 \sin^2 \theta - b^2 \cos^2 \theta)^{1/2}$$

Let D represent a rectangle whose center is at the origin, whose vertical edges are tangent to the hyperbola, and whose vertices (corners) lie on the *asymptotes* of the hyperbola (Fig. 1.25). Then a represents the distance from the origin to D as measured along the "horizontal" ray $\theta = 0$, and b represents the distance from the origin to D as measured along the "vertical" ray $\theta = \pi/2$. The values $2a$ and $2b$ represent the lengths of the *axes* of the hyperbola; the greater value is the length of the *major axis,* and the lesser value is the length of the *minor axis*.

Equation of lemniscate

The equation of a *lemniscate* centered at the origin in the polar plane is given by the following formula:

$$r = a\,(\cos 2\theta)^{1/2}$$

where a is a real number and $a > 0$. This is illustrated in Fig. 1.26. The area A of each loop of the figure is given by:

$$A = a^2$$

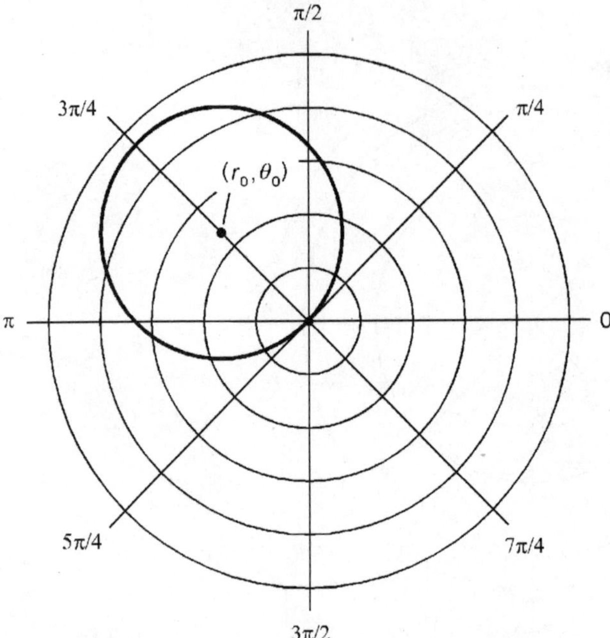

Figure 1.23 Polar graph of circle passing through the origin.

Equation of three-leafed rose

The equation of a *three-leafed rose* centered at the origin in the polar plane is given by either of the following two formulas:

$$r = a \cos 3\theta$$

$$r = a \sin 3\theta$$

where a is a real number and $a > 0$. The cosine curve is illustrated in Fig. 1.27A; the sine curve is illustrated in Fig. 1.27B.

Equation of four-leafed rose

The equation of a *four-leafed rose* centered at the origin in the polar plane is given by either of the following two formulas:

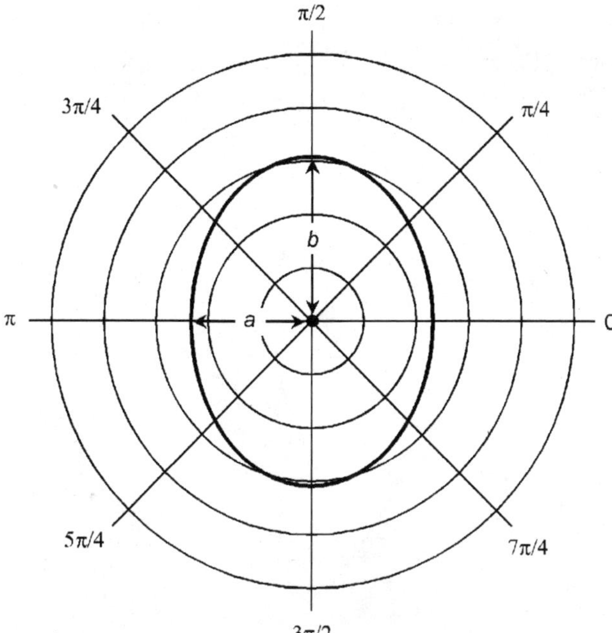

Figure 1.24 Polar graph of ellipse centered at the origin.

$$r = a \cos 2\theta$$

$$r = a \sin 2\theta$$

where a is a real number and $a > 0$. The cosine curve is illustrated in Fig. 1.28A; the sine curve is illustrated in Fig. 1.28B.

Equation of spiral

The equation of a *spiral* centered at the origin in the polar plane is given by the following formula:

$$r = a\theta$$

where a is a real number and $a > 0$. This is illustrated in Fig. 1.29.

Equation of cardioid

The equation of a *cardioid* centered at the origin in the polar plane is given by the following formula:

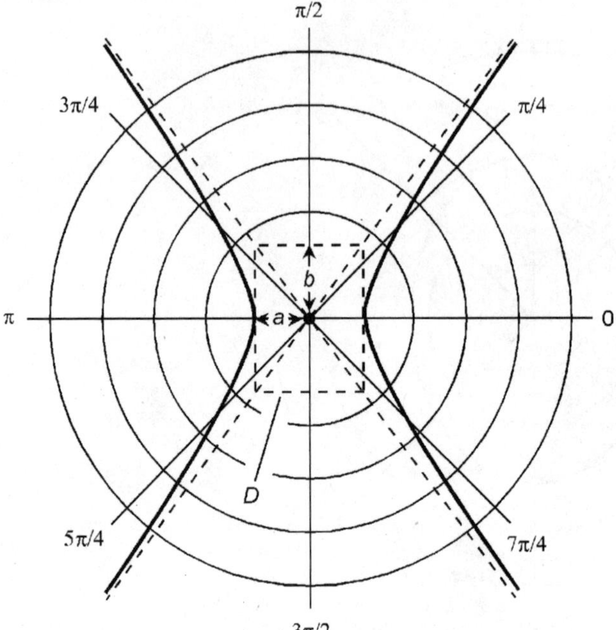

Figure 1.25 Polar graph of hyperbola centered at the origin.

$$r = 2a\ (1 + \cos\ \theta)$$

where a is a real number and $a > 0$. This is illustrated in Fig. 1.30.

Other Coordinate Systems

The following paragraphs describe other coordinate systems that are used in scientific and engineering applications.

Latitude and longitude

Latitude and *longitude* angles uniquely define the positions of points on the surface of a sphere or in the sky. The scheme for geographic locations on the earth is illustrated in Fig. 1.31A. The *polar axis* connects two specified points at antipodes on the sphere. These points are assigned latitude $\theta = 90°$ (north pole) and $\theta = -90°$ (south pole). The *equatorial axis* runs outward

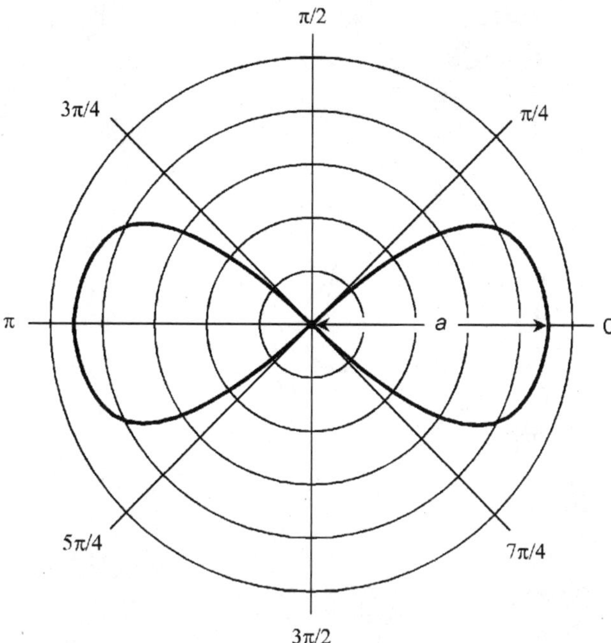

Figure 1.26 Polar graph of lemniscate centered at the origin.

from the center of the sphere at a 90° angle to the polar axis. It is assigned longitude $\phi = 0°$. Latitude θ is measured positively (north) and negatively (south) relative to the plane of the equator. Longitude ϕ is measured counterclockwise (east) and clockwise (west) relative to the equatorial axis. The angles are restricted as follows:

$$-90° \leq \theta \leq 90°$$

$$-180° < \phi \leq 180°$$

On the earth's surface, the half-circle connecting the 0° longitude line with the poles passes through Greenwich, England and is known as the *Greenwich meridian* or the *prime meridian*. Longitude angles are defined with respect to this meridian.

Celestial coordinates

Celestial latitude and *celestial longitude* are extensions of the earth's latitude and longitude into the heavens. Figure 1.31A,

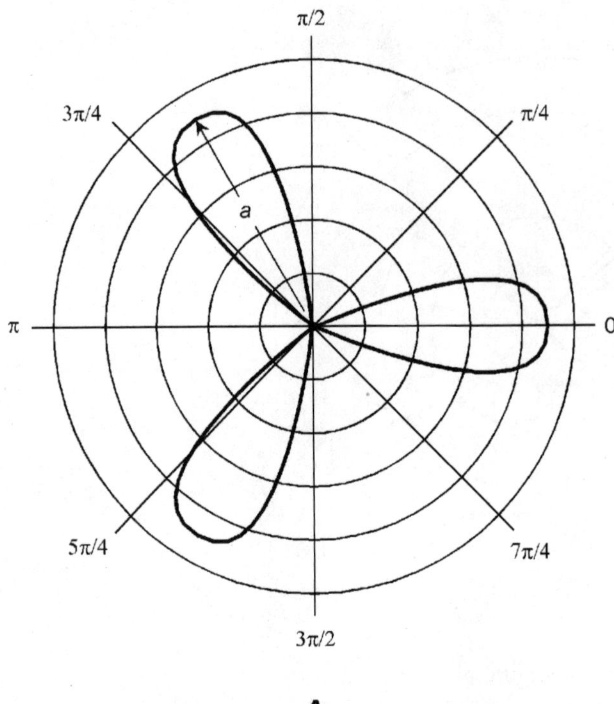

Figure 1.27A Polar graph of three-leafed rose $r = a \cos 3\theta$ centered at the origin.

the same set of coordinates used for geographic latitude and longitude, applies to this system. An object whose celestial latitude and longitude coordinates are (θ,ϕ) appears at the zenith in the sky from the point on the earth's surface whose latitude and longitude coordinates are (θ,ϕ).

Declination and right *ascension* define the positions of objects in the sky relative to the stars. Figure 1.31B applies to this system. Declination (θ) is identical to celestial latitude. Right ascension (ϕ) is measured eastward from the *vernal equinox* (the position of the sun in the heavens at the moment spring begins in the northern hemisphere). The angles are restricted as follows:

$$-90° \leq \theta \leq 90°$$

$$0° \leq \phi < 360°$$

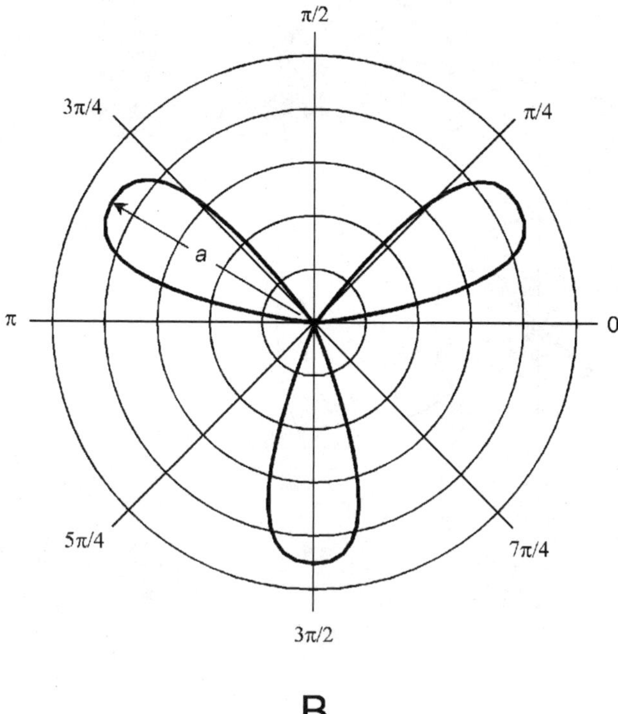

Figure 1.27B Polar graph of three-leafed rose $r = a \sin 3\theta$ centered at the origin.

Cartesian three-space

An extension of rectangular coordinates into three dimensions is *Cartesian three-space* (Fig. 1.32), also called *xyz-space*. Independent variables are usually plotted along the x and y axes; the dependent variable is plotted along the z axis. The scales are normally linear, although the divisions need not represent the same increments. Variations of this scheme can employ logarithmic graduations for one, two, or all three scales.

Cylindrical coordinates

Figure 1.33 shows a system of *cylindrical coordinates* for specifying the positions of points in three-space. Given a set of Cartesian coordinates or xyz-space, an angle ϕ is defined in the xy-plane, measured in radians counterclockwise from the x axis.

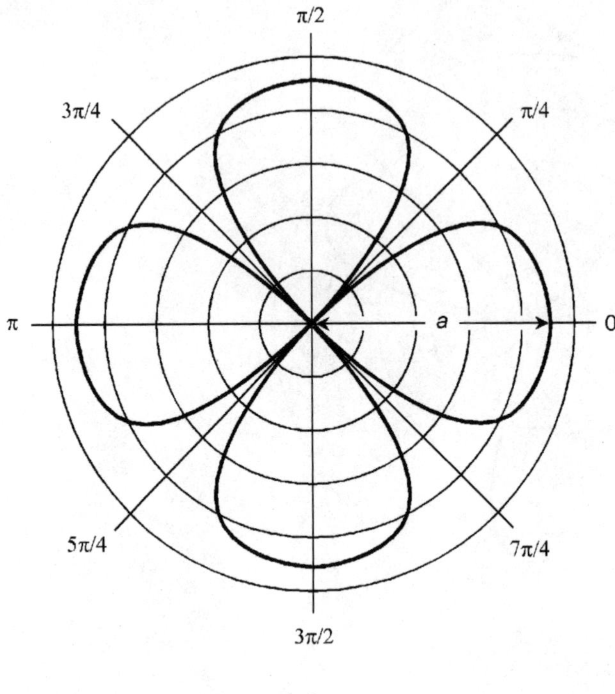

Figure 1.28A Polar graph of four-leafed rose $r = a \cos 2\theta$ centered at the origin.

Given a point P in space, consider its projection P' onto the xy-plane. The position of P is defined by the ordered triple (ϕ,r,z) such that:

$$\phi = \text{angle between } P' \text{ and the } x \text{ axis in the } xy\text{-plane}$$

$$r = \text{distance (radius) from } P \text{ to the origin}$$

$$z = \text{distance (altitude) of } P \text{ above the } xy\text{-plane}$$

Spherical coordinates

Figure 1.34 shows a system of *spherical coordinates* for defining points in space. This scheme is identical to the system for declination and right ascension, with the addition of a radius vector r representing the distance of point P from the origin. The location of a point P is defined by the ordered triple (θ,ϕ,r) such that:

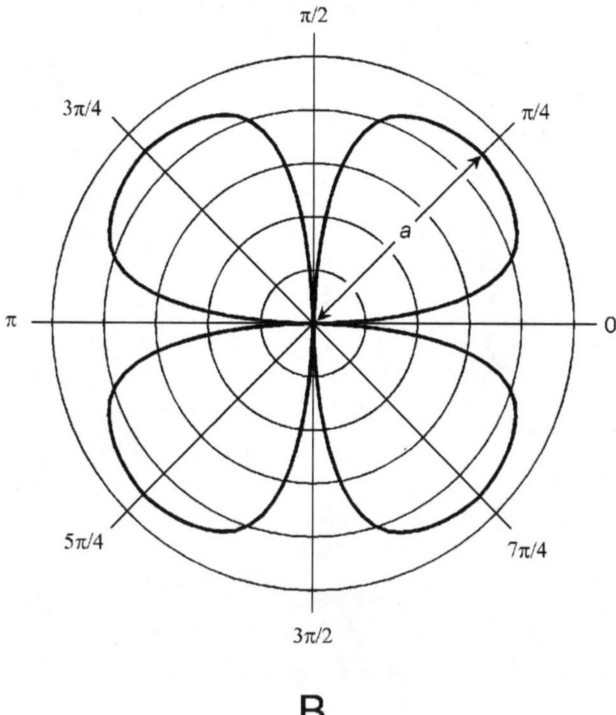

Figure 1.28B Polar graph of four-leafed rose $r = a \sin 2\theta$ centered at the origin.

$$\theta = \text{declination of } P$$

$$\phi = \text{right ascension of } P$$

$$r = \text{distance (radius) from } P \text{ to the origin}$$

In this example, angles are specified in degrees; alternatively they can be expressed in radians. There are several variations of this system, all of which are commonly called spherical coordinates.

Semilog (x-linear) coordinates

Figure 1.35 shows *semilogarithmic (semilog) coordinates* for defining points in a portion of the xy-plane. The independent-variable axis is linear, and the dependent-variable axis is logarithmic. The numerical values that can be depicted on the y

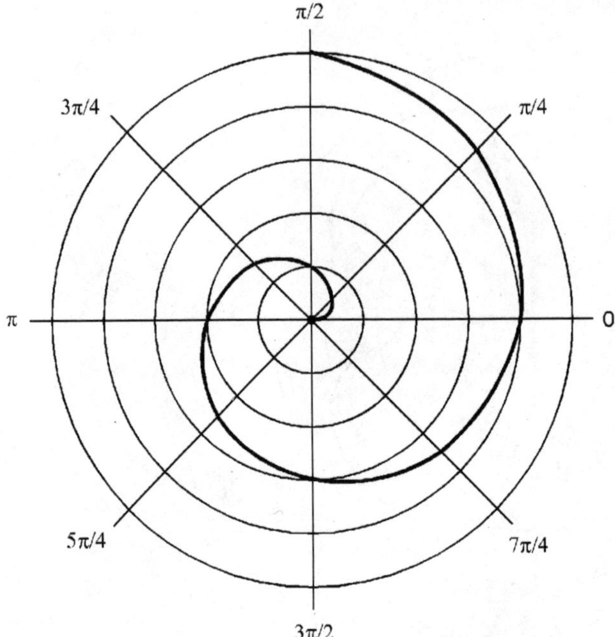

Figure 1.29 Polar graph of spiral centered at the origin.

axis are restricted to one sign or the other, positive or negative. In this example, functions can be plotted with domains and ranges as follows:

$$-1 \leq x \leq 1$$

$$0.1 \leq y \leq 10$$

The y axis in Fig. 1.35 spans two orders of magnitude (powers of 10). The span could be larger or smaller than this, but in any case the y values cannot extend to zero. In the example shown here, only portions of the first and second quadrants of the xy-plane can be depicted. If the y axis were inverted (its values made negative), the resulting plane would cover corresponding parts of the third and fourth quadrants.

Semilog (y-linear) coordinates

Figure 1.36 shows semilog coordinates for defining points in a portion of the xy-plane. The independent-variable axis is logarithmic, and the dependent-variable axis is linear. The numer-

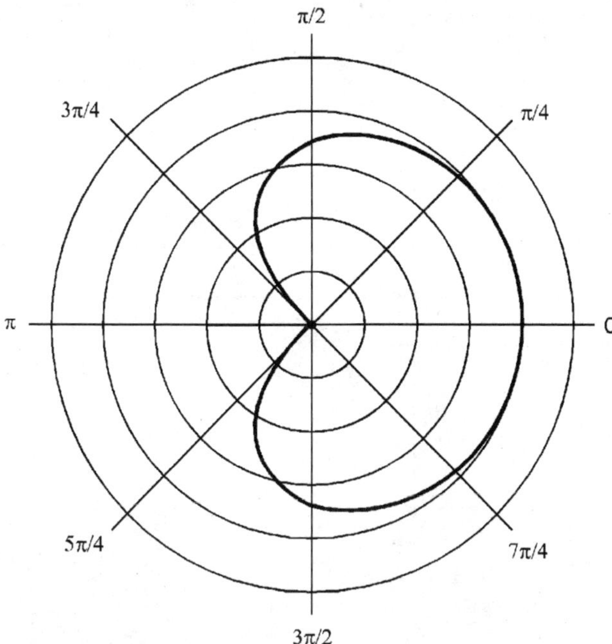

Figure 1.30 Polar graph of cardioid centered at the origin.

ical values that can be depicted on the x axis are restricted to one sign or the other (positive or negative). In this example, functions can be plotted with domains and ranges as follows:

$$0.1 \leq x \leq 10$$

$$-1 \leq y \leq 1$$

The x axis in Fig. 1.36 spans two orders of magnitude (powers of 10). The span could be larger or smaller, but in any case the x values cannot extend to zero. In the example shown here, only portions of the first and fourth quadrants of the xy-plane can be depicted. If the x axis were inverted (its values made negative), the resulting plane would cover corresponding parts of the second and third quadrants.

Log-log coordinates

Figure 1.37 shows *log-log coordinates* for defining points in a portion of the xy-plane. Both axes are logarithmic. The numer-

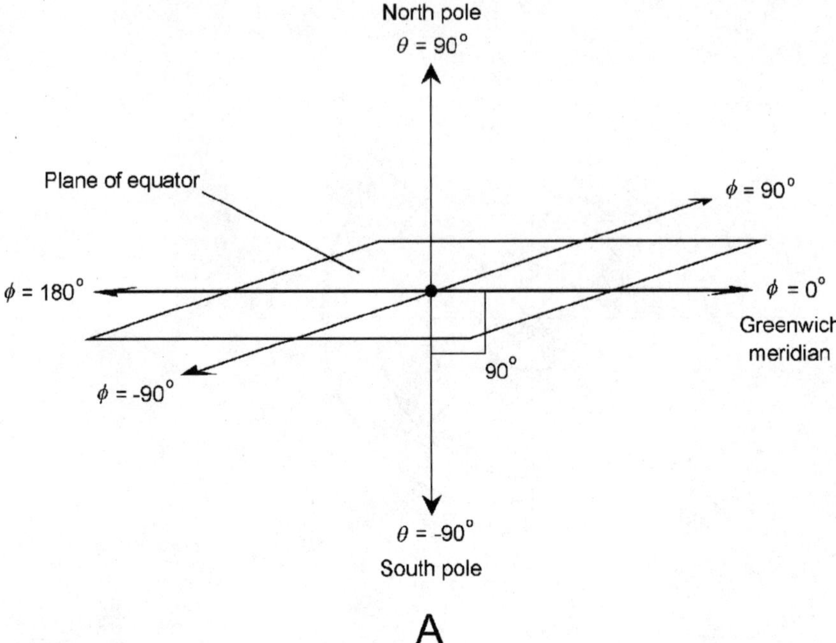

Figure 1.31A Scheme for latitude and longitude.

ical values that can be depicted on either axis are restricted to one sign or the other, positive or negative. In this example, functions can be plotted with domains and ranges as follows:

$$0.1 \leq x \leq 10$$

$$0.1 \leq y \leq 10$$

The axes in Fig. 1.37 span two orders of magnitude (powers of 10). The span of either axis could be larger or smaller, but in any case the values cannot extend to zero. In the example shown here, only a portion of the first quadrant of the *xy*-plane can be depicted. By inverting the signs of one or both axes, corresponding portions of any of the other three quadrants can be covered.

Geometric *xy*-coordinates

Figure 1.38 shows an *xy*-coordinate system on which both scales are graduated geometrically. The points corresponding to 1 on

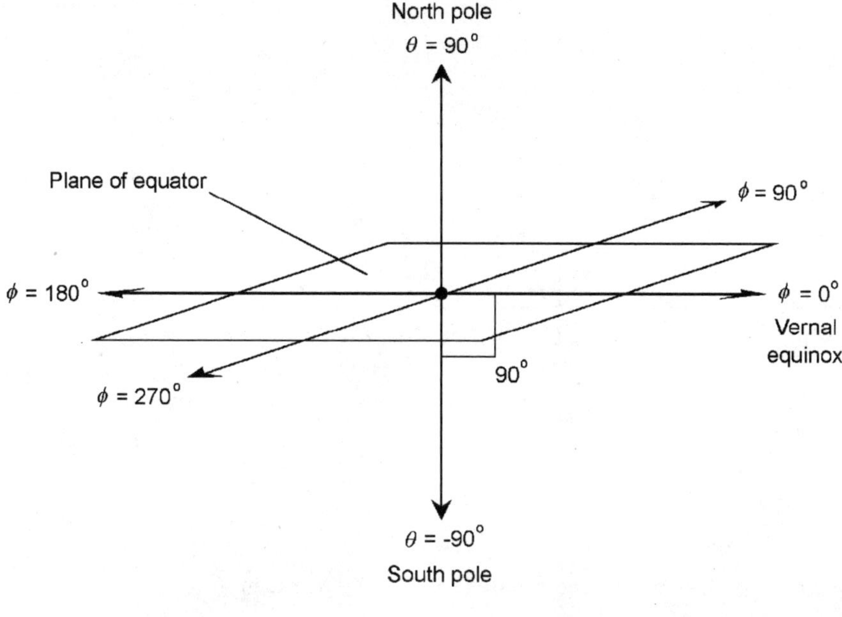

Figure 1.31B Scheme for declination and right ascension.

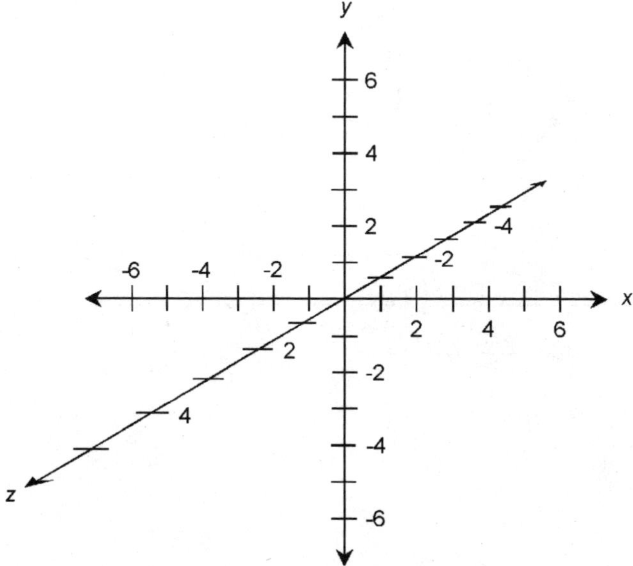

Figure 1.32 Cartesian three-space, also called xyz-space.

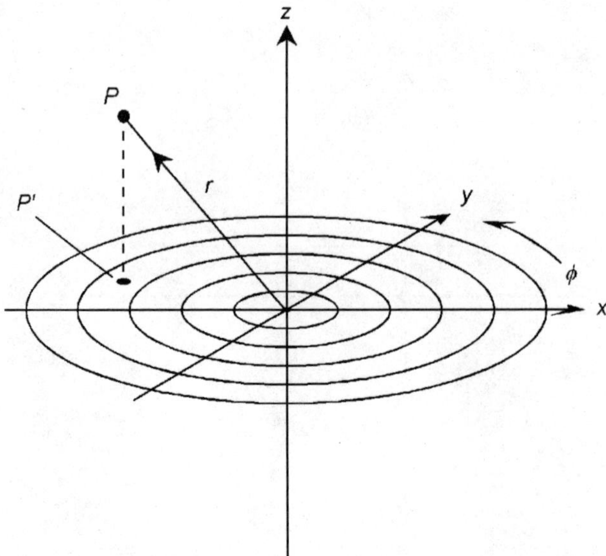

Figure 1.33 Cylindrical coordinates for defining points in three-space.

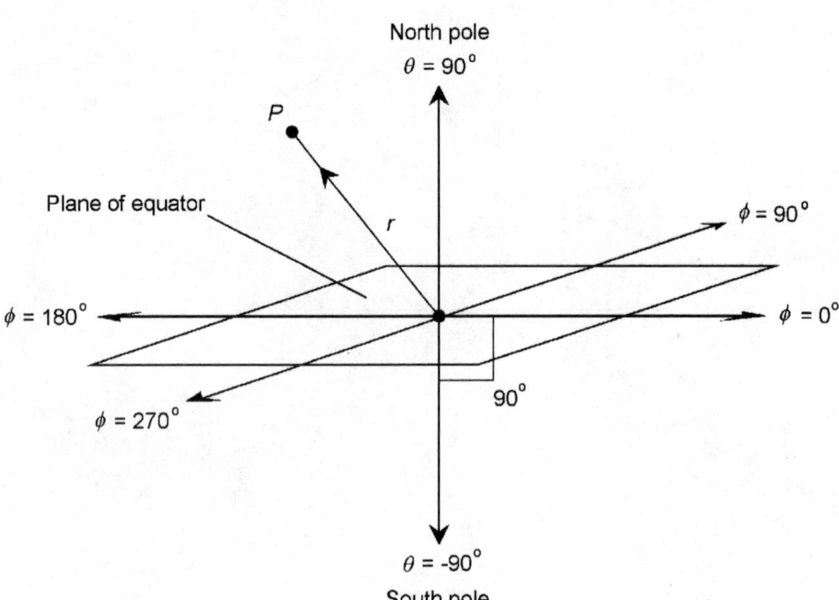

Figure 1.34 Spherical coordinates for defining points in three-space.

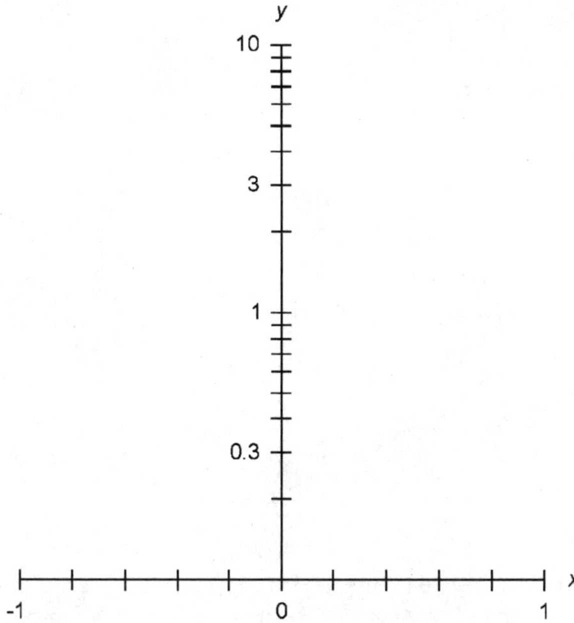

Figure 1.35 Semilog xy-plane with linear x axis and logarithmic y axis.

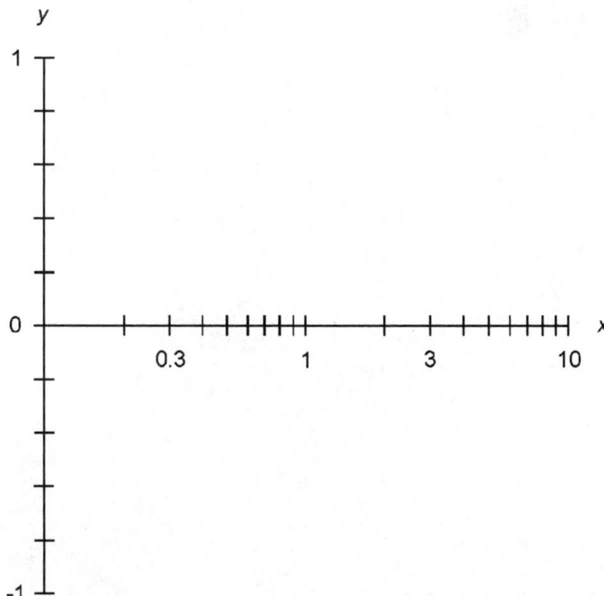

Figure 1.36 Semilog xy-plane with logarithmic x axis and linear y axis.

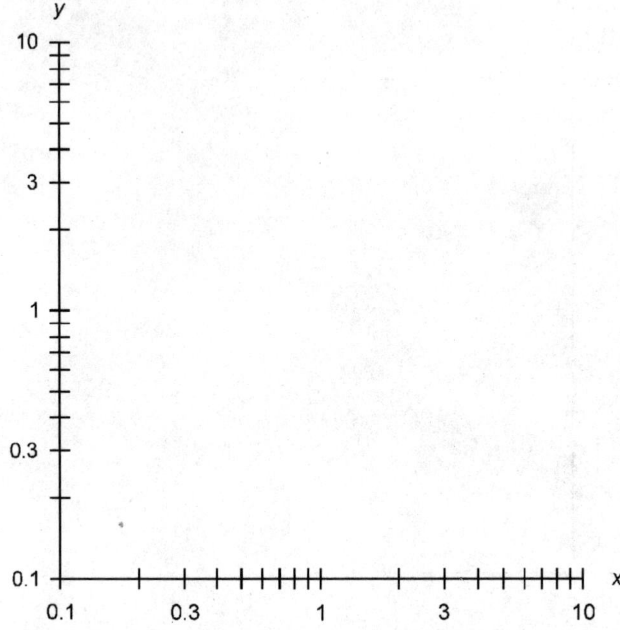

Figure 1.37 Log-log xy-plane.

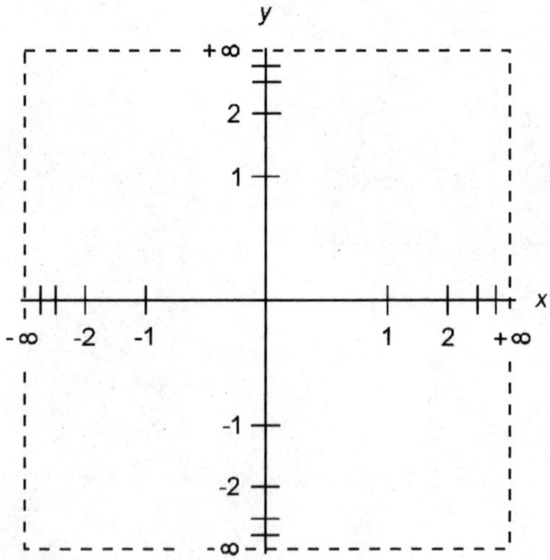

Figure 1.38 Geometric xy-plane.

the axes are halfway between the origin and the positive outer ends, which are labeled $+\infty$. The points corresponding to -1 on the axes are halfway between the origin and the negative outer ends, which are labeled $-\infty$. Succeeding integer points are placed halfway between previous integers (next closest to zero) and the outer ends. The result of this scheme is that the entire coordinate xy-plane is depicted within a finite open square.

The domain and/or range scales of this coordinate system can be expanded or compressed by multiplying all the values on either axis or both axes by a constant. This allows various relations and functions to be plotted, minimizing distortion in particular regions of interest. Distortion relative to the Cartesian (conventional rectangular) xy-plane is greatest near the periphery, and is least near the origin.

Geometric polar plane

Figure 1.39 shows a polar plane on which the radial scale is graduated geometrically. The point corresponding to 1 on the r

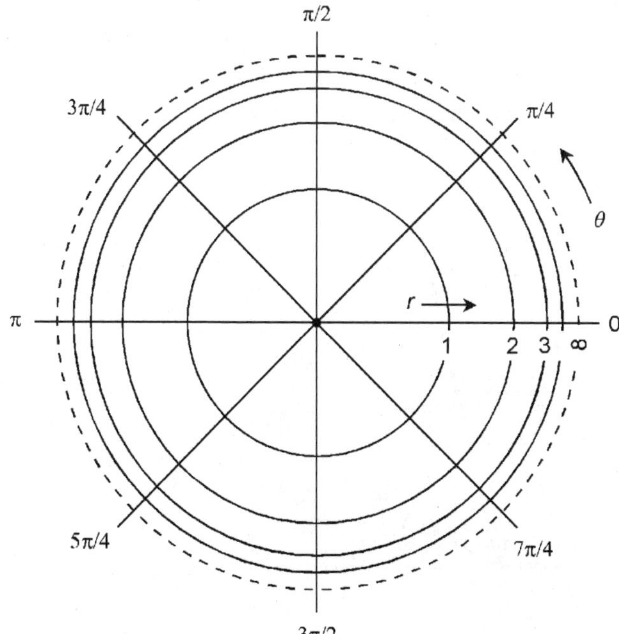

Figure 1.39 Geometric polar plane.

axis is halfway between the origin and the outer periphery, which is labeled ∞. Succeeding integer points are placed halfway between previous integer points and the outer periphery. The result of this scheme is that the entire polar coordinate polar plane is depicted within a finite open circle.

The radial scale of this coordinate system can be expanded or compressed by multiplying all the values on the r axis by a constant. This allows various relations and functions to be plotted, minimizing distortion in particular regions of interest. Distortion relative to the conventional polar coordinate plane is greatest near the periphery, and is least near the origin.

Vector Algebra

A *vector* is a mathematical expression for a quantity exhibiting two independently variable properties: *magnitude* and *direction*.

Vectors in the *xy*-plane

In the *xy*-plane, vectors **a** and **b** can be denoted as rays from the origin (0,0) to points (x_a,y_a) and (x_b,y_b) as shown in Fig. 1.40.

The magnitude of **a**, written |**a**|, is given by:

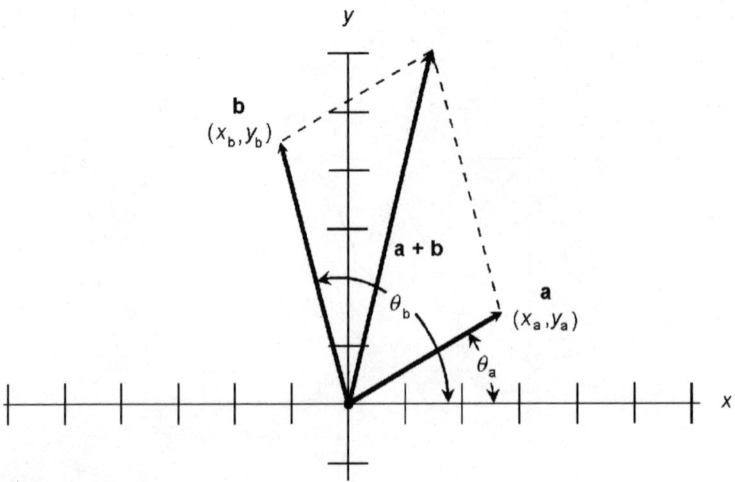

Figure 1.40 Vectors in the *xy*-plane.

$$|\mathbf{a}| = (x_a^2 + y_a^2)^{1/2}$$

The direction of **a**, written dir **a**, is the angle θ_a that a nonzero vector **a** subtends counterclockwise from the positive x axis:

$$\text{dir } \mathbf{a} = \theta_a = \arctan (y_a/x_a) = \tan^{-1} (y_a/x_a)$$

By convention, the following restrictions hold:

$$0 \leq \theta_a < 360 \text{ for } \theta_a \text{ in degrees}$$

$$0 \leq \theta_a < 2\pi \text{ for } \theta_a \text{ in radians}$$

The sum of vectors **a** and **b** is:

$$\mathbf{a} + \mathbf{b} = ((x_a + x_b),(y_a + y_b))$$

This sum can be found geometrically by constructing a parallelogram with **a** and **b** as adjacent sides; then **a + b** is the diagonal of this parallelogram.

The *dot product,* also known as the *scalar product* and written **a • b**, of vectors **a** and **b** is a real number given by the formula:

$$\mathbf{a} \bullet \mathbf{b} = x_a x_b + y_a y_b$$

The *cross product,* also known as the *vector product* and written **a × b**, of vectors **a** and **b** is a vector perpendicular to the plane containing **a** and **b**. Let θ be the angle between vectors **a** and **b**, as measured in the plane containing them both. The magnitude of **a × b** is given by the formula:

$$|\mathbf{a} \times \mathbf{b}| = |\mathbf{a}|\ |\mathbf{b}| \sin \theta$$

If the direction angle θ_b is greater than the direction angle θ_a (as shown in Fig. 1.40), then **a × b** points toward the observer. If $\theta_b < \theta_a$, then **a × b** points away from the observer.

Vectors in the Polar Plane

In the polar coordinate plane, vectors **a** and **b** can be denoted as rays from the origin (0,0) to points (r_a,θ_a) and (r_b,θ_b) as shown in Fig. 1.41.

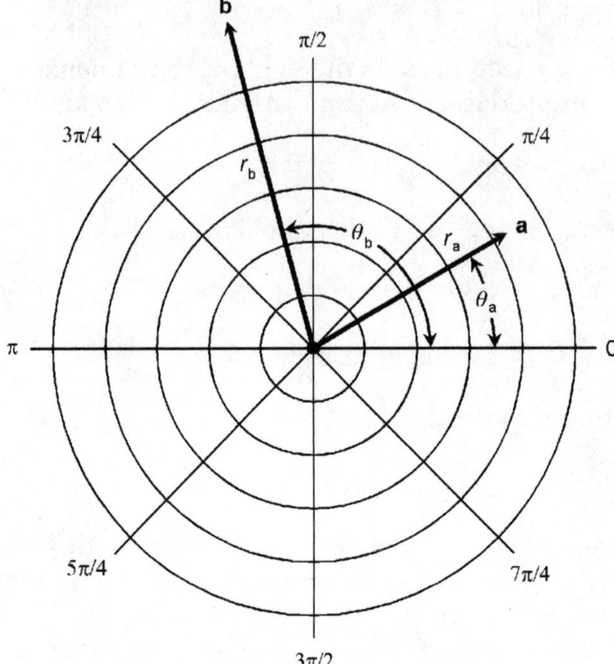

Figure 1.41 Vectors in the polar plane.

The magnitude and direction of vector **a** in the polar coordinate plane are defined directly:

$$|\mathbf{a}| = r_a$$

$$\text{dir } \mathbf{a} = \theta_a$$

By convention, the following restrictions hold:

$$r \geq 0$$

$$0 \leq \theta_a < 360 \text{ for } \theta_a \text{ in degrees}$$

$$0 \leq \theta_a < 2\pi \text{ for } \theta_a \text{ in radians}$$

The sum of **a** and **b** is best found by converting into rectangular (*xy*-plane) coordinates, adding the vectors according to the formula for the *xy*-plane, and then converting the resultant back to polar coordinates. To convert vector **a** from polar to rectangular coordinates, these formulas apply:

$$x_a = r_a \cos \theta_a$$

$$y_a = r_a \sin \theta_a$$

To convert vector **a** from rectangular coordinates to polar coordinates, these formulas apply:

$$r_a = (x_a^2 + y_a^2)^{1/2}$$

$$\theta_a = \arctan (y_a/x_a) = \tan^{-1} (y_a/x_a)$$

Let r_a be the radius of vector **a**, and r_b be the radius of vector **b** in the polar plane. Then the dot product of **a** and **b** is given by:

$$\mathbf{a} \cdot \mathbf{b} = |\mathrm{a}|\ |\mathrm{b}| \cos (\theta_b - \theta_a)$$

$$= r_a r_b \cos (\theta_b - \theta_a)$$

The cross product of **a** and **b** is perpendicular to the polar plane. Its magnitude is given by:

$$|\mathbf{a} \times \mathbf{b}| = |\mathbf{a}|\ |\mathbf{b}| \sin (\theta_b - \theta_a)$$

$$= r_a r_b \sin (\theta_b - \theta_a)$$

If $\theta_b > \theta_a$ (as is the case in Fig. 1.41), then **a** × **b** points toward the observer. If $\theta_b < \theta_a$, then **a** × **b** points away from the observer.

Vectors in *xyz*-space

In rectangular *xyz*-space, vectors **a** and **b** can be denoted as rays from the origin (0,0,0) to points (x_a,y_a,z_a) and (x_b,y_b,z_b) as shown in Fig. 1.42. The magnitude of **a**, written |**a**|, is given by:

$$|\mathbf{a}| = (x_a^2 + y_a^2 + z_a^2)^{1/2}$$

The direction of **a** is denoted by measuring the angles θ_x, θ_y, and θ_z that the vector **a** subtends relative to the positive x, y, and z axes respectively (Fig. 1.43). These angles, expressed in radians as an ordered triple $(\theta_x,\theta_y,\theta_z)$, are the *direction angles* of **a**. Often the cosines of these angles are specified. These are the *direction cosines* of **a**:

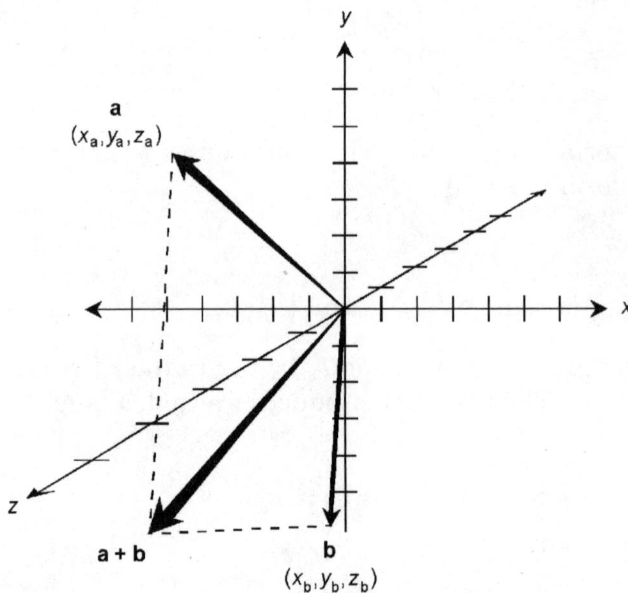

Figure 1.42 Vectors in *xyz*-space. This is a perspective drawing, so the vector-addition parallelogram appears distorted.

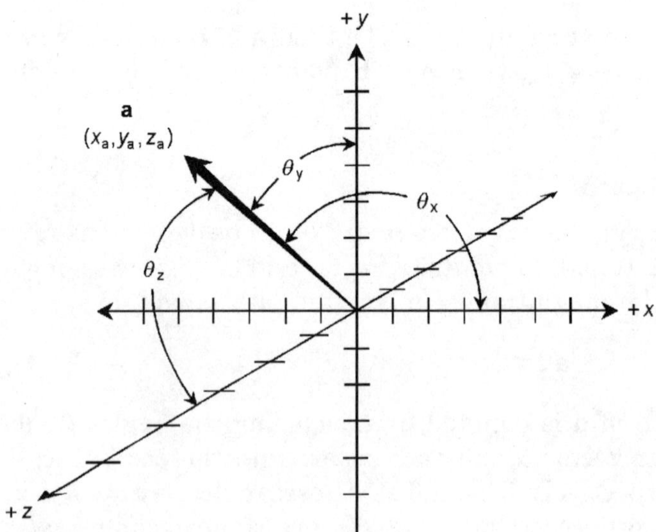

Figure 1.43 Direction angles of a vector in *xyz*-space.

$$\text{dir } \mathbf{a} = (\alpha,\beta,\gamma)$$

$$\alpha = \cos \theta_x$$

$$\beta = \cos \theta_y$$

$$\gamma = \cos \theta_z$$

The sum of vectors **a** and **b** is:

$$\mathbf{a} + \mathbf{b} = ((x_a + x_b),(y_a + y_b),(z_a + z_b))$$

This sum can, as in the two-dimensional case, be found geometrically by constructing a parallelogram with **a** and **b** as adjacent sides. The sum **a** + **b** is the diagonal.

The dot product **a** • **b** of two vectors **a** and **b** in *xyz*-space is a real number given by the formula:

$$\mathbf{a} \bullet \mathbf{b} = x_a x_b + y_a y_b + z_a z_b$$

The cross product **a** × **b** of vectors **a** and **b** in *xyz*-space is a vector perpendicular to the plane *P* containing both **a** and **b**, and whose magnitude is given by the formula:

$$|\mathbf{a} \times \mathbf{b}| = |\mathbf{a}|\ |\mathbf{b}| \sin \theta_{ab},$$

where θ_{ab} is the smaller angle between **a** and **b** as measured in *P*. Vector **a** × **b** is perpendicular to *P*. If **a** and **b** are observed from some point on a line perpendicular to *P* and intersecting *P* at the origin, and θ_{ab} is expressed counterclockwise from **a** to **b**, then **a** × **b** points toward the observer. If **a** and **b** are observed from some point on a line perpendicular to *P* and intersecting *P* at the origin, and θ_{ab} is expressed clockwise from **a** to **b**, then **a** × **b** points away from the observer.

Standard and nonstandard form

In most discussions, vectors are expressed as rays whose origins coincide with the origins of the coordinate systems in which they are denoted. This is the *standard form of a vector*. In standard form, a vector can be depicted as an ordered set of coordinates such as $(x,y,z) = (3,-5,5)$ or $(r,\theta) = (10,\pi/4)$. In rectangular coordinates, the origin of a vector does not have to coincide with the origin of the coordinate system, but it is customary to place it there unless there is a reason to place it elsewhere.

In the xy-plane. In two-dimensional Cartesian coordinates, suppose that the end points of a vector **a**′ are P_1 and P_2, defined as follows:

$$P_1 = (x_1,y_1)$$

$$P_2 = (x_2,y_2)$$

where (x_1,y_1) represents the origin of the vector; that is, the direction of the vector is from P_1 to P_2. Then the standard form of **a**′, denoted **a**, is defined by point P such that:

$$P = (x,y) = ((x_2 - x_1),(y_2 - y_1))$$

The two vectors **a** and **a**′ represent the same quantity. In effect they are identical because they have the same magnitude and the same direction, even though their end points differ (Fig. 1.44).

In the polar plane. Vectors in polar coordinates are always denoted in standard form, that is, with their origins at $(r,\theta) = (0,0)$.

In xyz-space. In three-dimensional Cartesian coordinates, suppose that the end points of a vector **a**′ are P_1 and P_2, defined as follows:

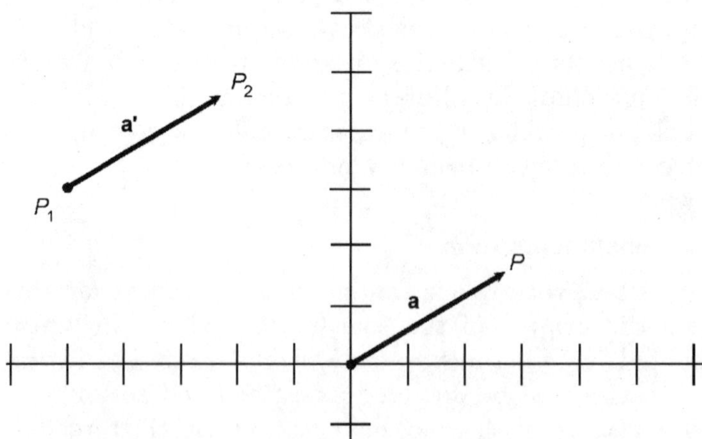

Figure 1.44 Standard and nonstandard forms of a vector in the *xy*-plane.

$$P_1 = (x_1,y_1,z_1)$$

$$P_2 = (x_2,y_2,z_2)$$

where (x_1,y_1,z_1) represents the origin of the vector; that is, the direction of the vector is from P_1 to P_2. Then the standard form of **a**′, call it **a**, is defined by point P such that:

$$P = (x,y,z) = ((x_2 - x_1),(y_2 - y_1),(z_2 - z_1))$$

The two vectors **a** and **a**′ represent the same quantity. In effect they are identical, because they have the same magnitude and the same direction, even though their end points differ (Fig. 1.45).

Equality of vectors

Two vectors **a** and **b** are *equal* (written **a** = **b**) if and only if they have the same magnitude and direction. The end points of the vectors need not coincide, unless both vectors are denoted in standard form.

In the xy-plane. In two-dimensional Cartesian coordinates, suppose the end points of vector **a** are (x_{a1},y_{a1}) and (x_{a2},y_{a2}). Suppose the end points of vector **b** are (x_{b1},y_{b1}) and (x_{b2},y_{b2}). Further

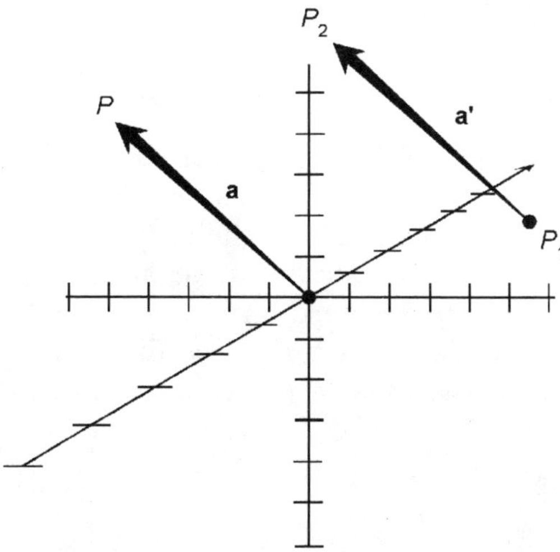

Figure 1.45 Standard and nonstandard forms of a vector in *xyz*-space.

suppose that no two of these points coincide. Then **a** = **b** if and only if the following equations hold:

$$x_{b1} - x_{b2} = x_{a1} - x_{a2}$$

$$y_{b1} - y_{b2} = y_{a1} - y_{a2}$$

In the polar plane. In two-dimensional polar coordinates, suppose the end points of vector **a** are (0,0) and (r_a,θ_a), and the end points of vector **b** are (0,0) and (r_b,θ_b). Then **a** = **b** if and only if the following equations hold:

$$r_a = r_b$$

$$\theta_a = \theta_b$$

In xyz-space. In three-dimensional Cartesian coordinates, suppose the end points of vector **a** are (x_{a1},y_{a1},z_{a1}) and (x_{a2},y_{a2},z_{a2}). Suppose the end points of vector **b** are (x_{b1},y_{b1},z_{b1}) and (x_{b2},y_{b2},z_{b2}). Further suppose that no two of these points coincide. Then **a** = **b** if and only if the following equations hold:

$$x_{b1} - x_{b2} = x_{a1} - x_{a2}$$

$$y_{b1} - y_{b2} = y_{a1} - y_{a2}$$

$$z_{b1} - z_{b2} = z_{a1} - z_{a2}$$

Multiplication of vector by scalar

When any vector is multiplied by a real-number scalar, the vector magnitude (length) is multiplied by that scalar. The direction (angle or angles) remain(s) unchanged if the scalar is positive, but is exactly reversed if the scalar is negative. The following rules apply to vectors in standard form.

In the xy-plane. In two-dimensional Cartesian coordinates, let vector **a** be defined by the coordinates (x,y) as shown in Fig. 1.46. Suppose **a** is multiplied by a positive real scalar k. Then the following equation holds:

$$k\mathbf{a} = k(x,y) = (kx,ky)$$

If **a** is multiplied by a negative real scalar $-k$, then:

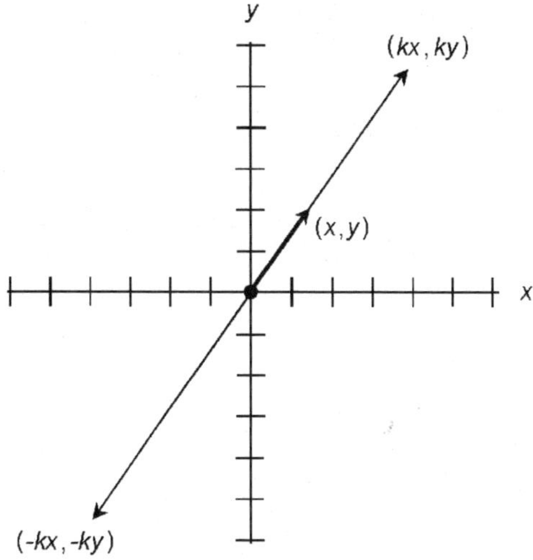

Figure 1.46 Multiplication of a vector by a positive real scalar k, and by a negative real scalar $-k$, in the xy-plane.

$$-k\mathbf{a} = -k(x,y) = (-kx,-ky)$$

In the polar plane. In two-dimensional polar coordinates, let vector **a** be defined by the coordinates (r,θ) as shown in Fig. 1.47. Suppose **a** is multiplied by a positive real scalar k. Then the following equation holds:

$$k\mathbf{a} = (kr,\theta)$$

If **a** is multiplied by a negative real scalar $-k$, then:

$$-k\mathbf{a} = (kr,\ \theta+\pi)$$

The addition of π (180 angular degrees) to θ reverses the direction of **a**. The same effect can be accomplished by adding or subtracting any odd integer multiple of π.

In xyz-space. In three-dimensional Cartesian coordinates, let vector **a** be defined by the coordinates (x,y,z) as shown in Fig. 1.48. Suppose **a** is multiplied by a positive real scalar k. Then the following equation holds:

$$k\mathbf{a} = k(x,y,z) = (kx,ky,kz)$$

If **a** is multiplied by a negative real scalar $-k$, then:

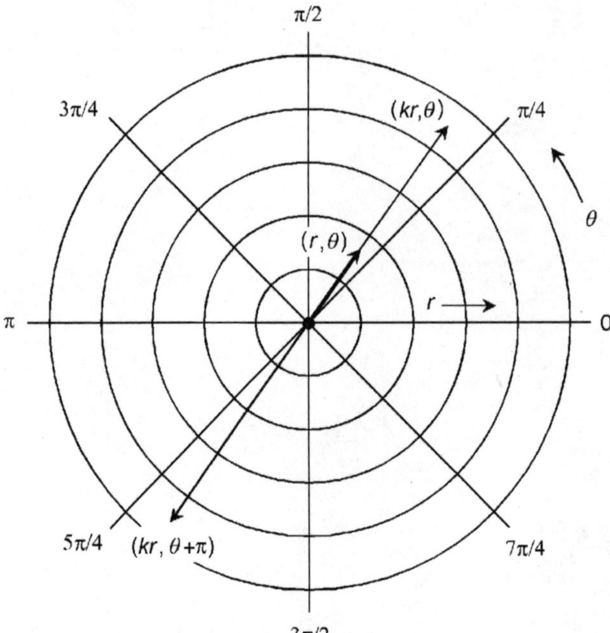

Figure 1.47 Multiplication of a vector by a positive real scalar k, and by a negative real scalar $-k$, in the polar plane.

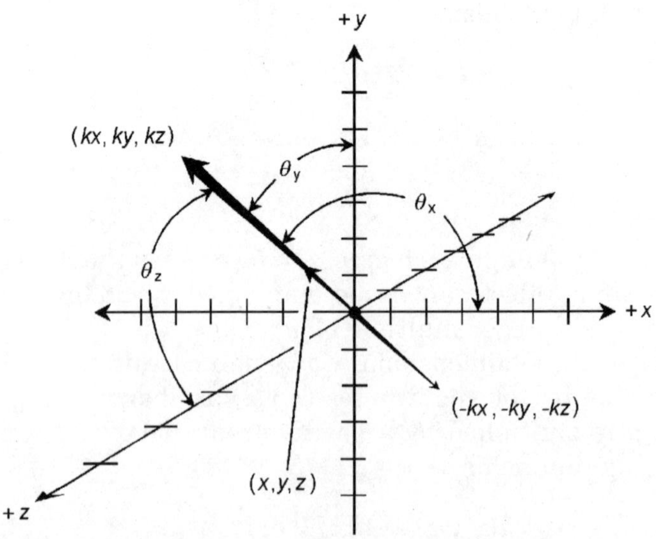

Figure 1.48 Multiplication of a vector by a positive real scalar k, and by a negative real scalar $-k$, in xyz-space.

$$-k\mathbf{a} = -k(x,y,z) = (-kx,-ky,-kz)$$

Suppose the direction angles of **a** are represented by $(\theta_x,\theta_y,\theta_z)$. Then the direction angles of k**a** are also given by $(\theta_x,\theta_y,\theta_z)$. The direction angles of $-k$**a** are all increased by π (180 angular degrees), so they are represented by $((\theta_x + \pi),(\theta_y + \pi),(\theta_z + \pi))$. The same effect can be accomplished by adding or subtracting any odd integer multiple of π to each of these direction angles.

Commutativity of vector addition

When summing any two vectors, it does not matter in which order the sum is performed. If **a** and **b** are vectors, then:

$$\mathbf{a} + \mathbf{b} = \mathbf{b} + \mathbf{a}$$

In the xy-plane. In two-dimensional Cartesian coordinates, let vector **a** be defined by the coordinates (x_a,y_a), and let vector **b** be defined by the coordinates (x_b,y_b). The commutativity of vector addition follows directly from the commutativity of the addition of real numbers:

$$\mathbf{a} + \mathbf{b} = (x_a,y_a) + (x_b,y_b)$$

$$= ((x_a + x_b),(y_a + y_b))$$

and

$$\mathbf{b} + \mathbf{a} = (x_b,y_b) + (x_a,y_a)$$

$$= ((x_b + x_a),(y_b + y_a))$$

$$= ((x_a + x_b),(y_a + y_b))$$

In the polar plane. The sum of **a** and **b** is best found by converting into rectangular (xy-plane) coordinates, adding the vectors according to the formula for the xy-plane, and then converting the resultant back to polar coordinates.

In xyz-space. In three-dimensional Cartesian coordinates, let vector **a** be defined by the coordinates (x_a,y_a,z_a), and let vector **b** be defined by the coordinates (x_b,y_b,z_b). The commutativity of vector addition follows directly from the commutativity of the addition of real numbers:

$$\mathbf{a} + \mathbf{b} = (x_a,y_a,z_a) + (x_b,y_b,z_b)$$

$$= ((x_a + x_b),(y_a + y_b),(z_a + z_b))$$

and

$$\mathbf{b} + \mathbf{a} = (x_b,y_b,z_b) + (x_a,y_a,z_a)$$

$$= ((x_b + x_a),(y_b + y_a),(z_b + z_a))$$

$$= ((x_a + x_b),(y_a + y_b),(z_a + z_b))$$

Commutativity of vector-scalar multiplication

When a vector is multiplied by a scalar, it does not matter in which order the product is performed. If **a** is a vector and k is a real number, then:

$$k\mathbf{a} = \mathbf{a}k$$

In the xy-plane. In two-dimensional Cartesian coordinates, let vector **a** be defined by the coordinates (x,y). Let k be any real number. The commutativity of vector-scalar multiplication follows directly from the commutativity of the multiplication of real numbers:

$$k\mathbf{a} = k(x,y) = (kx,ky)$$

and

$$\mathbf{a}k = (x,y)k = (xk,yk) = (kx,ky)$$

In the polar plane. In two-dimensional polar coordinates, let vector **a** be defined by the coordinates (r,θ). Let k be a real number. If $k > 0$, then:

$$k\mathbf{a} = k(r,\theta) = (kr,\theta)$$

and

$$\mathbf{a}k = (r,\theta)k = (rk,\theta) = (kr,\theta)$$

If $k = 0$, then:

$$k\mathbf{a} = 0(r,\theta) = (0r,\theta) = (0,\theta) = \text{zero vector}$$

and

$$\mathbf{a}k = (r,\theta)0 = (r0,\theta) = (0,\theta) = \text{zero vector}$$

Let $-k$ be a negative real; that is, suppose that $-k < 0$. Then:

$$-k\mathbf{a} = -k(r,\theta) = (kr,\ \theta+\pi)$$

and

$$\mathbf{a}(-k) = (r,\theta)(-k) = (r(-k),\theta + \pi) = (kr,\ \theta+\pi)$$

In xyz-space. In three-dimensional Cartesian coordinates, let vector **a** be defined by the coordinates (x,y,z). Let k be any real number. The commutativity of vector-scalar multiplication follows directly from the commutativity of the multiplication of real numbers:

$$k\mathbf{a} = k(x,y,z) = (kx,ky,kz)$$

and

$$\mathbf{a}k = (x,y,z)k = (xk,yk,zk) = (kx,ky,kz)$$

Commutativity of dot product

When the dot product of two vectors is found, it does not matter in which order the vectors are placed. If **a** and **b** are vectors, then:

$$\mathbf{a} \bullet \mathbf{b} = \mathbf{b} \bullet \mathbf{a}$$

In the xy-plane. In two-dimensional Cartesian coordinates, let vector **a** be defined by the coordinates (x_a,y_a), and let vector **b** be defined by the coordinates (x_b,y_b). The commutativity of the dot product follows directly from the commutativity of the multiplication of real numbers:

$$\mathbf{a} \bullet \mathbf{b} = x_a x_b + y_a y_b$$

and

$$\mathbf{b} \bullet \mathbf{a} = x_b x_a + y_b y_a$$

$$= x_a x_b + y_a y_b$$

In the polar plane. In two-dimensional polar coordinates, let vector **a** be defined by (r_a, θ_a); let vector **b** be defined by (r_b, θ_b). Note that, for any angles θ_a, θ_b, and ϕ, the following equations hold:

$$(\theta_a - \theta_b) = -(\theta_b - \theta_a)$$

$$\cos(-\phi) = \cos \phi$$

The commutativity of the dot product is demonstrated as follows:

$$\mathbf{a} \cdot \mathbf{b} = |a|\,|b| \cos(\theta_b - \theta_a)$$

$$= r_a r_b \cos(\theta_b - \theta_a)$$

and

$$\mathbf{b} \cdot \mathbf{a} = |b|\,|a| \cos(\theta_a - \theta_b)$$

$$= r_b r_a \cos(\theta_a - \theta_b)$$

$$= r_a r_b \cos -(\theta_b - \theta_a)$$

$$= r_a r_b \cos(\theta_b - \theta_a)$$

In xyz-space. In three-dimensional Cartesian coordinates, let vector **a** be defined by the coordinates (x_a, y_a, z_a), and let vector **b** be defined by the coordinates (x_b, y_b, z_b). The commutativity of the dot product follows directly from the commutativity of the multiplication of real numbers:

$$\mathbf{a} \cdot \mathbf{b} = x_a x_b + y_a y_b + z_a z_b$$

and

$$\mathbf{b} \cdot \mathbf{a} = x_b x_a + y_b y_a + z_b z_a$$

$$= x_a x_b + y_a y_b + z_a z_b$$

Negative commutativity of cross product

Let θ be the angle between two vectors **a** and **b** as measured in the plane containing **a** and **b**. Suppose that θ is measured by rotating from **a** to **b**, in the direction such that the angle traversed is less than or equal to π radians. The magnitude of the cross-product vector is independent of the order in which the cross product is performed. This can be derived from the commutativity of the multiplication of scalars:

$$|\mathbf{a} \times \mathbf{b}| = |\mathbf{a}|\ |\mathbf{b}| \sin \theta$$

and

$$|\mathbf{b} \times \mathbf{a}| = |\mathbf{b}|\ |\mathbf{a}| \sin \theta = |\mathbf{a}|\ |\mathbf{b}| \sin \theta$$

The direction of **b** × **a** is opposite that of **a** × **b**. The sense of these vectors, and their relationship with **a** and **b**, is shown in the example of Fig. 1.49. From this, it follows that:

$$\mathbf{b} \times \mathbf{a} = (-1)(\mathbf{a} \times \mathbf{b}) = -(\mathbf{a} \times \mathbf{b})$$

Associativity of vector addition

When summing any three vectors, it makes no difference how the sum is grouped. If **a**, **b**, and **c** are vectors, then:

$$(\mathbf{a} + \mathbf{b}) + \mathbf{c} = \mathbf{a} + (\mathbf{b} + \mathbf{c})$$

In the xy-plane. In two-dimensional Cartesian coordinates, let vector **a** be defined by the coordinates (x_a, y_a), let vector **b** be

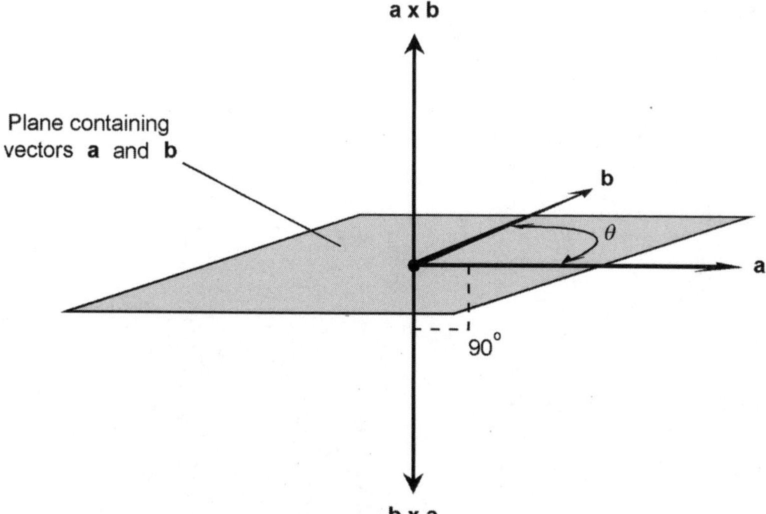

Figure 1.49 The vector **b** × **a** has the same magnitude as vector **a** × **b**, but points in the opposite direction.

defined by the coordinates (x_b, y_b), and let vector **c** be defined by the coordinates (x_c, y_c). The associativity of vector addition follows directly from the associativity of the addition of real numbers:

$$(\mathbf{a} + \mathbf{b}) + \mathbf{c} = ((x_a, y_a) + (x_b, y_b)) + (x_c, y_c)$$

$$= (((x_a + x_b) + x_c), ((y_a + y_b) + y_c))$$

and

$$\mathbf{a} + (\mathbf{b} + \mathbf{c}) = (x_a, y_a) + ((x_b, y_b) + (x_c, y_c))$$

$$= ((x_a + (x_b + x_c)), (y_a + (y_b + y_c)))$$

$$= (((x_a + x_b) + x_c), ((y_a + y_b) + y_c))$$

This situation is shown in Figs. 1.50 and 1.51. Fig. 1.50 is an illustration of summed vectors (**a** + **b**) + **c**; Fig. 1.51 is an illustration of summed vectors **a** + (**b** + **c**).

In the polar plane. The sum of vectors is best found by converting into rectangular (*xy*-plane) coordinates, adding the vec-

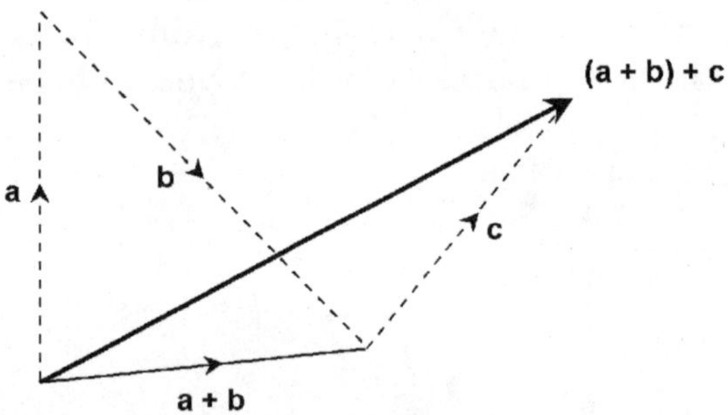

Figure 1.50 Sum of three vectors: (**a** + **b**) + **c**.

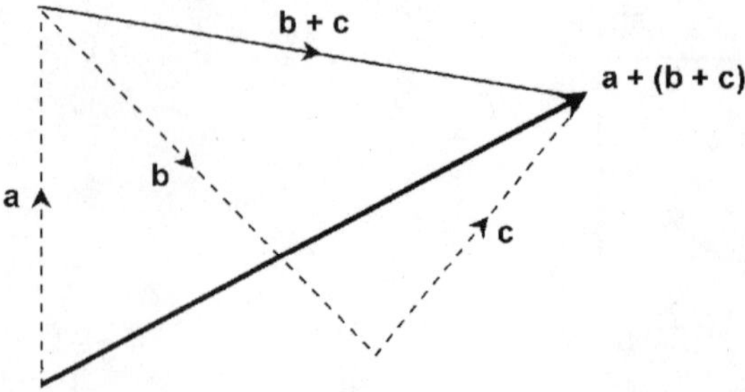

Figure 1.51 Sum of three vectors: **a** + (**b** + **c**).

tors according to the formula for the xy-plane, and then converting the resultant back to polar coordinates.

In xyz-space. In three-dimensional Cartesian coordinates, let vector **a** be defined by the coordinates (x_a,y_a,z_a), let vector **b** be defined by the coordinates (x_b,y_b,z_b), and let vector **c** be defined by the coordinates (x_c,y_c,z_c). The associativity of vector addition follows directly from the associativity of the addition of real numbers:

$$(\mathbf{a} + \mathbf{b}) + \mathbf{c} = ((x_a,y_a,z_a) + (x_b,y_b,z_b)) + (x_c,y_c,z_c)$$

$$= (((x_a + x_b) + x_c),((y_a + y_b) + y_c),((z_a + z_b) + z_c))$$

and

$$\mathbf{a} + (\mathbf{b} + \mathbf{c}) = (x_a,y_a,z_a) + ((x_b,y_b,z_b) + (x_c,y_c,z_c))$$

$$= ((x_a + (x_b + x_c)),(y_a + (y_b + y_c)),(z_a + (z_b + z_c)))$$

$$= (((x_a + x_b) + x_c),((y_a + y_b) + y_c),((z_a + z_b) + z_c))$$

Associativity of vector-scalar multiplication

Let **a** be a vector, and let k_1 and k_2 be real-number scalars. Then the following equation holds:

$$k_1(k_2\mathbf{a}) = (k_1k_2)\mathbf{a}$$

In the xy-plane. In two-dimensional Cartesian coordinates, let vector **a** be defined by the coordinates (x,y). Let k_1 *and* k_2 be real numbers. The associativity of vector-scalar multiplication follows directly from the associativity of the multiplication of real numbers:

$$k_1(k_2\mathbf{a}) = k_1(k_2(x,y)) = k_1(k_2x,k_2y) = (k_1k_2x,k_1k_2y)$$

and

$$(k_1k_2)\mathbf{a} = ((k_1k_2)x,(k_1k_2)y) = (k_1k_2x,k_1k_2y)$$

In the polar plane. The products in this case are best found by converting into rectangular (xy-plane) coordinates, multiplying the vector by scalars according to the formulas in the preceding paragraph, and then converting the resultant back to polar coordinates.

In xyz-space. In three-dimensional Cartesian coordinates, let vector **a** be defined by the coordinates (x,y,z). Let k_1 *and* k_2 be real numbers. The associativity of vector-scalar multiplication follows directly from the associativity of the multiplication of real numbers:

$$k_1(k_2\mathbf{a}) = k_1(k_2(x,y,z)) = k_1(k_2x,k_2y,k_2z) = (k_1k_2x,k_1k_2y,k_1k_2z)$$

and

$$(k_1k_2)\mathbf{a} = ((k_1k_2)x,(k_1k_2)y,(k_1k_2)z) = (k_1k_2x,k_1k_2y,k_1k_2z)$$

Other properties of vector operations

The following theorems apply to vectors and real-number scalars in the *xy*-plane, in the polar plane, or in *xyz*-space.

Distributivity of scalar multiplication over scalar addition. Let **a** be a vector, and let k_1 and k_2 be real-number scalars. Then the following equations hold:

$$(k_1 + k_2)\mathbf{a} = k_1\mathbf{a} + k_2\mathbf{a}$$

$$\mathbf{a}(k_1 + k_2) = \mathbf{a}k_1 + \mathbf{a}k_2 = k_1\mathbf{a} + k_2\mathbf{a}$$

Distributivity of scalar multiplication over vector addition. Let **a** and **b** be vectors, and let k be a real-number scalar. Then the following equations hold:

$$k(\mathbf{a} + \mathbf{b}) = k\mathbf{a} + k\mathbf{b}$$

$$(\mathbf{a} + \mathbf{b})k = \mathbf{a}k + \mathbf{b}k = k\mathbf{a} + k\mathbf{b}$$

Distributivity of dot product over vector addition. Let **a**, **b**, and **c** be vectors. Then the following equations hold:

$$\mathbf{a} \cdot (\mathbf{b} + \mathbf{c}) = \mathbf{a} \cdot \mathbf{b} + \mathbf{a} \cdot \mathbf{c}$$

$$(\mathbf{b} + \mathbf{c}) \cdot \mathbf{a} = \mathbf{b} \cdot \mathbf{a} + \mathbf{c} \cdot \mathbf{a} = \mathbf{a} \cdot \mathbf{b} + \mathbf{a} \cdot \mathbf{c}$$

Distributivity of cross product over vector addition. Let **a**, **b**, and **c** be vectors. Then the following equations hold:

$$\mathbf{a} \times (\mathbf{b} + \mathbf{c}) = \mathbf{a} \times \mathbf{b} + \mathbf{a} \times \mathbf{c}$$

$$\begin{aligned}(\mathbf{b} + \mathbf{c}) \times \mathbf{a} &= \mathbf{b} \times \mathbf{a} + \mathbf{c} \times \mathbf{a} \\ &= -(\mathbf{a} \times \mathbf{b}) - (\mathbf{a} \times \mathbf{c}) \\ &= -(\mathbf{a} \times \mathbf{b} + \mathbf{a} \times \mathbf{c})\end{aligned}$$

Dot product of cross products. Let **a**, **b**, **c**, and **d** be vectors. Then the following equation holds:

$$(\mathbf{a} \times \mathbf{b}) \cdot (\mathbf{c} \times \mathbf{d}) = (\mathbf{a} \cdot \mathbf{c})(\mathbf{b} \cdot \mathbf{d}) - (\mathbf{a} \cdot \mathbf{d})(\mathbf{b} \cdot \mathbf{c})$$

Chapter

2

Geometry, Trigonometry, Logarithms, and Exponential Functions

This chapter outlines basic principles and formulas relevant to Euclidean geometry, circular and hyperbolic trigonometric functions, common and natural logarithmic functions, and exponential functions.

Principles of Geometry

The fundamental rules of geometry are widely used in Newtonian (non-relativistic) physics and engineering.

Two point principle

Let P and Q be two distinct points. Then the following statements hold true, as shown in Fig. 2.1:

- P and Q lie on a common line L
- L is the only line on which both points lie

Three point principle

Let P, Q, and R be three distinct points, not all of which lie on a straight line. Then the following statements hold true:

- P, Q, and R all lie in a common Euclidean plane S
- S is the only Euclidean plane in which all three points lie

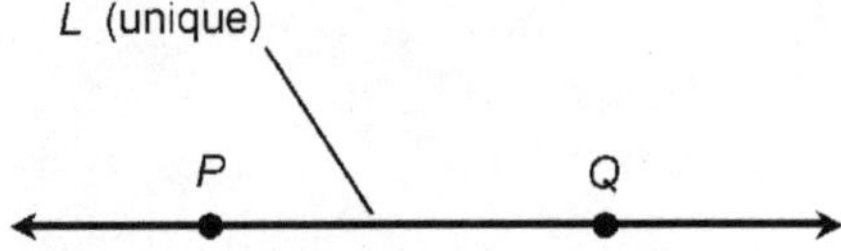

Figure 2.1 Two point principle.

Principle of *n* points

Let P_1, P_2, P_3, . . . , and P_n be n distinct points, not all of which lie in the same Euclidean space of $n - 1$ dimensions. Then the following statements hold true:

- P_1, P_2, P_3, . . . , and P_n all lie in a common Euclidean space U of n dimensions.
- U is the only n-dimensional Euclidean space in which all n points lie.

Distance notation

The distance between any two points P and Q, as measured from P towards Q along the straight line connecting them, is symbolized by writing PQ.

Midpoint principle

Given a line segment connecting two points P and R, there exists one and only one point Q on the line segment, between P and R, such that $PQ = QR$. This is illustrated in Fig. 2.2.

Angle notation

Let P, Q, and R be three distinct points. Let L be the line segment connecting P and Q; let M be the line segment connecting R and Q. Then the smaller of the 2 angles angle θ between L and M, as measured at the vertex point Q in the plane defined by the three points, can be written either as $\angle PQR$ or as $\angle RQP$.

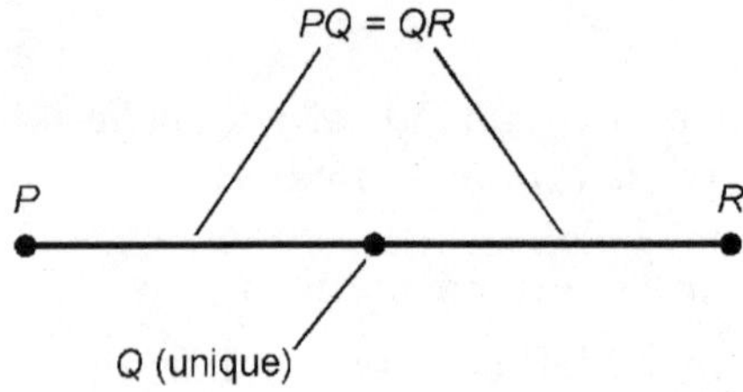

Figure 2.2 Midpoint principle.

If the rotational sense of measurement is specified, then $\angle PQR$ indicates the angle as measured from L to M, and $\angle RQP$ indicates the angle as measured from M to L (Fig. 2.3.) These notations can also stand for the measures of angles in degrees or radians.

Angle bisection

Given an angle $\angle PQR$ measuring less than 180 degrees and defined by three points P, Q, and R, there exists exactly one ray M that bisects $\angle PQR$. If S is any point on M and $S \neq Q$, then the following statement is always true:

$$\angle PQS = \angle SQR$$

That is, ray M is uniquely defined by points Q and S. Every angle has one and only one ray that bisects it. This is illustrated in Fig. 2.4.

Perpendicularity

Let L be a line through points P and Q, and let R be a point not on L. Then there exists one and only one perpendicular line M through point R, intersecting line L at a point S, such that the following holds (Fig. 2.5):

$$\angle PSR = \angle QSR = 90° = \pi/2 \text{ radians}$$

Perpendicular bisector

Let L be a line segment connecting two points P and R. Then there exists one and only one perpendicular line M that inter-

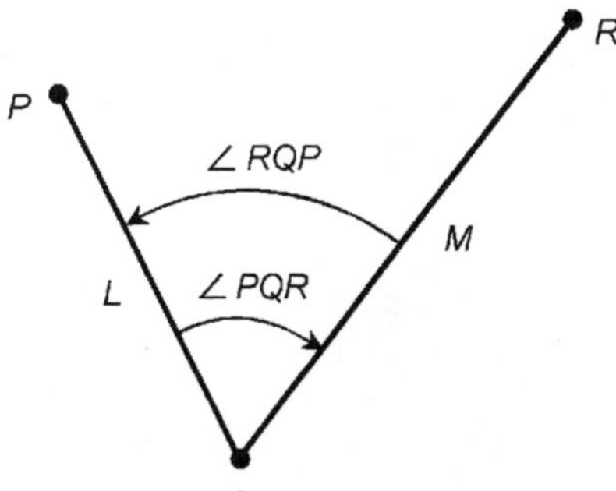

Figure 2.3 Angle notation and measurement.

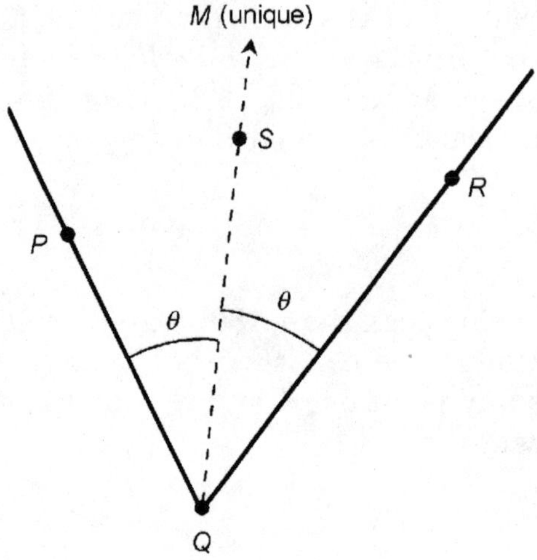

Figure 2.4 Angle bisection principle.

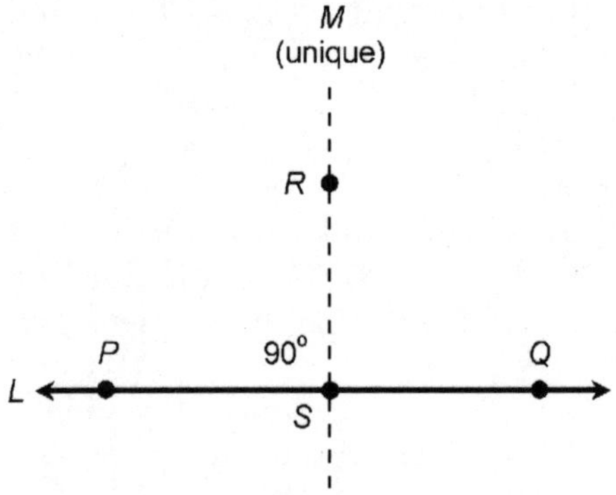

Figure 2.5 Perpendicular principle.

sects line segment L in a point Q, such that the distance from P to Q is equal to the distance from Q to R. That is, every line segment has exactly one perpendicular bisector. This is illustrated in Fig. 2.6.

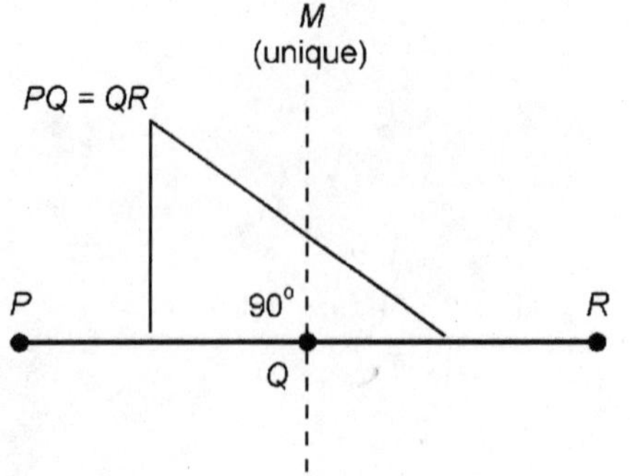

Figure 2.6 Perpendicular bisector principle.

Distance addition and subtraction

Let P, Q, and R be points on a line L, such that Q is between P and R. Then the following equations hold concerning distances as measured along L (Fig. 2.7).

$$PQ + QR = PR$$

$$PR - PQ = QR$$

$$PR - QR = PQ$$

Angle addition and subtraction

Let P, Q, R, and S be four points that lie in a common plane. Let Q be the vertex of three angles $\angle PQR$, $\angle PQS$, and $\angle SQR$ as shown in Fig. 2.8. Then the following equations hold concerning the angular measures:

$$\angle PQS + \angle SQR = \angle PQR$$

$$\angle PQR - \angle PQS = \angle SQR$$

$$\angle PQR - \angle SQR = \angle PQS$$

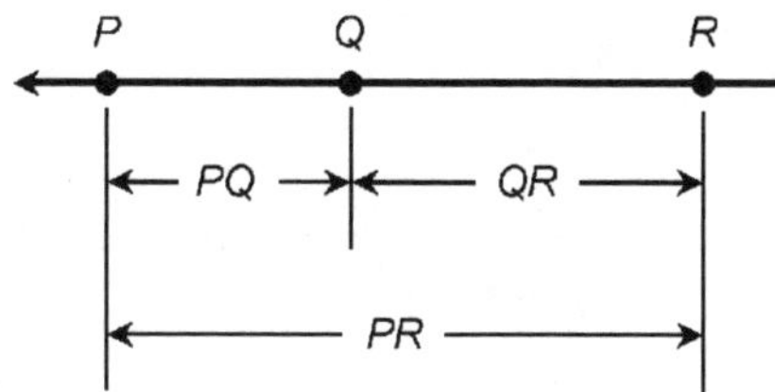

Figure 2.7 Distance addition and subtraction.

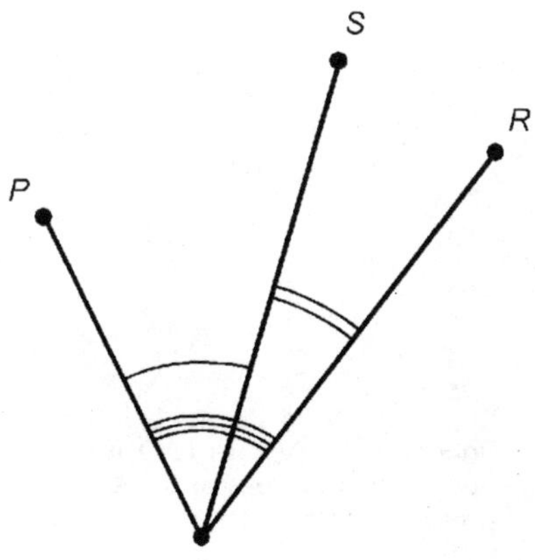

Figure 2.8 Angular addition and subtraction.

Vertical angles

Let L and M be two lines that intersect at a point P. Opposing pairs of angles, defined by the intersection of the lines, are known as *vertical angles* and always have equal measure (Fig. 2.9). In the example shown, $\theta \neq \phi$. Lines L and M are perpendicular if and only if $\theta = \phi$.

Alternate interior angles

Let L and M be parallel lines. Let N be a transversal line that intersects L and M at points P and Q, respectively. In Fig. 2.10, angles labeled θ are *alternate interior angles*; the same holds true for angles labeled ϕ. Alternate interior angles always have equal measure. The transversal line N is perpendicular to lines L and M if and only if $\theta = \phi$.

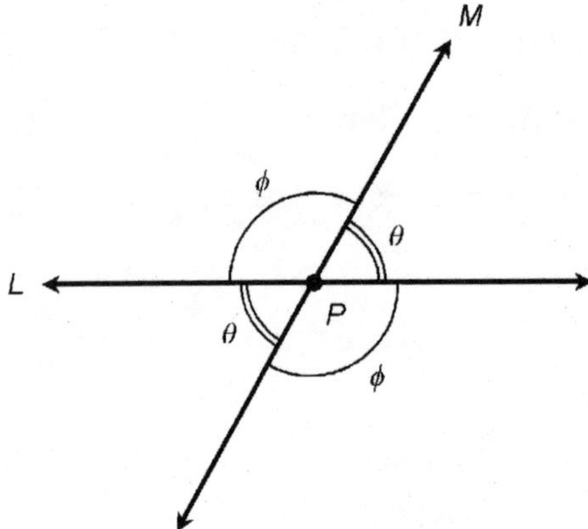

Figure 2.9 Vertical angles have equal measure.

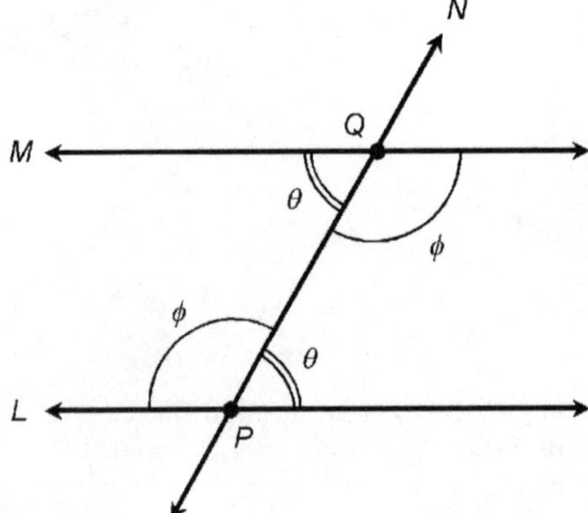

Figure 2.10 Alternate interior angles have equal measure.

Alternate exterior angles

Let L and M be parallel lines. Let N be a transversal line that intersects L and M at points P and Q, respectively. In Fig. 2.11, angles labeled θ are *alternate exterior angles*; the same holds true for angles labeled ϕ. Alternate exterior angles always have equal measure. The transversal line N is perpendicular to lines L and M if and only if $\theta = \phi$.

Corresponding angles

Let L and M be parallel lines. Let N be a transversal line that intersects L and M at points P and Q, respectively. In Fig. 2.12, angles labeled θ_1 are *corresponding angles;* the same holds true for angles labeled θ_2, θ_3, and θ_4. Corresponding angles always have equal measure. The transversal line N is perpendicular to lines L and M if and only if the following equation holds:

$$\theta_1 = \theta_2 = \theta_3 = \theta_4 = 90° = \pi/2 \text{ radians}$$

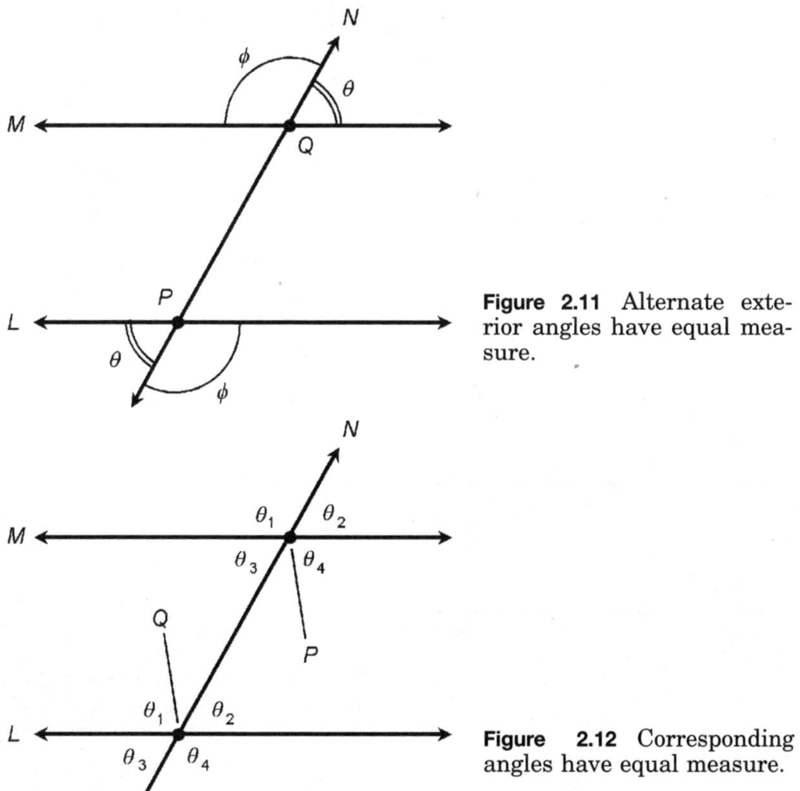

Figure 2.11 Alternate exterior angles have equal measure.

Figure 2.12 Corresponding angles have equal measure.

Parallel principle

Let L be a line; let P be a point not on L. Then there exists one and only one line M through P, such that line M is parallel to line L (Fig. 2.13). This is one of the most important postulates in Euclidean geometry. Its negation can take two forms: either there exists no such line M, or there exists more than one such line M. Either form of the negation of this principle constitutes a cornerstone of non-Euclidean geometry.

Mutual perpendicularity

Let L and M be coplanar lines (lines that lie in a common plane), both of which intersect a third line N, and both of which are perpendicular to line N. Then lines L and M are parallel (Fig. 2.14).

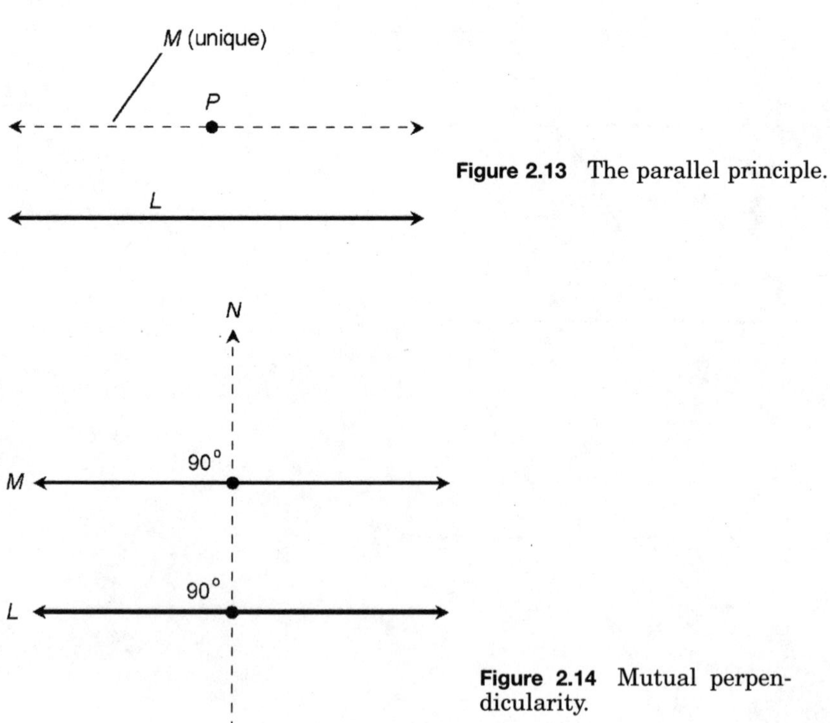

Figure 2.13 The parallel principle.

Figure 2.14 Mutual perpendicularity.

Point-point-point triangle

Let P, Q, and R be three distinct points, not all of which lie on the same straight line. Then the following statements hold true (Fig. 2.15):

- P, Q, and R lie at the vertices of a triangle T.
- T is the only triangle having vertices P, Q, and R.

Side-side-side triangles

Let S_1, S_2, and S_3 be line segments. Let s_1, s_2, and s_3 be the lengths of S_1, S_2, and S_3 respectively. Suppose S_1, S_2, and S_3 are joined at their end points P, Q, and R (Fig. 2.15). Then the following statements hold true:

- Line segments S_1, S_2, and S_3 determine a triangle T.
- T is the only triangle having sides S_1, S_2, and S_3.
- All triangles having sides of lengths s_1, s_2, and s_3 are congruent.

Side-angle-side triangles

Let S_1 and S_2 be distinct line segments. Let P be a point that lies at the ends of both S_1 and S_2. Let s_1 and s_2 be the lengths of S_1 and S_2, respectively. Suppose line segments S_1 and S_2 subtend an angle θ at point P (Fig. 2.16). Then the following statements hold true:

- S_1, S_2, and θ determine a triangle T.
- T is the only triangle having sides S_1 and S_2 subtending an angle θ at point P.

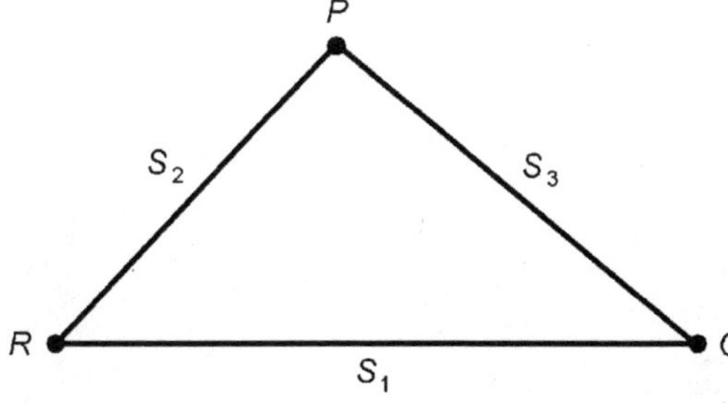

Figure 2.15 The three-point principle; side-side-side triangles.

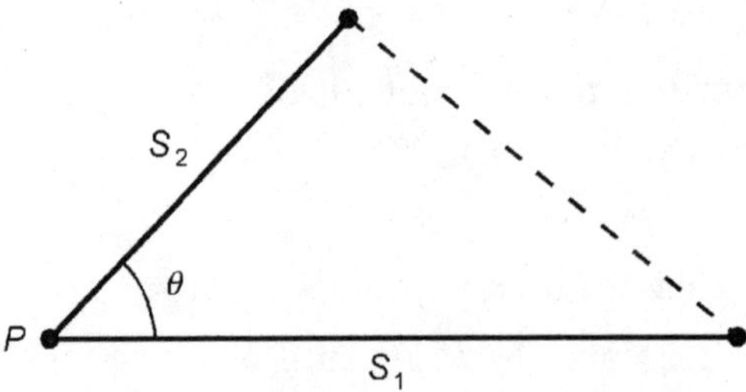

Figure 2.16 Side-angle-side triangles.

- All triangles containing two sides of lengths s_1 and s_2 that subtend an angle θ are congruent.

Angle-side-angle triangles

Let S be a line segment having length s, and whose end points are P and Q. Let θ_1 and θ_2 be angles relative to S subtended by nonparallel lines L_1 and L_2 that run through P and Q, respectively (Fig. 2.17). Then the following statements hold true:

- S, θ_1, and θ_2 determine a triangle T.
- T is the only triangle determined by S, θ_1, and θ_2.
- All triangles containing one side of length s, and whose other two sides subtend angles of θ_1 and θ_2 relative to the side whose length is s, are congruent.

Angle-angle-angle triangles

Let L_1, L_2, and L_3 be lines that lie in a common plane and intersect in three points as illustrated in Fig. 2.18. Let the an-

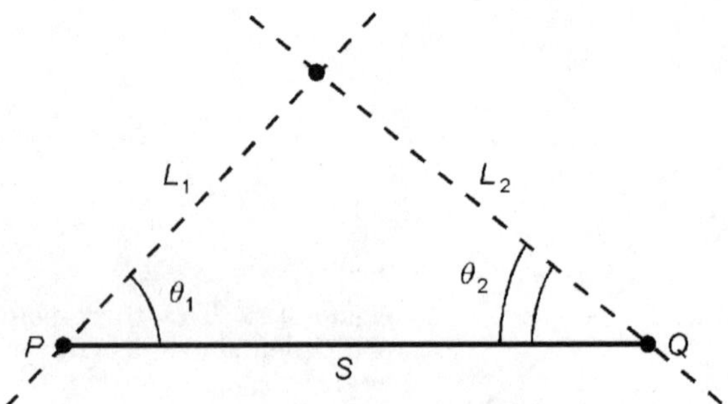

Figure 2.17 Angle-side-angle triangles.

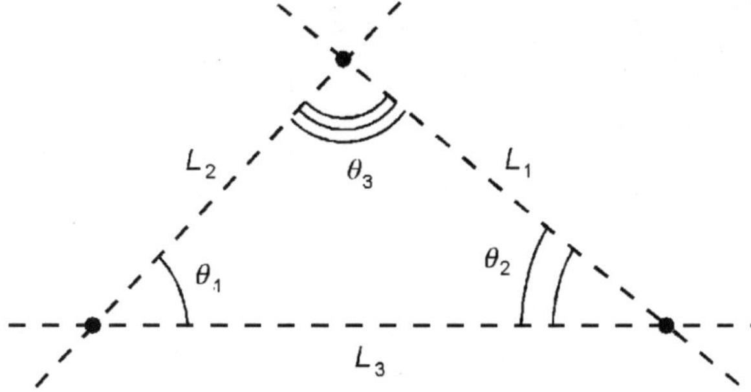

Figure 2.18 Angle-angle-angle triangles.

gles at these points be θ_1, θ_2, and θ_3. Then the following statements hold true:

- There are infinitely many triangles with interior angles θ_1, θ_2, and θ_3 in the sense shown.
- All triangles with interior angles θ_1, θ_2, and θ_3 in the sense shown are similar to each other.

Isosceles triangle

Let T be a triangle with sides S_1, S_2, and S_3 having lengths s_1, s_2, and s_3. Let θ_1, θ_2, and θ_3 be the angles opposite S_1, S_2, and S_3 respectively (Fig. 2.19). Suppose any of the following equations hold:

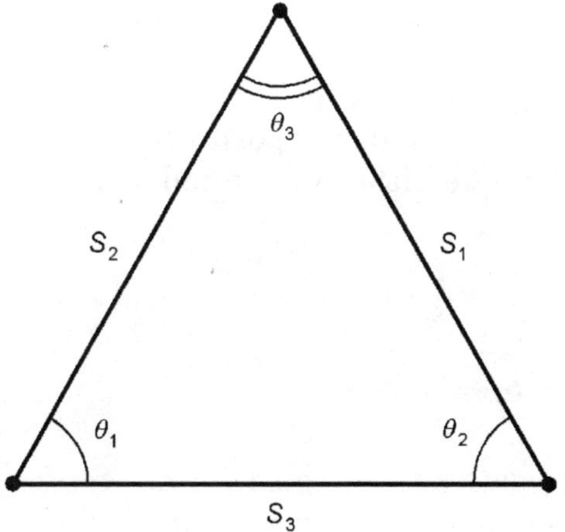

Figure 2.19 Isosceles and equilateral triangles.

$$s_1 = s_2$$

$$s_2 = s_3$$

$$s_1 = s_3$$

$$\theta_1 = \theta_2$$

$$\theta_2 = \theta_3$$

$$\theta_1 = \theta_3$$

Then T is known as an *isosceles triangle,* and the following logical statements are valid:

$$s_1 = s_2 \rightarrow \theta_1 = \theta_2$$

$$s_2 = s_3 \rightarrow \theta_2 = \theta_3$$

$$s_1 = s_3 \rightarrow \theta_1 = \theta_3$$

$$\theta_1 = \theta_2 \rightarrow s_1 = s_2$$

$$\theta_2 = \theta_3 \rightarrow s_2 = s_3$$

$$\theta_1 = \theta_3 \rightarrow s_1 = s_3$$

Equilateral triangle

Let T be a triangle with sides S_1, S_2, and S_3 having lengths s_1, s_2, and s_3. Let θ_1, θ_2, and θ_3 be the angles opposite S_1, S_2, and S_3 respectively (Fig. 2.19). Suppose either of the following are true:

$$s_1 = s_2 = s_3$$

$$\theta_1 = \theta_2 = \theta_3$$

Then T is a special type of isosceles triangle called an *equilateral triangle*, and the following logical statements are valid:

$$s_1 = s_2 = s_3 \rightarrow \theta_1 = \theta_2 = \theta_3$$

$$\theta_1 = \theta_2 = \theta_3 \rightarrow s_1 = s_2 = s_3$$

That is, all equilateral triangles have precisely the same shape; if T and U are any two equilateral triangles, then T and U are similar.

Isosceles triangle bisection

Let T be an isosceles triangle with sides S_1, S_2, and S_3 having lengths s_1, s_2, and s_3. Suppose $s_1 = s_2$, so that $\theta_1 = \theta_2$. Then S_3 is the *base* of the triangle, and the point P opposite S_3, whose angle is θ_3, is the *vertex*. Let L be a line through P that cuts the vertex angle θ_3 into two smaller angles θ_{3a} and θ_{3b}, and that cuts the base S_3 into two shorter line segments S_{3a} and S_{3b} whose lengths are s_{3a} and s_{3b} as shown in Fig. 2.20. Then the following logical statements are valid:

$$s_{3a} = s_{3b} \rightarrow \theta_{3a} = \theta_{3b}$$

$$\theta_{3a} = \theta_{3b} \rightarrow s_{3a} = s_{3b}$$

$$L \perp S_3$$

That is, if L bisects the base, L also bisects the vertex angle; and if L bisects the vertex angle, L also bisects the base. Also, if L bisects either the base or the vertex angle, then L is perpendicular to the base.

Parallelogram diagonals

Let V be a parallelogram defined by four points P, Q, R, and S. Let D_1 be a line segment connecting P and R as shown in Fig. 2.21A. Then D_1 is a *minor diagonal* of the parallelogram, and the triangles defined by D_1 are congruent:

$$\Delta PQR \cong \Delta RSP$$

Let D_2 be a line segment connecting Q and S (Fig. 2.21B). Then D_2 is a *major diagonal* of the parallelogram, and the triangles defined by D_2 are congruent:

$$\Delta QRS \cong \Delta SPQ$$

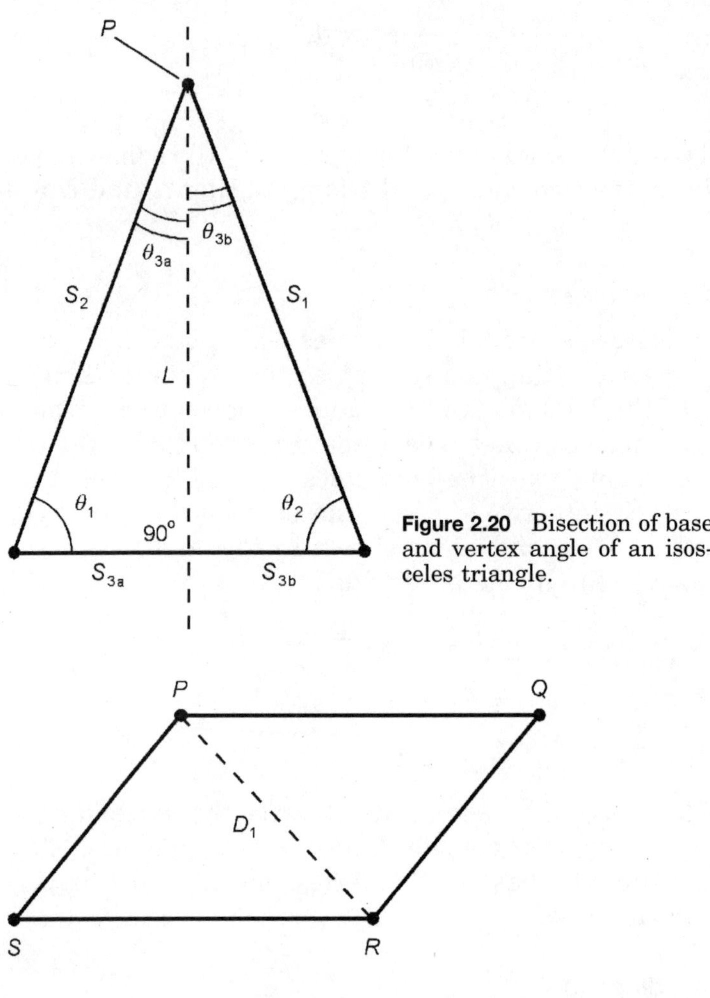

Figure 2.20 Bisection of base and vertex angle of an isosceles triangle.

Figure 2.21a The triangles defined by the minor diagonal of a parallelogram are congruent.

Bisection of parallelogram diagonals

Let V be a parallelogram defined by four points P, Q, R, and S. Let D_1 be the diagonal connecting P and R; let D_2 be the diagonal connecting Q and S (Fig. 2.22). Then D_1 and D_2 bisect each

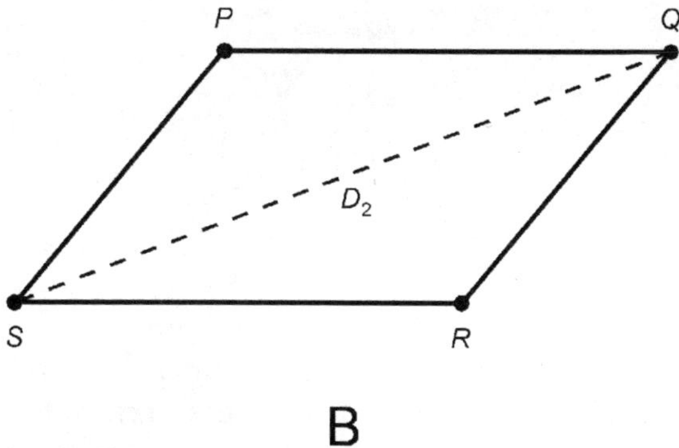

Figure 2.21b The triangles defined by the major diagonal of a parallelogram are congruent.

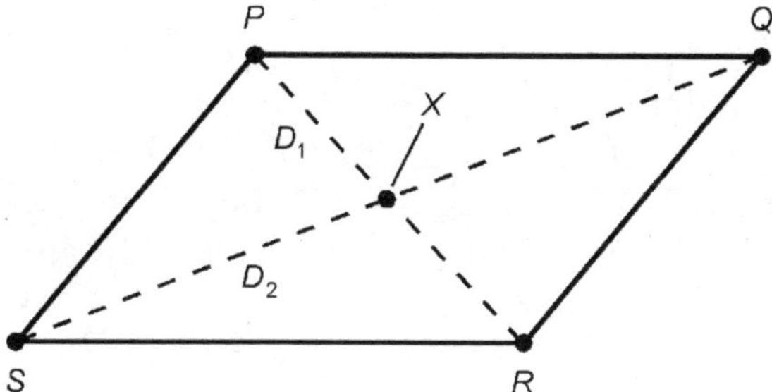

Figure 2.22 The diagonals of a parallelogram bisect each other.

other at their intersection point X. In addition, the following pairs of triangles are congruent:

$$\Delta PQX \cong \Delta RSX$$

$$\Delta QRX \cong \Delta SPX$$

The converse of the foregoing is also true: if V is a convex plane quadrilateral whose diagonals bisect each other, then V is a parallelogram.

Rectangle

Let V be a parallelogram defined by four points P, Q, R, and S. Suppose any of the following statements is true for angles in degrees:

$$\angle PQR = 90° = \pi/2 \text{ radians}$$

$$\angle QRS = 90° = \pi/2 \text{ radians}$$

$$\angle RSP = 90° = \pi/2 \text{ radians}$$

$$\angle SPQ = 90° = \pi/2 \text{ radians}$$

Then all four interior angles measure 90 degrees, and V is a *rectangle:* a four-sided plane polygon whose interior angles are all congruent (Fig. 2.23). The converse of this is also true: if V is a rectangle, then any given interior angle has a measure of 90 degrees.

Rectangle diagonals

Let V be a parallelogram defined by four points P, Q, R, and S. Let D_1 be the diagonal connecting P and R; let D_2 be the diagonal connecting Q and S. Let the length of D_1 be d_1; let the length of D_2 be d_2 (Fig. 2.24). If $d_1 = d_2$, then V is a rectangle. The converse is also true: if V is a rectangle, then $d_1 = d_2$. That is, a parallelogram is a rectangle if and only if its diagonals have equal lengths.

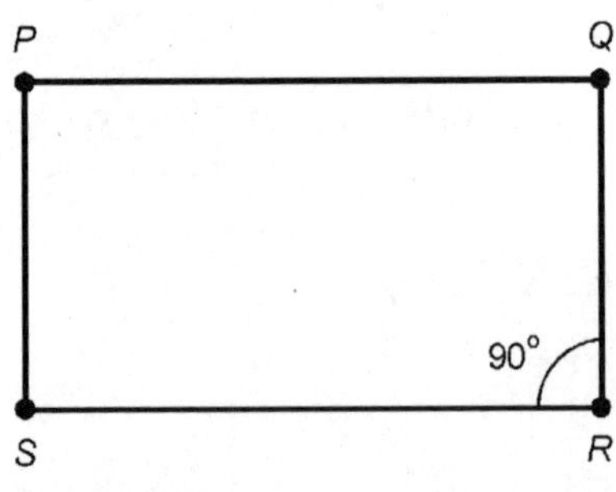

Figure 2.23 If a parallelogram has one right interior angle, then the parallelogram is a rectangle.

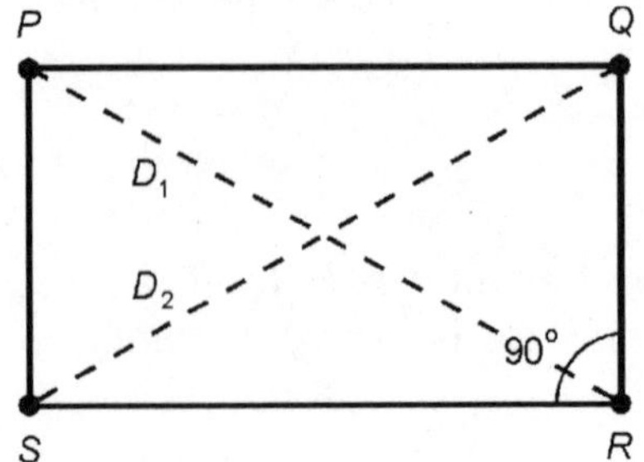

Figure 2.24 The diagonals of a rectangle have equal length.

Rhombus diagonals

Let V be a parallelogram defined by four points P, Q, R, and S. Let D_1 be the diagonal connecting P and R; let D_2 be the diagonal connecting Q and S. If $D_1 \perp D_2$, then V is a *rhombus*: a four-sided plane polygon whose sides are all congruent (Fig. 2.25). The converse is also true: if V is a rhombus, then $D_1 \perp D_2$. That is, a parallelogram is a rhombus if and only if its diagonals are perpendicular.

Trapezoid within triangle

Let T be a triangle defined by three points P, Q, and R. Let X be the midpoint of side PR, and let Y be the midpoint of side PQ. Then line segments XY and RQ are parallel, and the figure defined by $XYQR$ is a *trapezoid*: a four-sided plane polygon with one pair of parallel sides (Fig. 2.26). In addition, the length of line segment XY is half the length of line segment RQ.

Median of trapezoid

Let V be a trapezoid defined by four points P, Q, R, and S. Let X be the midpoint of side PS, and let Y be the midpoint of side

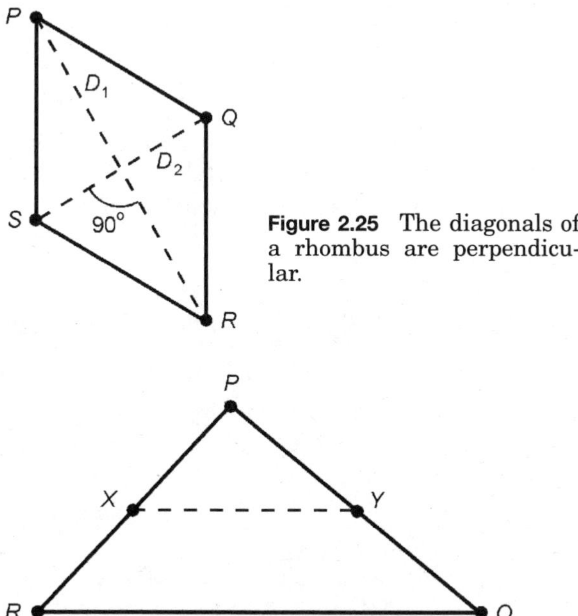

Figure 2.25 The diagonals of a rhombus are perpendicular.

Figure 2.26 A trapezoid within a triangle.

QR. Line segment XY is called the *median* of trapezoid $PQRS$. Let W_1 be the polygon defined by P, Q, Y, and X; let W_2 be the polygon defined by X, Y, R, and S. Then W_1 and W_2 are trapezoids (Fig. 2.27). In addition, the length of line segment XY is half the sum of the lengths of line segments PQ and SR:

$$XY = (PQ + SR)/2$$

Formulas for Plane Figures

The following formulas apply to common geometric figures in Euclidean two-space (a "flat" plane).

Sum of interior angles of triangle

Let T be a triangle, and let the interior angles be θ_1, θ_2, and θ_3. Then the following equation holds if the angular measures are given in degrees:

$$\theta_1 + \theta_2 + \theta_3 = 180$$

If the angular measures are given in radians, then the following holds:

$$\theta_1 + \theta_2 + \theta_3 = \pi$$

Theorem of Pythagoras

Let T be a right triangle whose sides are S_1, S_2, and S_3 having lengths s_1, s_2, and s_3 respectively. Let S_3 be the side opposite the right angle (Fig. 2.28). Then the following equation holds:

$$s_1^2 + s_2^2 = s_3^2$$

The converse of this is also true: If T is a triangle whose sides

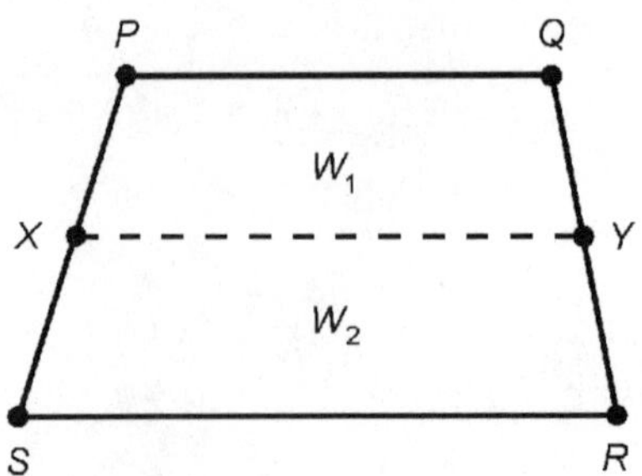

Figure 2.27 The median of a trapezoid.

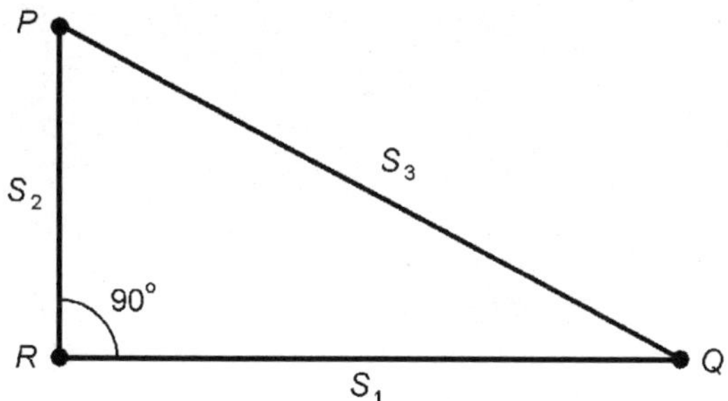

Figure 2.28 The Theorem of Pythagoras.

have lengths s_1, s_2, and s_3, and the above formula holds, then T is a right triangle.

Sum of interior angles of plane quadrilateral

Let V be a plane quadrilateral, and let the interior angles be θ_1, θ_2, θ_3, and θ_4 (Fig. 2.29). Then the following equation holds if the angular measures are given in degrees:

$$\theta_1 + \theta_2 + \theta_3 + \theta_4 = 360$$

If the angular measures are given in radians, then the following holds:

$$\theta_1 + \theta_2 + \theta_3 + \theta_4 = 2\pi$$

Sum of interior angles of plane polygon

Let V be a plane polygon having n sides. Let the interior angles be θ_1, θ_2, θ_3, . . . , θ_n (Fig. 2.30). Then the following equation holds if the angular measures are given in degrees:

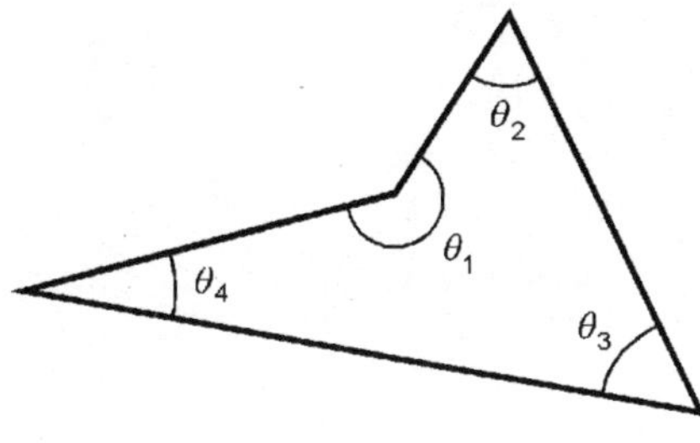

Figure 2.29 Interior angles of a plane quadrilateral.

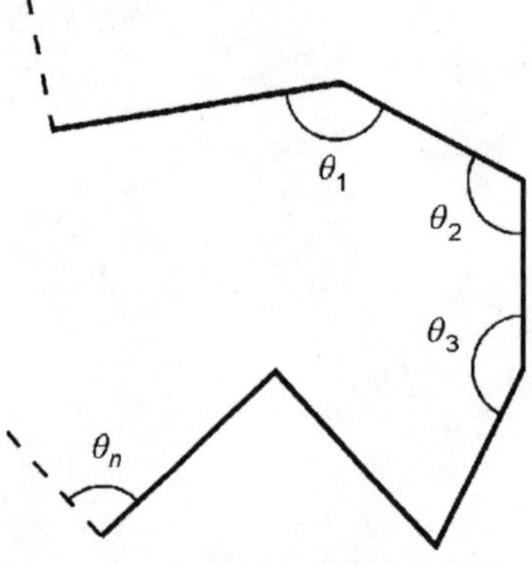

Figure 2.30 Interior angles of a plane polygon.

$$\theta_1 + \theta_2 + \theta_3 + \ldots + \theta_n = 180n - 360 = 180\,(n - 2)$$

If the angular measures are given in radians, then the following holds:

$$\theta_1 + \theta_2 + \theta_3 + \ldots + \theta_n = \pi n - 2\pi = \pi\,(n - 2)$$

Individual interior angles of regular plane polygon

Let V be a plane polygon having n sides whose interior angles all have equal measure given by s, and whose sides all have equal length (Fig. 2.31). Then V is a *regular polygon*, and the measure of each interior angle, θ, in degrees is given by the following formula:

$$\theta = (180n - 360)/n$$

If the angular measures are given in radians, then the following holds:

$$\theta = (\pi n - 2\pi)/n$$

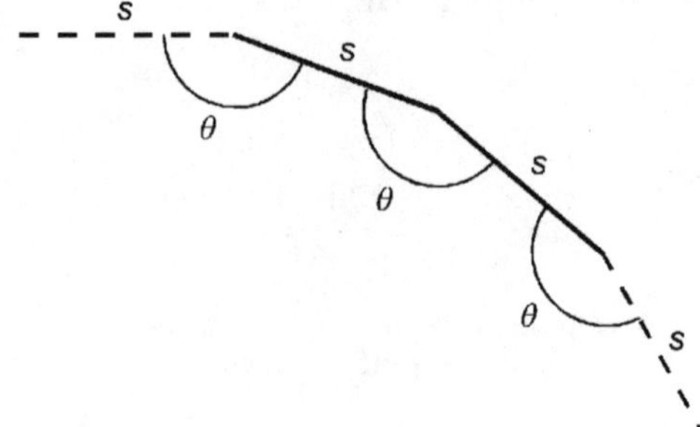

Figure 2.31 Interior angles of a regular plane polygon.

Positive and negative exterior angles

An *exterior angle* of a polygon is measured counterclockwise between a specific side and the extension of a side next to it (Fig. 2.32). If the arc lies outside the polygon, then the resulting angle θ has a measure between, but not including, 0 and +180 degrees:

$$0 < \theta < +180$$

If the arc lies inside the polygon, then the angle is measured clockwise ("negatively counterclockwise"). This results in an angle ϕ with a measure between, but not including, −180 and 0 degrees:

$$-180 < \phi < 0$$

Exterior angles of plane polygon

Let V be a plane polygon having n sides. Let the exterior angles be $\theta_1, \theta_2, \theta_3, \ldots, \theta_n$. Then the following equation holds if the angular measures are given in degrees:

$$\theta_1 + \theta_2 + \theta_3 + \ldots + \theta_n = 360$$

If the angular measures are given in radians, then the following holds:

$$\theta_1 + \theta_2 + \theta_3 + \ldots + \theta_n = 2\pi$$

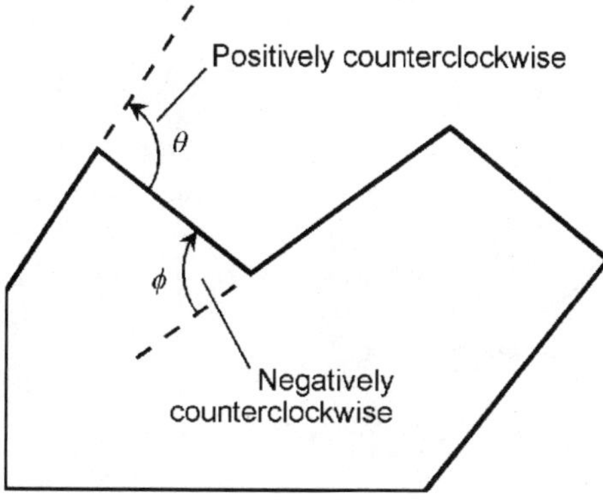

Figure 2.32 In this polygon, θ is a positive exterior angle, as are all the others except ϕ, which is a negative exterior angle.

Perimeter of triangle

Let T be a triangle defined by points P, Q, and R, and having sides of lengths s_1, s_2, and s_3 as shown in Fig. 2.33. Let s_1 be the base length, h be the height, and θ be the angle between the sides having length s_1 and s_2. Then the perimeter, B, of the triangle is given by the following formula:

$$B = s_1 + s_2 + s_3$$

Interior area of triangle

Let T be a triangle as defined above and in Fig. 2.33. The interior area, A, is given by either of the following formulas:

$$A = s_1 h \;/\; 2$$

$$A = (s_1 s_2 \sin\,\theta)/2$$

Perimeter of parallelogram

Let V be a parallelogram defined by points P, Q, R, and S, and having sides of lengths s_1 and s_2 as shown in Fig. 2.34. Let s_1 be the base length, h be the height, and θ be the angle between the sides having lengths s_1 and s_2. Then the perimeter, B, of the parallelogram is given by the following formula:

$$B = 2s_1 + 2s_2$$

Interior area of parallelogram

Let V be a parallelogram as defined above and in Fig. 2.34. The interior area, A, is given by either of the following formulas:

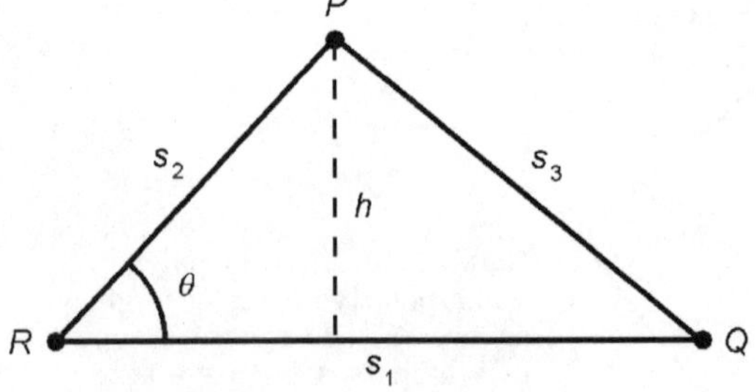

Figure 2.33 Perimeter and area of triangle.

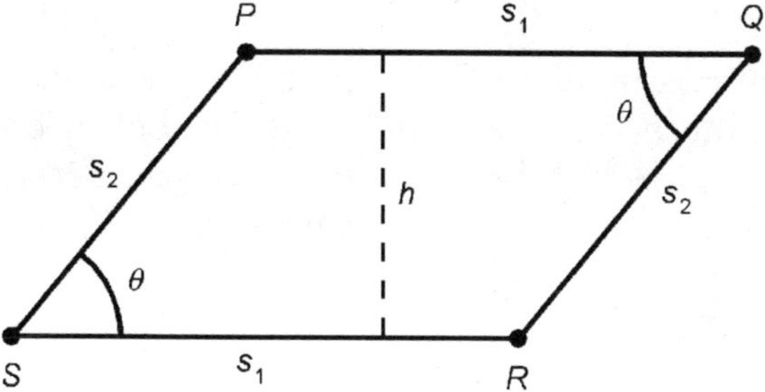

Figure 2.34 Perimeter and area of parallelogram. If $s_1 = s_2$, the figure is a rhombus.

$$A = s_1 h$$

$$A = s_1 s_2 \sin \theta$$

Perimeter of rhombus

Let V be a rhombus defined by points P, Q, R, and S, and having sides all of which have length s. Let the lengths of the diagonals of the rhombus be d_1 and d_2. The rhombus is a special case of the parallelogram (Fig. 2.34) in which the following holds:

$$s_1 = s_2 = s$$

The perimeter, B, of the rhombus is given by the following formula:

$$B = 4s$$

Interior area of rhombus

Let V be a rhombus as defined above. The interior area, A, of the rhombus is given by any of the following formulas:

$$A = sh$$

$$A = s^2 \sin \theta$$

$$A = d_1 d_2 / 2$$

Perimeter of rectangle

Let V be a rectangle defined by points P, Q, R, and S, and having sides of lengths s_1 and s_2 as shown in Fig. 2.35. Let s_1 be the base length, and let s_2 be the height. Then the perimeter, B, of the rectangle is given by the following formula:

$$B = 2s_1 + 2s_2$$

Interior area of rectangle

Let V be a rectangle as defined above and in Fig. 2.35. The interior area, A, is given by:

$$A = s_1 s_2$$

Perimeter of square

Let V be a square defined by points P, Q, R, and S, and having sides all of which have length s. This is a special case of the rectangle (Fig. 2.35) in which the following holds:

$$s_1 = s_2 = s$$

The perimeter, B, of the square is given by the following formula:

$$B = 4s$$

Interior area of square

Let V be a square as defined above. The interior area, A, is given by:

$$A = s^2$$

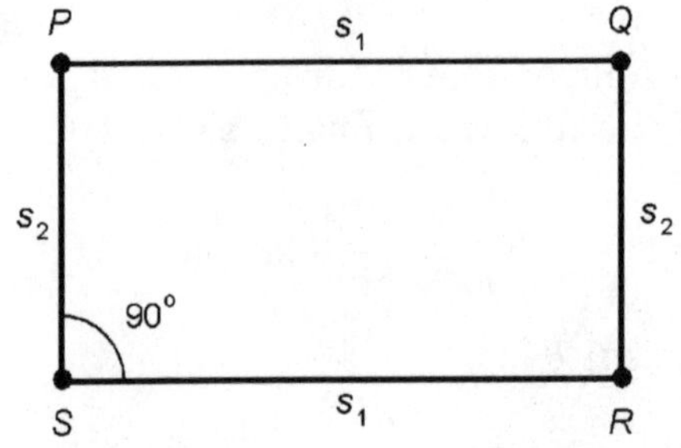

Figure 2.35 Perimeter and area of rectangle. If $s_1 = s_2$, the figure is a square.

Perimeter of trapezoid

Let V be a trapezoid defined by points P, Q, R, and S, and having sides of lengths s_1, s_2, s_3, and s_4 as shown in Fig. 2.36. Let s_1 be the base length, h be the height, θ be the angle between the sides having length s_1 and s_2, and ϕ be the angle between the sides having length s_1 and s_4. Let sides having lengths s_1 and s_3 (line segments PQ and RS) be parallel. Then the perimeter, B, of the trapezoid is given by either of the following formulas:

$$B = s_1 + s_2 + s_3 + s_4$$

$$B = s_1 + s_3 + h \csc \theta + h \csc \phi$$

Interior area of trapezoid

Let V be a trapezoid as defined above and in Fig. 2.36. The interior area, A, is given by:

$$A = (s_1 h + s_3 h)/2$$

Perimeter of regular polygon

Let V be a regular plane polygon having n sides of length s, and whose vertices are P_1, P_2, P_3, . . . , P_n as shown in Fig. 2.37. Then the perimeter, B, of the polygon is given by the following formula:

$$B = ns$$

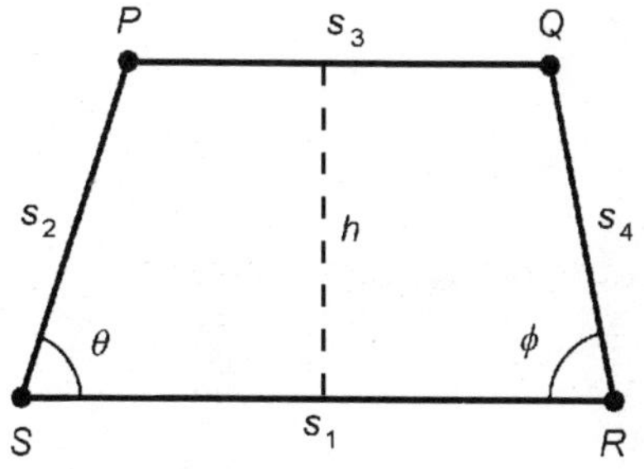

Figure 2.36 Perimeter and area of trapezoid.

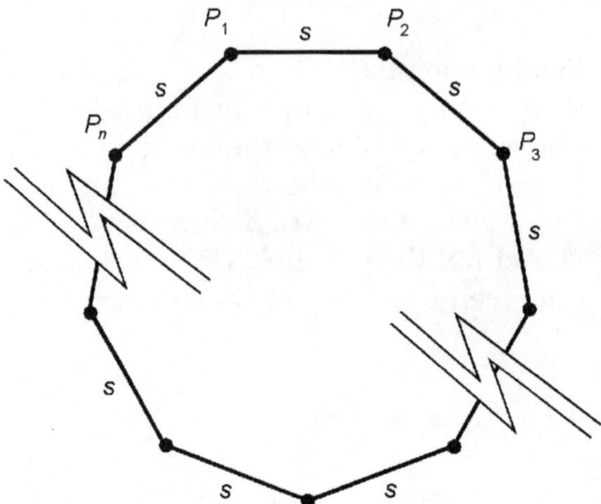

Figure 2.37 Perimeter and area of regular polygon.

Interior area of regular polygon

Let V be a regular polygon as defined above and in Fig. 2.37. The interior area, A, is given by the following formula if angles are specified in degrees:

$$A = (ns^2/4) \cot (180/n)$$

If angles are specified in radians, then:

$$A = (ns^2/4) \cot (\pi/n)$$

Perimeter of circle

Let C be a circle having radius r as shown in Fig. 2.38. Then the perimeter, B, of the circle is given by the following formula:

$$B = 2\pi r$$

Interior area of circle

Let C be a circle as defined above and in Fig. 2.38. The interior area, A, of the circle is given by:

$$A = \pi r^2$$

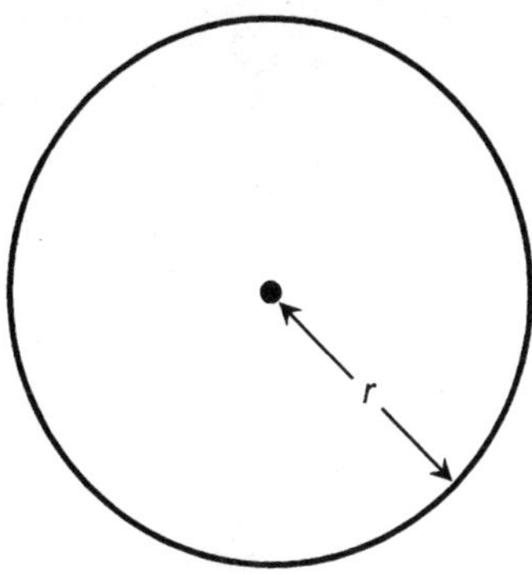

Figure 2.38 Perimeter and area of circle.

Perimeter of ellipse

Let E be an ellipse whose major semi-axis measures r_1, and whose minor semi-axis measures r_2 as shown in Fig. 2.39. Then the perimeter, B, of the ellipse is given approximately by the following formula:

$$B \approx 2\pi((r_1{}^2 + r_2{}^2)/2)^{1/2}$$

Interior area of ellipse

Let E be an ellipse as defined above and in Fig. 2.39. The interior area, A, of the ellipse is given by:

$$A = \pi r_1 r_2$$

Perimeter of regular polygon in circle

Let V be a regular plane polygon having n sides, and whose vertices $P_1, P_2, P_3, \ldots, P_n$ lie on a circle of radius r (Fig. 2.40). Then the perimeter, B, of the polygon is given by the following formula when angles are specified in degrees:

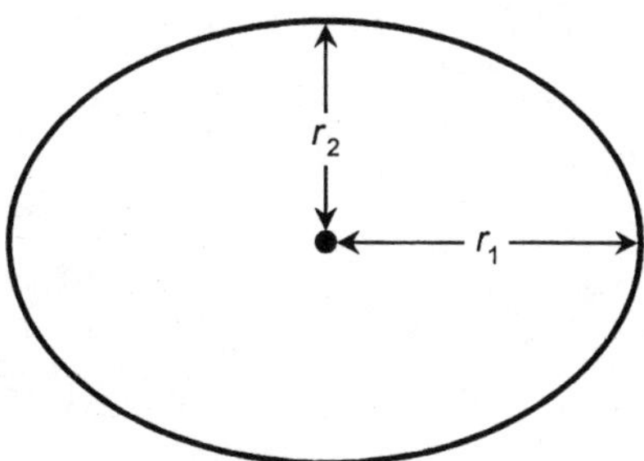

Figure 2.39 Perimeter and area of ellipse.

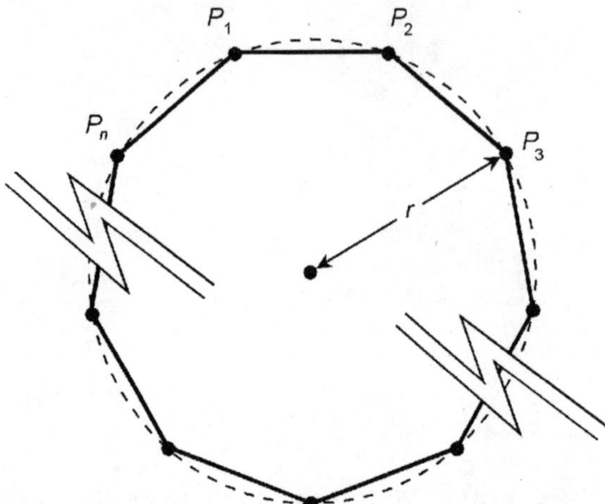

Figure 2.40 Perimeter and area of regular polygon inscribed within a circle.

$$B = 2nr \sin (180/n)$$

If angles are specified in radians, then:

$$B = 2nr \sin (\pi/n)$$

Interior area of regular polygon in circle

Let V be a regular polygon as defined above and in Fig. 2.40. The interior area, A, of the polygon is given by the following formula if angles are specified in degrees:

$$A = (nr^2/2) \sin (360/n)$$

If angles are specified in radians, then:

$$A = (nr^2/2) \sin (2\pi/n)$$

Perimeter of regular polygon around circle

Let V be a regular plane polygon having n sides whose center points $P_1, P_2, P_3, \ldots, P_n$ lie on a circle of radius r (Fig. 2.41). Then the perimeter, B, of the polygon is given by the following formula when angles are specified in degrees:

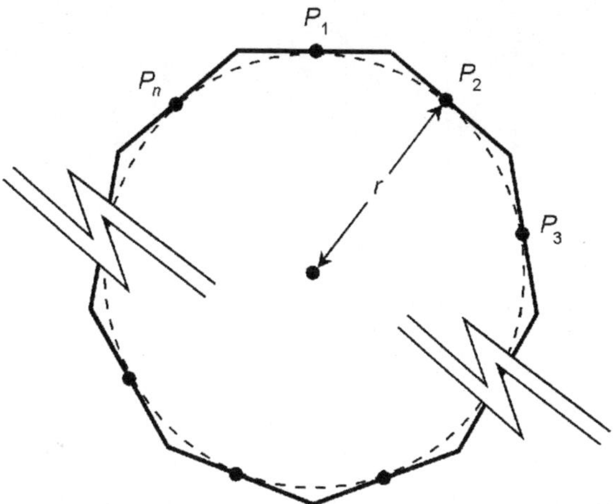

Figure 2.41 Perimeter and area of regular polygon that circumscribes a circle.

$$B = 2nr \tan (180/n)$$

If angles are specified in radians, then:

$$B = 2nr \tan (\pi/n)$$

Interior area of regular polygon around circle

Let V be a regular polygon as defined above and in Fig. 2.41. The interior area, A, of the polygon is given by the following formula if angles are specified in degrees:

$$A = nr^2 \tan (180/n)$$

If angles are specified in radians, then:

$$A = nr^2 \tan (\pi/n)$$

Perimeter of circular sector

Let S be a sector of a circle whose radius is r (Fig. 2.42). Let θ be the apex angle in radians. Then the perimeter, B, of the sector is given by the following formula:

$$B = r(2 + \theta)$$

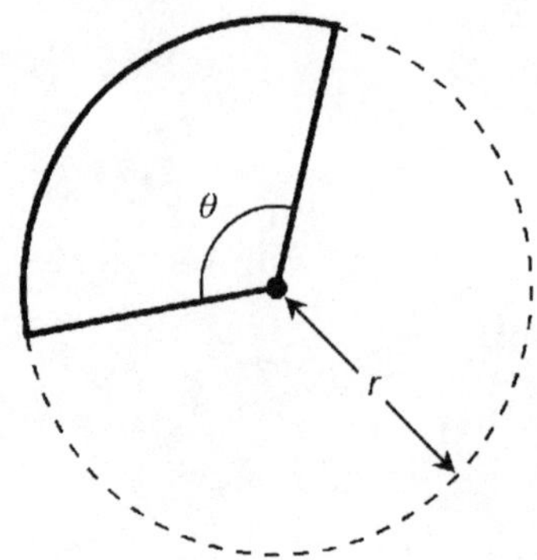

Figure 2.42 Perimeter and area of circular sector.

Interior area of circular sector

Let S be a sector of a circle as defined above and in Fig. 2.42. If θ is in radians, then the interior area, A, of the sector is given by:

$$A = r^2\theta/2$$

If θ is specified in degrees, then the perimeter, B, of the sector is given by:

$$B = (\pi r\theta/180) + 2r$$

If θ is specified in degrees, then the interior area, A, of the sector is given by:

$$A = \pi r^2\theta/360$$

Formulas for Solids

The following formulas apply to common geometric solids in Euclidean three-space.

Volume of pyramid

Let W be a pyramid whose base is a polygon with area A, and whose height is h (Fig. 2.43). The volume, V, of the pyramid is given by:

$$V = Ah/3$$

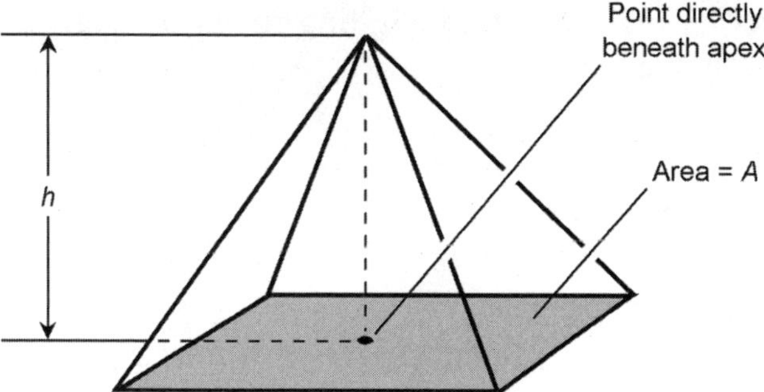

Figure 2.43 Volume of pyramid.

Surface area of right circular cone

Let W be a cone whose base is a circle. Let P be the apex of the cone, and let Q be the center of the base (Fig. 2.44). Suppose line segment PQ is perpendicular to the base, so that W is a *right circular cone*. Let r be the radius of the base, let h be the height of the cone (the length of line segment PQ), and let s be the slant height of the cone as measured from any point on the edge of the circle to the apex P. Then the surface area S_1 of the cone (including the base) is given by either of the following formulas:

$$S_1 = \pi r^2 + \pi rs$$

$$S_1 = \pi r^2 + \pi r(r^2 + h^2)^{1/2}$$

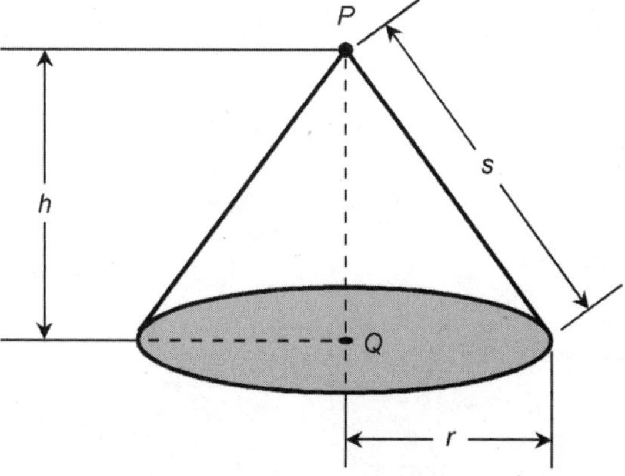

Figure 2.44 Surface area and volume of right circular cone and enclosed solid.

The surface area S_2 of the cone (not including the base) is given by either of the following:

$$S_2 = \pi rs$$

$$S_2 = \pi r(r^2 + h^2)^{1/2}$$

Volume of right circular conical solid

Let W be a right circular cone as defined above and in Fig. 2.44. The volume, V, of the corresponding *right circular conical solid* is given by:

$$V = \pi r^2 h/3$$

Surface area of frustum of right circular cone

Let W be a cone whose base is a circle, and which is truncated by a plane parallel to the base. Let P be the center of the circle defined by the truncation, and let Q be the center of the base (Fig. 2.45). Suppose line segment PQ is perpendicular to the base, so that W is a frustum of a right circular cone. Let r_1 be the radius of the top, let r_2 be the radius of the base, let h be

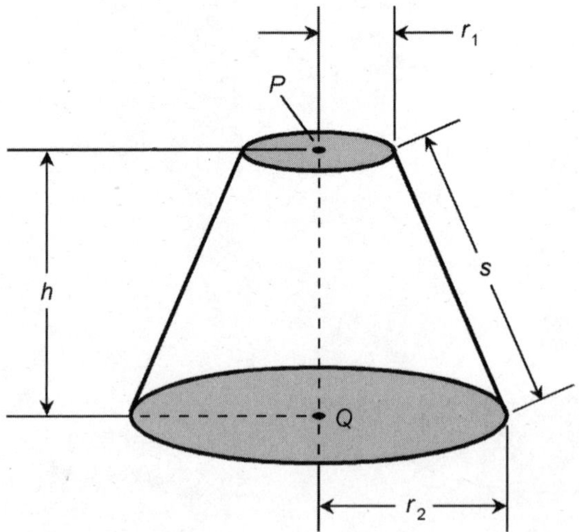

Figure 2.45 Surface area and volume of frustum of right circular cone and enclosed solid.

the height of the object (the length of line segment PQ), and let s be the slant height. Then the surface area S_1 of the object (including the base and the top) is given by either of the following formulas:

$$S_1 = \pi(r_1 + r_2)(s^2 + (r_2 - r_1)^2)^{1/2} + \pi(r_1^{\,2} + r_2^{\,2})$$

$$S_1 = \pi s(r_1 + r_2) + \pi(r_1^{\,2} + r_2^{\,2})$$

The surface area S_2 of the object (not including the base or the top) is given by either of the following:

$$S_2 = \pi(r_1 + r_2)(s^2 + (r_2 - r_1)^2)^{1/2}$$

$$S_2 = \pi s(r_1 + r_2)$$

Volume of frustum of right circular conical solid

Let W be a cone as defined above and in Fig. 2.45. The volume, V, of the corresponding *right circular conical solid* is given by:

$$V = \pi r^2 h/3$$

Volume of general conical solid

Let W be a cone whose base is any enclosed plane curve. Let A be the interior area of the base of the cone. Let P be the apex of the cone, and let Q be a point in the plane X containing the base such that line segment PQ is perpendicular to X (Fig. 2.46).

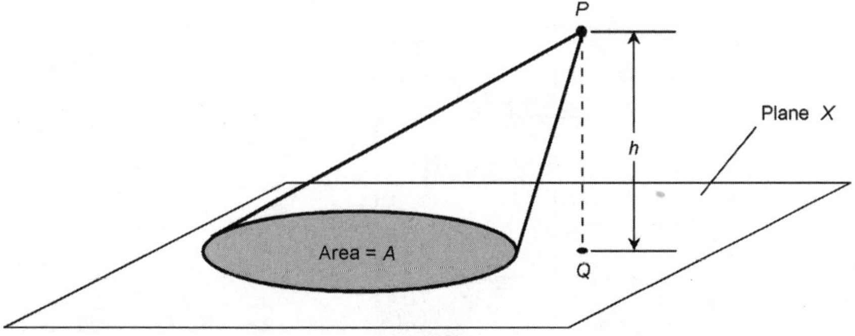

Figure 2.46 Volume of general conical solid.

Let h be the height of the cone (the length of line segment PQ). Then the volume, V, of the corresponding conical solid is given by:

$$V = Ah/3$$

Surface area of right circular cylinder

Let W be a cylinder whose base is a circle. Let P be the center of the top of the cylinder, and let Q be the center of the base (Fig. 2.47). Suppose line segment PQ is perpendicular to both the top and the base, so that W is a *right circular cylinder*. Let r be the radius of the cylinder, and let h be the height (the length of line segment PQ). Then the surface area S_1 of the cylinder (including the base) is given by:

$$S_1 = 2\pi rh + 2\pi r^2 = 2\pi r(h + r)$$

The surface area S_2 of the cylinder (not including the base) is given by:

$$S_2 = 2\pi rh$$

Volume of right circular cylindrical solid

Let W be a cylinder as defined above (Fig. 2.47). The volume, V, of the corresponding *right circular cylindrical solid* is given by:

$$V = \pi r^2 h$$

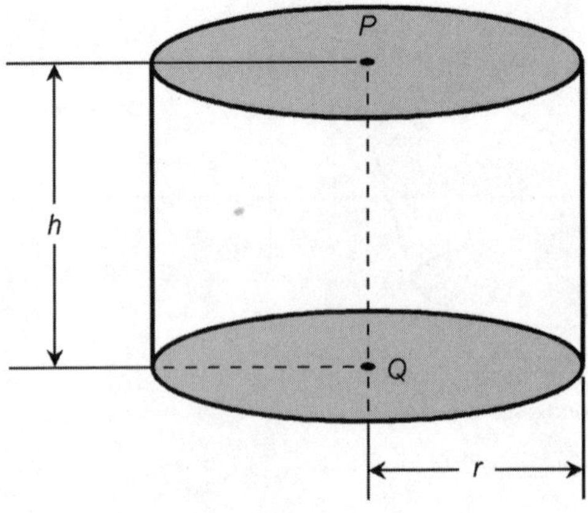

Figure 2.47 Surface area and volume of right circular cylinder and enclosed solid.

Surface area of general cylinder

Let W be a *general cylinder* whose base is any enclosed plane curve. Let A be the interior area of the base of the cylinder (thus also the interior area of the top). Let B be the perimeter of the base (thus also the perimeter of the top). Let h be height of the cylinder, or the perpendicular distance separating the planes containing the top and the base. Let θ be the angle between the plane containing the base and any line segment PQ connecting corresponding points P and Q in the top and the base, respectively. Let s be the slant height of the cylinder, or the length of line segment PQ (Fig. 2.48). Then the surface area S_1 of the cylinder (not including the base or the top) is given by either of the the following formulas:

$$S_1 = 2A + Bh$$

$$S_1 = 2A + Bs \sin \theta$$

The surface area S_2 of the cylinder (not including the base) is given by either of the following:

$$S_2 = Bh$$

$$S_2 = Bs \sin \theta$$

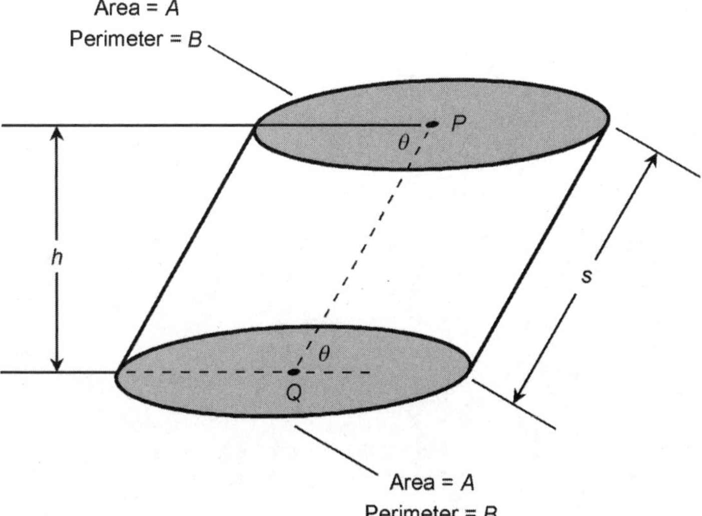

Figure 2.48 Surface area and volume of general cylinder and enclosed solid.

Volume of general cylindrical solid

Let W be a general cylinder as defined above (Fig. 2.48). The volume, V, of the corresponding *general cylindrical solid* is given by either of the following formulas:

$$V = Ah$$

$$V = As \sin \theta$$

Surface area of sphere

Let S be a sphere having radius r as shown in Fig. 2.49. The surface area, A, of the sphere is given by:

$$A = 4\pi r^2$$

Volume of spherical solid

Let S be a sphere as defined above and in Fig. 2.49. The volume, V, of the solid enclosed by the sphere is given by:

$$V = 4\pi r^3/3$$

Volume of ellipsoidal solid

Let E be an ellipsoid whose semi-axes are r_1, r_2, and r_3 (Fig. 2.50). The volume, V, of the enclosed solid is given by:

$$V = 4\pi r_1 r_2 r_3/3$$

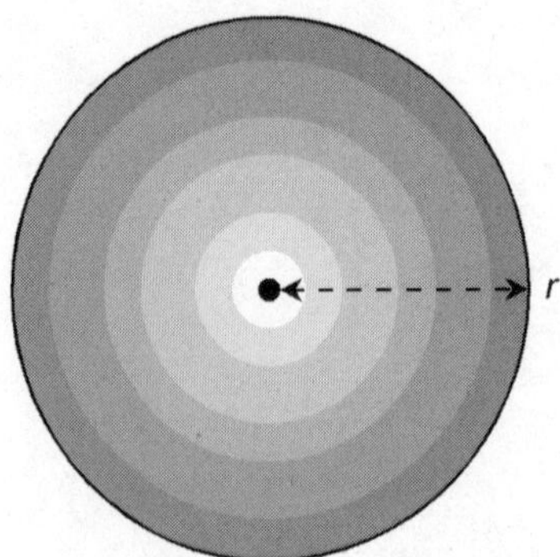

Figure 2.49 Surface area and volume of sphere and enclosed solid.

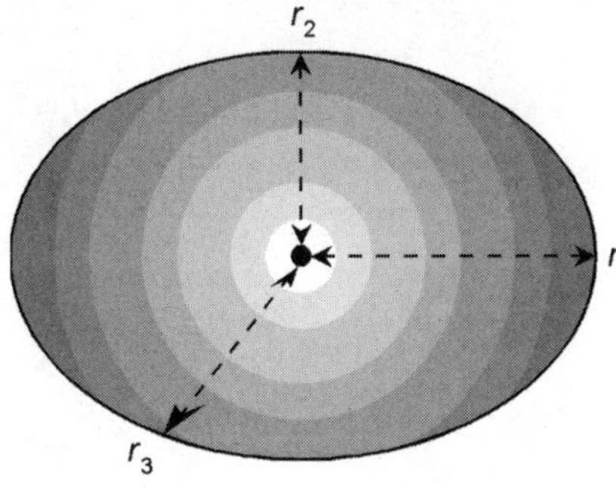

Figure 2.50 Volume of solid enclosed by ellipsoid.

Surface area of cube

Let K be a cube whose edges each have length s, as shown in Fig. 2.51. The surface area, A, of the cube is given by:

$$A = 6s^2$$

Volume of cubical solid

Let K be a cube as defined above and in Fig. 2.51. The volume, V, of the solid enclosed by the cube is given by:

$$V = s^3$$

Surface area of rectangular prism

Let W be a rectangular prism whose edges have lengths s_1, s_2, and s_3 as shown in Fig. 2.52. The surface area, A, of the prism is given by:

$$A = 2s_1s_2 + 2s_1s_3 + 2s_2s_3$$

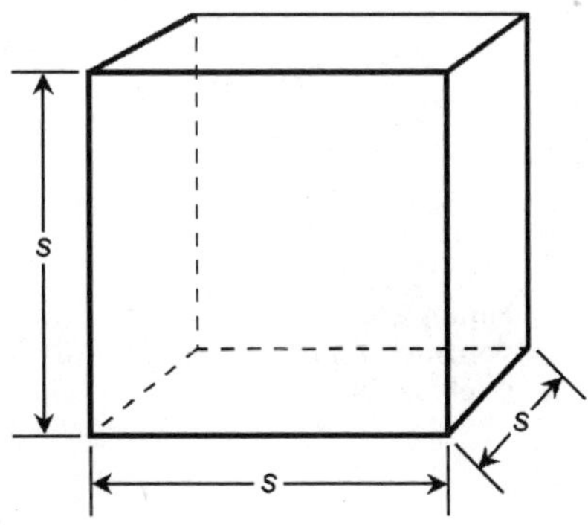

Figure 2.51 Surface area and volume of cube and enclosed solid.

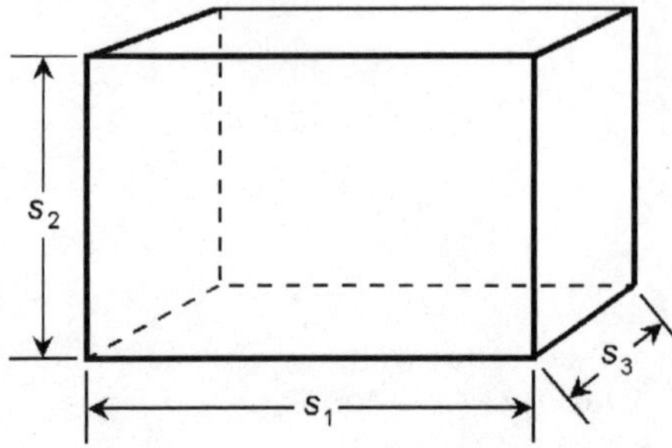

Figure 2.52 Surface area and volume of rectangular prism and enclosed solid.

Volume of rectangular prism solid

Let W be a rectangular prism as defined above and in Fig. 2.52. The volume, V, of the enclosed solid is given by:

$$V = s_1 s_2 s_3$$

Surface area of parallelepiped

Let W be a parallelepiped whose edges have lengths s_1, s_2, and s_3. Suppose the acute angle between edges s_1 and s_2 is θ as shown in Fig. 2.53. The surface area, A, of the parallelepiped is given by:

$$A = 2s_1 s_2 \sin\theta + 2s_1 s_3 + 2s_2 s_3$$

Volume of parallelepiped solid

Let W be a parallelepiped as defined above and in Fig. 2.53. The volume, V, of the enclosed solid is given by:

$$V = s_1 s_2 s_3 \sin\theta$$

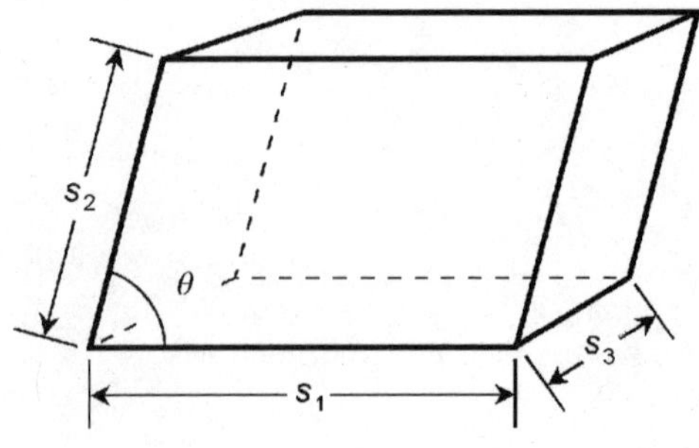

Figure 2.53 Surface area and volume of parallelepiped and enclosed solid.

Surface area of torus

Let T be a torus with a circular cross section, an inner radius of r_1, and an outer radius of r_2 as shown in Fig. 2.54. The surface area, A, of the torus is given by:

$$A = \pi^2(r_2{}^2 - r_1{}^2)$$

Volume of toroidal solid

Let T be a torus as defined above and in Fig. 2.54. The volume, V, of the enclosed toroidal solid is given by:

$$V = \pi^2(r_2 + r_1)(r_2 - r_1)^2/4$$

Circular Functions

There are six *circular trigonometric functions*. They operate on angles to yield real numbers, and are known as *sine, cosine, tangent, cosecant, secant*, and *cotangent*. In formulas and equations, they are abbreviated sin, cos, tan, csc, sec, and cot respectively.

Basic circular functions

Consider a circle in rectangular coordinates with the following equation:

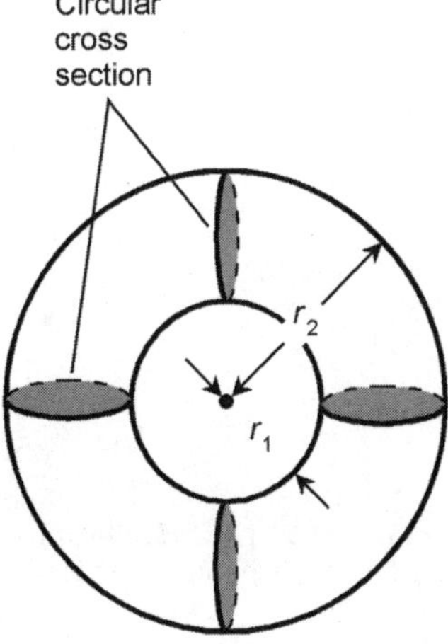

Figure 2.54 Surface area and volume of torus and enclosed solid.

$$x^2 + y^2 = 1$$

This is called the *unit circle* because its radius is one unit, and it is centered at the origin (0,0), as shown in Fig. 2.55. Let θ be an angle whose apex is at the origin, and that is measured counterclockwise from the abscissa (x axis). Suppose this angle corresponds to a ray that intersects the unit circle at some point $P = (x_0, y_0)$. Then

$$y_0 = \sin \theta$$

$$x_0 = \cos \theta$$

$$y_0 / x_0 = \tan \theta$$

Secondary circular functions

Three more circular trigonometric functions are derived from those defined above. They are the *cosecant* function, the *secant* function, and the *cotangent* function. In formulas and equations, they are abbreviated csc θ, sec θ, and cot θ. They are defined as follows:

$$\csc \theta = 1 \,/\, (\sin \theta) = 1/y_0$$

$$\sec \theta = 1 \,/\, (\cos \theta) = 1/x_0$$

$$\cot \theta = 1 \,/\, (\tan \theta) = x_0/y_0$$

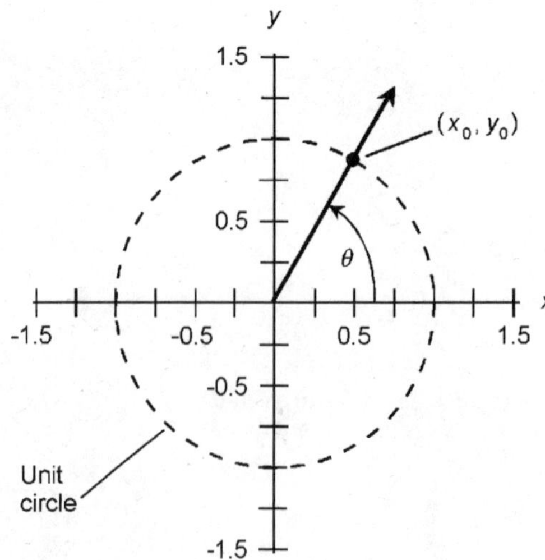

Figure 2.55 Unit-circle model for defining circular trigonometric functions.

Right-triangle model

Consider a right triangle ΔPQR, such that $\angle PQR$ is the right angle. Let a be the length of line segment RQ, b be the length of line segment QP, and c be the length of line segment RP as shown in Fig. 2.56. Let θ be the angle between line segments RQ and RP. The six circular trigonometric functions can be defined as ratios between the lengths of the sides, as follows:

$$\sin \theta = b/c$$

$$\cos \theta = a/c$$

$$\tan \theta = b/a$$

$$\csc \theta = c/b$$

$$\sec \theta = c/a$$

$$\cot \theta = a/b$$

Circular functions as imaginary powers of e

Let θ be a real-number angle in radians. The values of the circular functions of θ can be defined in exponential terms as imaginary-number powers of e, where e is the natural logarithm base and is equal to approximately 2.71828. The symbol j represents the unit imaginary number, which is the positive square root of -1. As long as denominators are nonzero, the following equations hold:

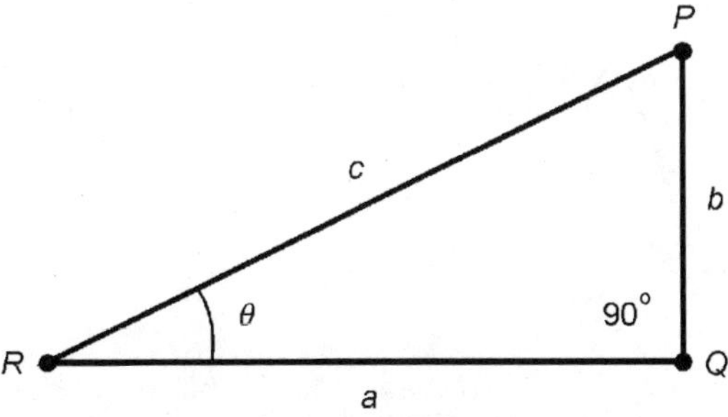

Figure 2.56 Right-triangle model for defining circular trigonometric functions.

$$\sin\theta = (e^{j\theta} - e^{-j\theta})/j2$$

$$\cos\theta = (e^{j\theta} + e^{-j\theta})/2$$

$$\tan\theta = (e^{j\theta} - e^{-j\theta})/j(e^{j\theta} + e^{-j\theta})$$

$$\csc\theta = j2/(e^{j\theta} - e^{-j\theta})$$

$$\sec\theta = 2/(e^{j\theta} + e^{-j\theta})$$

$$\cot\theta = j(e^{j\theta} + e^{-j\theta})/(e^{j\theta} - e^{-j\theta})$$

Graph of sine function

Figure 2.57 is a graph of the function $y = \sin x$ for values of the domain x in radians from -3π to 3π (-540 to 540 degrees). The range of the sine function is limited to values between, and including, -1 and 1.

Graph of cosine function

Figure 2.58 is a graph of the function $y = \cos x$ for values of the domain x in radians from -3π to 3π (-540 to 540 degrees). The range of the cosine function is limited to values between, and including, -1 and 1.

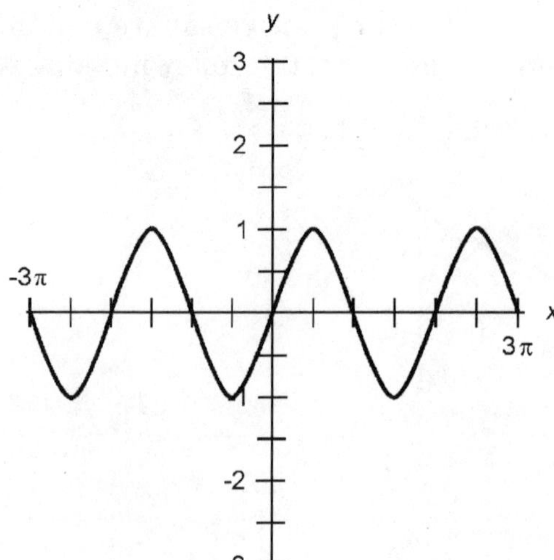

Figure 2.57 Approximate graph of the sine function.

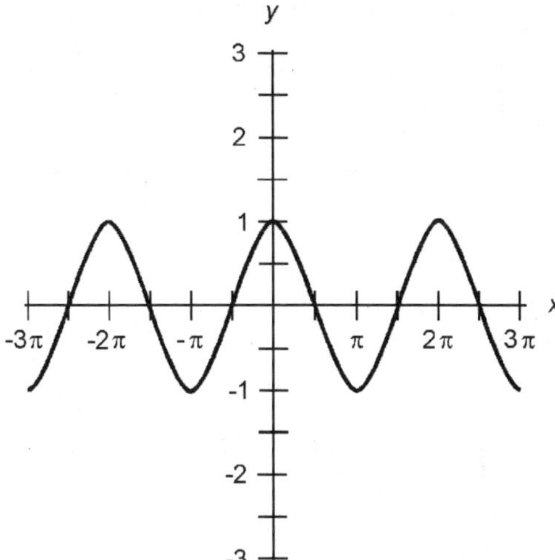

Figure 2.58 Approximate graph of the cosine function.

Graph of tangent function

Figure 2.59 is a graph of the function $y = \tan x$ for values of the domain x in radians from -3π to 3π (-540 to 540 degrees). The range of the tangent function encompasses the entire set of real numbers. Asymptotes are shown as dotted lines. For values of x where these asymptotes intersect the x axis, the function is undefined.

Graph of cosecant function

Figure 2.60 is a graph of the function $y = \csc x$ for values of the domain x in radians from -3π to 3π (-540 to 540 degrees). The range of the cosecant function encompasses all real numbers greater than or equal to 1, and all real numbers less than or equal to -1.

Graph of secant function

Figure 2.61 is a graph of the function $y = \sec x$ for values of the domain x in radians from -3π to 3π (-540 to 540 degrees). The range of the secant function encompasses all real numbers greater than or equal to 1, and all real numbers less than or equal to -1.

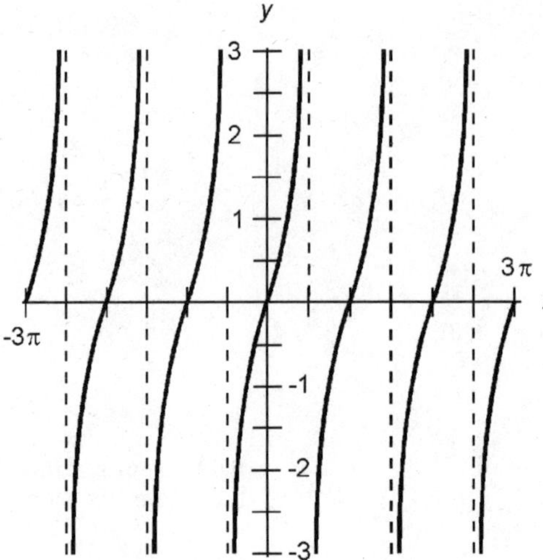

Figure 2.59 Approximate graph of the tangent function.

Graph of cotangent function

Figure 2.62 is a graph of the function $y = \cot x$ for values of the domain x in radians from -3π to 3π (-540 to 540 degrees). The range of the cotangent function encompasses the entire set of real numbers. Asymptotes are shown as dotted lines. For values of x where these asymptotes intersect the x axis, the function is undefined.

Inverse circular functions

Each of the six circular trigonometric functions has an inverse. These inverse relations operate on real numbers to yield angles, and are known as *arcsine, arccosine, arctangent, arccosecant, arcsecant*, and *arccotangent*. In formulas and equations, they are abbreviated arcsin or $\sin^{-1}$, arccos or $\cos^{-1}$, arctan or $\tan^{-1}$, arccsc or $\csc^{-1}$, arcsec or $\sec^{-1}$, and arccot or $\cot^{-1}$ respectively. They are functions when their domains are restricted as shown in the graphs of Figs. 2.63 through 2.68.

Graph of arcsine function

Figure 2.63 is an approximate graph of the function $y = \sin^{-1} x$ with its domain limited to values of x such that $-1 \leq x \leq 1$. The range of the arcsine function is limited to values of y between, and including, $-\pi/2$ and $\pi/2$ radians (-90 and 90 degrees).

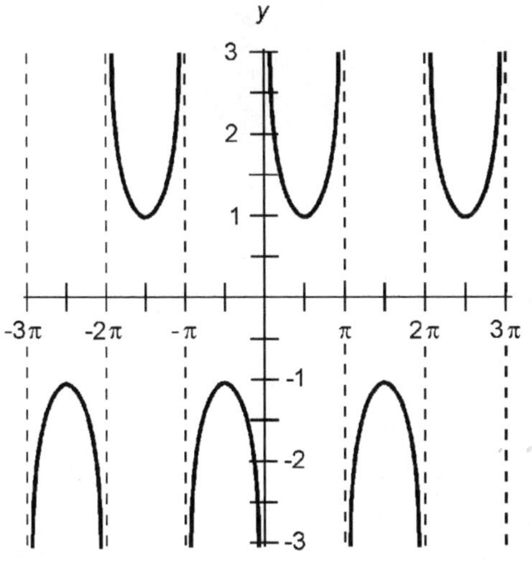

Figure 2.60 Approximate graph of the cosecant function.

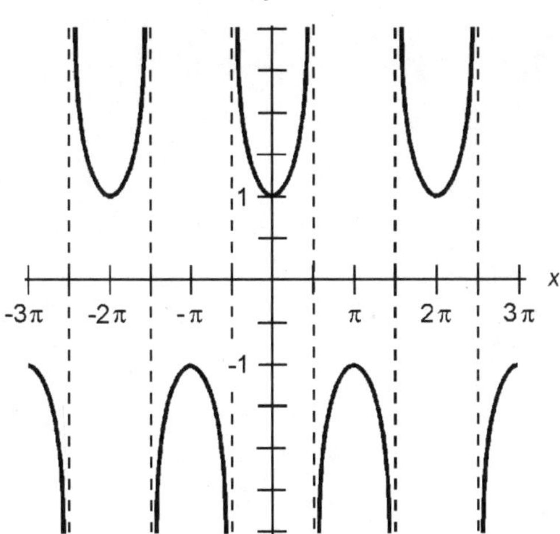

Figure 2.61 Approximate graph of the secant function.

Graph of arccosine function

Figure 2.64 is an approximate graph of the function $y = \cos^{-1} x$ with its domain limited to values of x such that $-1 \leq x \leq 1$. The range of the arccosine function is limited to values of y between, and including, 0 and π radians (0 and 180 degrees).

Graph of arctangent function

Figure 2.65 is an approximate graph of the function $y = \tan^{-1} x$; its domain encompasses the entire set of real num-

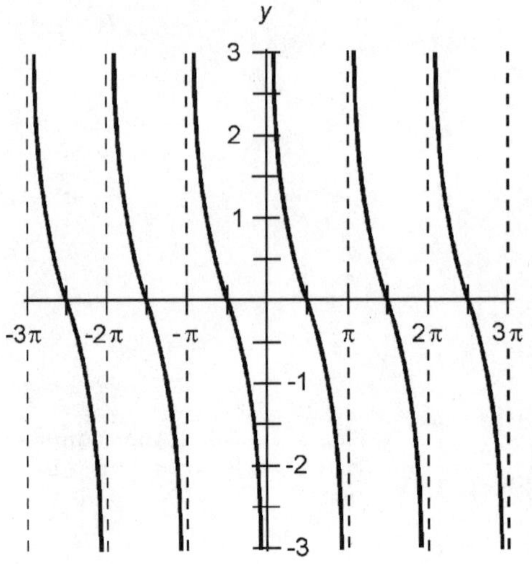

Figure 2.62 Approximate graph of the cotangent function.

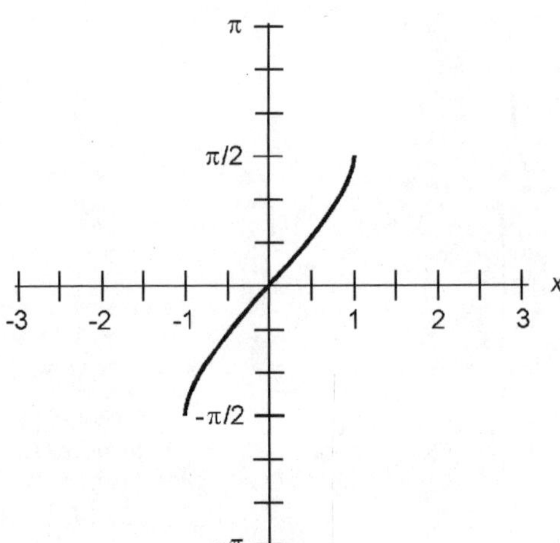

Figure 2.63 Approximate graph of the arcsine function.

bers. The range of the arctangent function is limited to values of y between, but not including, $-\pi/2$ and $\pi/2$ radians (-90 and 90 degrees).

Graph of arccosecant function

Figure 2.66 is an approximate graph of the function $y = \csc^{-1} x$ with its domain limited to values of x such that $x \leq -1$ or $x \geq 1$. The range of the arccosecant function is limited to

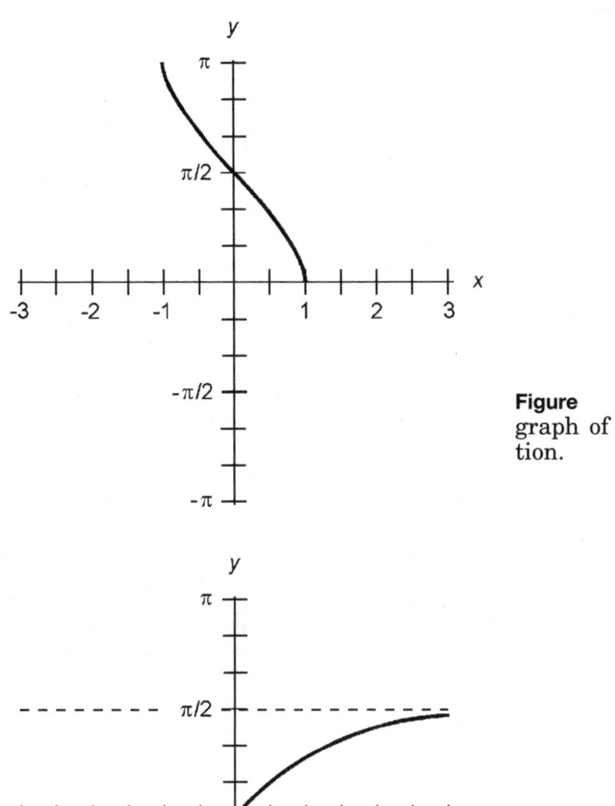

Figure 2.64 Approximate graph of the arccosine function.

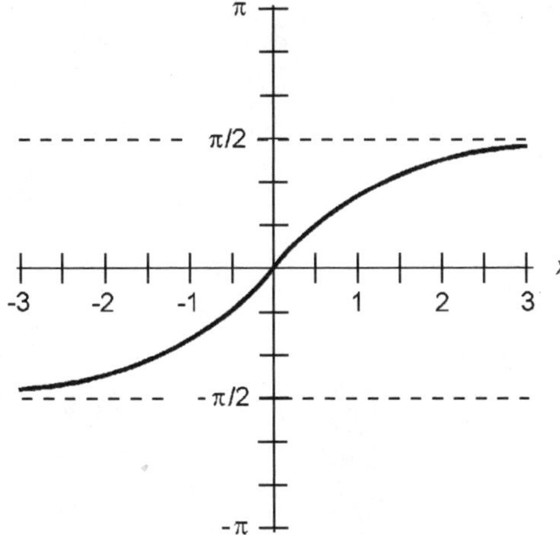

Figure 2.65 Approximate graph of the arctangent function.

values of y in radians such that $-\pi/2 \leq y < 0$ or $0 < y \leq \pi/2$. This corresponds to values of y in degrees such that $-90 \leq y < 0$ or $0 < y \leq 90$.

Graph of arcsecant function

Figure 2.67 is an approximate graph of the function $y = \sec^{-1} x$ with its domain limited to values of x such that $x \leq -1$ or $x \geq 1$. The range of the arcsecant function is limited to values

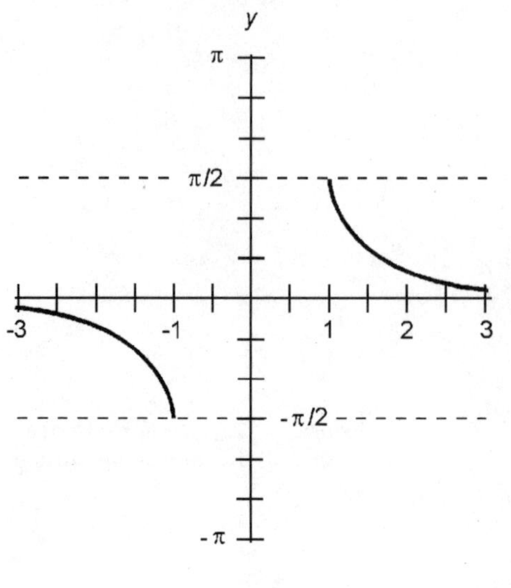

Figure 2.66 Approximate graph the arccosecant function.

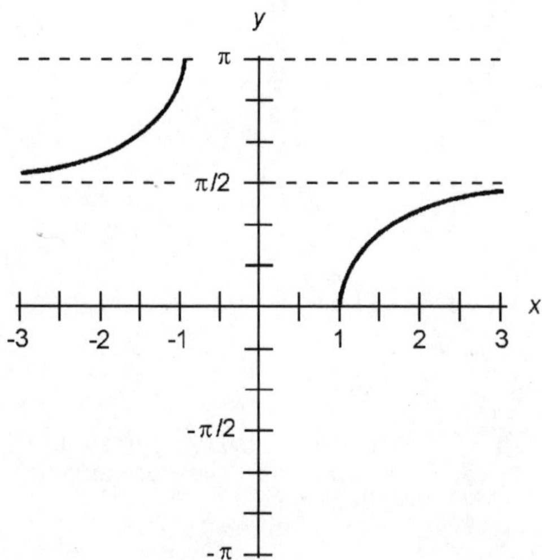

Figure 2.67 Approximate graph of the arcsecant function.

of y in radians such that $0 \leq y < \pi/2$ or $\pi/2 < y \leq \pi$. This corresponds to values of y in degrees such that $0 \leq y < 90$ or $90 < y \leq 180$.

Graph of arccotangent function

Figure 2.68 is an approximate graph of the function $y = \cot^{-1} x$; its domain encompasses the entire set of real numbers. The range of the arccotangent function is limited to values of y between, but not including, 0 and π radians (0 and 180 degrees).

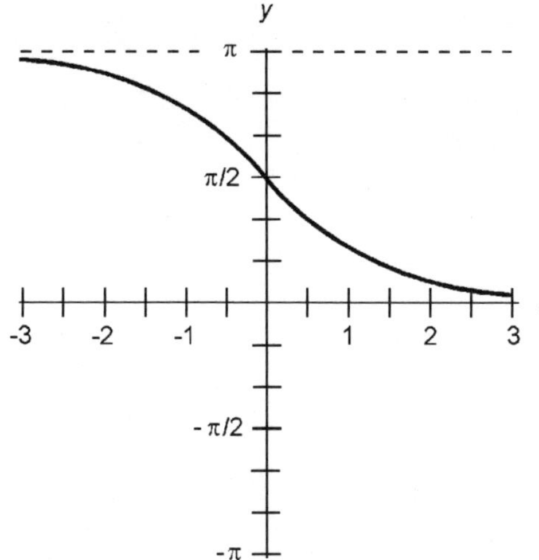

Figure 2.68 Approximate graph of the arccotangent function.

Circular Identities

The following paragraphs depict common *trigonometric identities* for the circular functions. Unless otherwise specified, these formulas apply to angles θ and ϕ in the standard range, as follows:

$$0 \leq \theta < 2\pi \text{ (in radians)}$$

$$0 \leq \theta < 360 \text{ (in degrees)}$$

$$0 \leq \phi < 2\pi \text{ (in radians)}$$

$$0 \leq \phi < 360 \text{ (in degrees)}$$

Angles outside the standard range should be converted to values within the standard range by adding or subtracting the appropriate integral multiple of 2π radians (360 degrees).

Pythagorean theorem for sine and cosine

The sum of the squares of the sine and cosine of an angle is always equal to 1. The following formula holds:

$$\sin^2 \theta + \cos^2 \theta = 1$$

Pythagorean theorem for secant and tangent

The difference between the squares of the secant and tangent of an angle is always equal to either 1 or -1. The following formulas apply for all angles except $\theta = \pi/2$ radians (90 degrees) and $\theta = 3\pi/2$ radians (270 degrees):

$$\sec^2 \theta - \tan^2 \theta = 1$$

$$\tan^2 \theta - \sec^2 \theta = -1$$

Sine of negative angle

The sine of the negative of an angle is equal to the negative (additive inverse) of the sine of the angle. The following formula holds:

$$\sin -\theta = -\sin \theta$$

Cosine of negative angle

The cosine of the negative of an angle is equal to the cosine of the angle. The following formula holds:

$$\cos -\theta = \cos \theta$$

Tangent of negative angle

The tangent of the negative of an angle is equal to the negative (additive inverse) of the tangent of the angle. The following formula applies for all angles except $\theta = \pi/2$ radians (90 degrees) and $\theta = 3\pi/2$ radians (270 degrees):

$$\tan -\theta = -\tan \theta$$

Cosecant of negative angle

The cosecant of the negative of an angle is equal to the negative (additive inverse) of the cosecant of the angle. The following formula applies for all angles except $\theta = 0$ radians (0 degrees) and $\theta = \pi$ radians (180 degrees):

$$\csc -\theta = -\csc \theta$$

Secant of negative angle

The secant of the negative of an angle is equal to the secant of the angle. The following formula applies for all angles except $\theta = \pi/2$ radians (90 degrees) and $\theta = 3\pi/2$ radians (270 degrees):

$$\sec -\theta = \sec \theta$$

Cotangent of negative angle

The cotangent of the negative of an angle is equal to the negative (additive inverse) of the cotangent of the angle. The following formula applies for all angles except $\theta = 0$ radians (0 degrees) and $\theta = \pi$ radians (180 degrees):

$$\cot -\theta = -\cot \theta$$

Periodicity of sine

The sine of an angle is equal to the sine of any integral multiple of 2π radians added to, or subtracted from, that angle. If k is an integer, then the following formulas hold for angles θ in radians:

$$\sin \theta = \sin (\theta + 2\pi k)$$

$$\sin \theta = \sin (\theta - 2\pi k)$$

For angles θ in degrees, the equivalent formulas are:

$$\sin \theta = \sin (\theta + 360k)$$

$$\sin \theta = \sin (\theta - 360k)$$

Periodicity of cosine

The cosine of an angle is equal to the cosine of any integral multiple of 2π radians added to, or subtracted from, that angle. If k is an integer, then the following formulas hold for angles θ in radians:

$$\cos\theta = \cos(\theta + 2\pi k)$$

$$\cos\theta = \cos(\theta - 2\pi k)$$

For angles θ in degrees, the equivalent formulas are:

$$\cos\theta = \cos(\theta + 360k)$$

$$\cos\theta = \cos(\theta - 360k)$$

Periodicity of tangent

The tangent of an angle is equal to the tangent of any integral multiple of π radians added to, or subtracted from, that angle. If k is an integer, the following formulas apply for all angles except $\theta = \pi/2$ radians and $\theta = 3\pi/2$ radians:

$$\tan\theta = \tan(\theta + \pi k)$$

$$\tan\theta = \tan(\theta - \pi k)$$

For angles θ in degrees, the following formulas apply for all angles except $\theta = 90$ and $\theta = 270$:

$$\tan\theta = \tan(\theta + 180k)$$

$$\tan\theta = \tan(\theta - 180k)$$

Periodicity of cosecant

The cosecant of an angle is equal to the cosecant of any integral multiple of 2π radians added to, or subtracted from, that angle. If k is an integer, the following formulas apply for all angles except $\theta = 0$ radians and $\theta = \pi$ radians:

$$\csc\theta = \csc(\theta + 2\pi k)$$

$$\csc\theta = \csc(\theta - 2\pi k)$$

For angles θ in degrees, the following formulas apply for all angles except $\theta = 0$ and $\theta = 180$:

$$\csc\theta = \csc(\theta + 360k)$$

$$\csc\theta = \csc(\theta - 360k)$$

Periodicity of secant

The secant of an angle is equal to the secant of any integral multiple of 2π radians added to, or subtracted from, that angle. If k is an integer, the following formulas apply for all angles except $\theta = \pi/2$ radians and $\theta = 3\pi/2$ radians:

$$\sec \theta = \sec (\theta + 2\pi k)$$

$$\sec \theta = \sec (\theta - 2\pi k)$$

For angles θ in degrees, the following formulas apply for all angles except $\theta = 90$ and $\theta = 270$:

$$\sec \theta = \sec (\theta + 360k)$$

$$\sec \theta = \sec (\theta - 360k)$$

Periodicity of cotangent

The cotangent of an angle is equal to the cotangent of any integral multiple of π radians added to, or subtracted from, that angle. If k is an integer, the following formulas apply for all angles except $\theta = 0$ radians and $\theta = \pi$ radians:

$$\cot \theta = \cot (\theta + \pi k)$$

$$\cot \theta = \cot (\theta - \pi k)$$

For angles θ in degrees, the following formulas apply for all angles except $\theta = 0$ and $\theta = 180$:

$$\cot \theta = \cot (\theta + 180k)$$

$$\cot \theta = \cot (\theta - 180k)$$

Sine of double angle

The sine of twice any given angle is equal to twice the sine of the original angle times the cosine of the original angle:

$$\sin 2\theta = 2 \sin \theta \ \cos \theta$$

Cosine of double angle

The cosine of twice any given angle can be found according to either of the following two formulas:

$$\cos 2\theta = 1 - (2 \sin^2 \theta)$$

$$\cos 2\theta = (2 \cos^2 \theta) - 1$$

Tangent of double angle

The tangent of twice any given angle can be found according to the following formula:

$$\tan 2\theta = (2 \tan \theta)/(1 - \tan^2 \theta)$$

One or both of the above tangent functions is/are undefined, and therefore the formula does not apply in the following cases:

$$\theta = \pi/4 \text{ radians (45 degrees)}$$

$$\theta = \pi/2 \text{ radians (90 degrees)}$$

$$\theta = 3\pi/4 \text{ radians (135 degrees)}$$

$$\theta = \pi \text{ radians (180 degrees)}$$

$$\theta = 5\pi/4 \text{ radians (225 degrees)}$$

$$\theta = 3\pi/2 \text{ radians (270 degrees)}$$

$$\theta = 7\pi/4 \text{ radians (315 degrees)}$$

Sine of half angle

The sine of half any given angle can be found according to the following formula when $0 \leq \theta < \pi$ radians ($0 \leq \theta < 180$ degrees):

$$\sin (\theta/2) = ((1 - \cos \theta)/2)^{1/2}$$

When $\pi \leq \theta < 2\pi$ radians ($180 \leq \theta < 360$ degrees), the formula is:

$$\sin (\theta/2) = -((1 - \cos \theta)/2)^{1/2}$$

Cosine of half angle

The cosine of half any given angle can be found according to the following formula when $0 \leq \theta < \pi/2$ radians ($0 \leq \theta < 90$ degrees) or $3\pi/2 \leq \theta < 2\pi$ radians ($270 \leq \theta < 360$ degrees):

$$\cos (\theta/2) = ((1 + \cos \theta)/2)^{1/2}$$

When $\pi/2 \leq \theta < 3\pi/2$ radians ($90 \leq \theta < 270$ degrees) the formula is:

$$\cos (\theta/2) = -((1 + \cos \theta)/2)^{1/2}$$

Tangent of half angle

The tangent of half any given angle can be found according to the following formula when $0 \leq \theta < \pi/2$ radians ($0 \leq \theta < 90$ degrees) or $\pi < \theta < 3\pi/2$ radians ($180 < \theta < 270$ degrees):

$$\tan (\theta/2) = ((1 - \cos \theta)/(1 + \cos \theta))^{1/2}$$

When $\pi/2 \leq \theta < \pi$ radians ($90 \leq \theta < 180$ degrees) or $3\pi/2 \leq \theta < 2\pi$ radians ($270 \leq \theta < 360$ degrees), the formula is:

$$\tan (\theta/2) = -((1 - \cos \theta)/(1 + \cos \theta))^{1/2}$$

The following formula can be used for all angles except π radians (180 degrees):

$$\tan (\theta/2) = (\sin \theta)/(1 + \cos \theta)$$

Either of the following two formulas hold for all angles except 0 and π radians (0 and 180 degrees):

$$\tan (\theta/2) = (1 - \cos \theta)/(\sin \theta)$$

$$\tan (\theta/2) = \csc \theta - \cot \theta$$

Sine of angular sum

The sine of the sum of two angles θ and ϕ can be found according to the following formula:

$$\sin (\theta + \phi) = (\sin \theta)(\cos \phi) + (\cos \theta)(\sin \phi)$$

Cosine of angular sum

The cosine of the sum of two angles θ and ϕ can be found according to the following formula:

$$\cos (\theta + \phi) = (\cos \theta)(\cos \phi) - (\sin \theta)(\sin \phi)$$

Tangent of angular sum

The tangent of the sum of two angles θ and ϕ can be found according to the following formula:

$$\tan(\theta + \phi) = (\tan\theta + \tan\phi)/(1 - (\tan\theta)(\tan\phi))$$

One or more of the functions within the above equation is/are undefined, and therefore the formula does not apply in the following cases:

$$\theta = \pi/2 \text{ radians (90 degrees)}$$

$$\phi = \pi/2 \text{ radians (90 degrees)}$$

$$\theta + \phi = \pi/2 \text{ radians (90 degrees)}$$

$$\theta = 3\pi/2 \text{ radians (270 degrees)}$$

$$\phi = 3\pi/2 \text{ radians (270 degrees)}$$

$$\theta + \phi = 3\pi/2 \text{ radians (270 degrees)}$$

$$(\tan\theta)(\tan\phi) = 1$$

Sine of angular difference

The sine of the difference between two angles θ and ϕ can be found according to the following formula:

$$\sin(\theta - \phi) = (\sin\theta)(\cos\phi) - (\cos\theta)(\sin\phi)$$

Cosine of angular difference

The cosine of the difference between two angles θ and ϕ can be found according to the following formula:

$$\cos(\theta - \phi) = (\cos\theta)(\cos\phi) + (\sin\theta)(\sin\phi)$$

Tangent of angular difference

The tangent of the difference between two angles θ and ϕ can be found according to the following formula:

$$\tan(\theta - \phi) = (\tan\theta - \tan\phi)/((1 + (\tan\theta)(\tan\phi))$$

One or more of the functions within the above equation is/are undefined, and therefore the formula does not apply in the following cases:

$$\theta = \pi/2 \text{ radians (90 degrees)}$$

$$\phi = \pi/2 \text{ radians (90 degrees)}$$

$$\theta - \phi = \pi/2 \text{ radians (90 degrees)}$$

$$\theta = 3\pi/2 \text{ radians (270 degrees)}$$

$$\phi = 3\pi/2 \text{ radians (270 degrees)}$$

$$\theta - \phi = 3\pi/2 \text{ radians (270 degrees)}$$

$$(\tan \theta)(\tan \phi) = -1$$

Sine of complementary angle

The sine of the complement of an angle is equal to the cosine of the angle. The following formula holds for angles in radians:

$$\sin (\pi/2 - \theta) = \cos \theta$$

For angles in degrees, the equivalent formula is:

$$\sin (90 - \theta) = \cos \theta$$

Cosine of complementary angle

The cosine of the complement of an angle is equal to the sine of the angle. The following formula holds:

$$\cos (\pi/2 - \theta) = \sin \theta$$

For angles in degrees, the equivalent formula is:

$$\cos (90 - \theta) = \sin \theta$$

Tangent of complementary angle

The tangent of the complement of an angle is equal to the cotangent of the angle. The following formula holds:

$$\tan(\pi/2 - \theta) = \cot\theta$$

For angles in degrees, the equivalent formula is:

$$\tan(90 - \theta) = \cot\theta$$

The functions within the above equations are undefined, and therefore the formulas do not apply in the following cases:

$$\theta = 0 \text{ radians (0 degrees)}$$

$$\theta = \pi \text{ radians (180 degrees)}$$

Cosecant of complementary angle

The cosecant of the complement of an angle is equal to the secant of the angle. The following formula holds:

$$\csc(\pi/2 - \theta) = \sec\theta$$

For angles in degrees, the equivalent formula is:

$$\csc(90 - \theta) = \sec\theta$$

The functions within the above equations are undefined, and therefore the formulas do not apply in the following cases:

$$\theta = \pi/2 \text{ radians (90 degrees)}$$

$$\theta = 3\pi/2 \text{ radians (270 degrees)}$$

Secant of complementary angle

The secant of the complement of an angle is equal to the cosecant of the angle. The following formula holds:

$$\sec(\pi/2 - \theta) = \csc\theta$$

For angles in degrees, the equivalent formula is:

$$\sec(90 - \theta) = \csc\theta$$

The functions within the above equations are undefined, and therefore the formulas do not apply in the following cases:

$$\theta = 0 \text{ radians (0 degrees)}$$

$$\theta = \pi \text{ radians (180 degrees)}$$

Cotangent of complementary angle

The cotangent of the complement of an angle is equal to the tangent of the angle. The following formula holds:

$$\cot (\pi/2 - \theta) = \tan \theta$$

For angles in degrees, the equivalent formula is:

$$\cot (90 - \theta) = \tan \theta$$

The functions within the above equations are undefined, and therefore the formulas do not apply in the following cases:

$$\theta = \pi/2 \text{ radians (90 degrees)}$$

$$\theta = 3\pi/2 \text{ radians (270 degrees)}$$

Sine of supplementary angle

The sine of the supplement of an angle is equal to the sine of the angle. The following formula holds:

$$\sin (\pi - \theta) = \sin \theta$$

For angles in degrees, the equivalent formula is:

$$\sin (180 - \theta) = \sin \theta$$

Cosine of supplementary angle

The cosine of the supplement of an angle is equal to the negative (additive inverse) of the cosine of the angle. The following formula holds:

$$\cos (\pi - \theta) = -\cos \theta$$

For angles in degrees, the equivalent formula is:

$$\cos (180 - \theta) = -\cos \theta$$

Tangent of supplementary angle

The tangent of the supplement of an angle is equal to the negative (additive inverse) of the tangent of the angle. The following formula holds:

$$\tan(\pi - \theta) = -\tan\theta$$

For angles in degrees, the equivalent formula is:

$$\tan(180 - \theta) = -\tan\theta$$

The functions within the above equations are undefined, and therefore the formulas do not apply in the following cases:

$$\theta = \pi/2 \text{ radians (90 degrees)}$$

$$\theta = 3\pi/2 \text{ radians (270 degrees)}$$

Cosecant of supplementary angle

The cosecant of the supplement of an angle is equal to the cosecant of the angle. The following formula holds:

$$\csc(\pi - \theta) = \csc\theta$$

For angles in degrees, the equivalent formula is:

$$\csc(180 - \theta) = \csc\theta$$

The functions within the above equations are undefined, and therefore the formulas do not apply in the following cases:

$$\theta = 0 \text{ radians (0 degrees)}$$

$$\theta = \pi \text{ radians (180 degrees)}$$

Secant of supplementary angle

The secant of the supplement of an angle is equal to the negative (additive inverse) of the secant of the angle. The following formula holds:

$$\sec(\pi - \theta) = -\sec\theta$$

For angles in degrees, the equivalent formula is:

$$\sec (180 - \theta) = -\sec \theta$$

The functions within the above equations are undefined, and therefore the formulas do not apply in the following cases:

$$\theta = \pi/2 \text{ radians (90 degrees)}$$

$$\theta = 3\pi/2 \text{ radians (270 degrees)}$$

Cotangent of supplementary angle

The cotangent of the supplement of an angle is equal to the negative (additive inverse) of the cotangent of the angle. The following formula holds:

$$\cot (\pi - \theta) = -\cot \theta$$

For angles in degrees, the equivalent formula is:

$$\cot (180 - \theta) = -\cot \theta$$

The functions within the above equations are undefined, and therefore the formulas do not apply in the following cases:

$$\theta = 0 \text{ radians (0 degrees)}$$

$$\theta = \pi \text{ radians (180 degrees)}$$

Hyperbolic Functions

There are six *hyperbolic functions* that are analogous in some ways to the circular trigonometric functions. They are known as *hyperbolic sine, hyperbolic cosine, hyperbolic tangent, hyperbolic cosecant, hyperbolic secant,* and *hyperbolic cotangent.* In formulas and equations, they are abbreviated sinh, cosh, tanh, csch, sech, and coth respectively.

Hyperbolic functions as powers of e

Let x be a real number. The values of the hyperbolic functions of x can be defined in exponential terms as powers of e, where e is the natural logarithm base and is equal to approximately 2.71828. As long as denominators are nonzero, the following equations hold:

$$\sinh x = (e^x - e^{-x})/2$$

$$\cosh x = (e^x + e^{-x})/2$$

$$\tanh x = \sinh x/\cosh x = (e^x - e^{-x})/(e^x + e^{-x})$$

$$\text{csch } x = 1/\sinh x = 2/(e^x - e^{-x})$$

$$\text{sech } x = 1/\cosh x = 2/(e^x + e^{-x})$$

$$\coth x = \cosh x/\sinh x = (e^x + e^{-x})/(e^x - e^{-x})$$

Hyperbolic functions as series

Let x be a real number. The exclamation symbol (!) denotes factorial. The following equations hold:

$$\sinh x = x + x^3/3! + x^5/5! + x^7/7! + \ldots$$

$$\cosh x = 1 + x^2/2! + x^4/4! + x^6/6! + \ldots$$

Graph of hyperbolic sine function

Figure 2.69 is an approximate graph of the function $y = \sinh x$; both its domain and range extend over the entire set of real numbers.

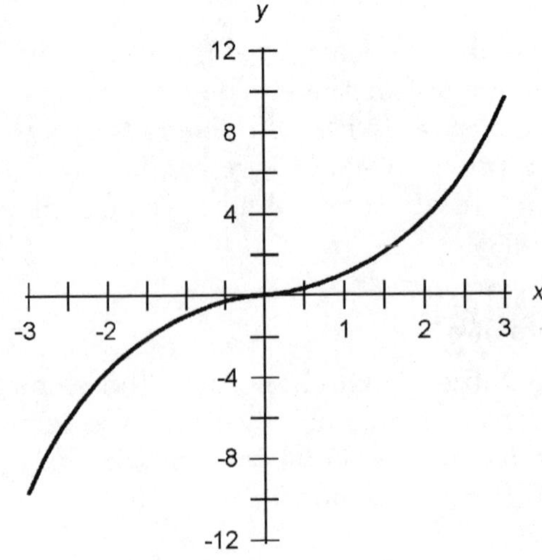

Figure 2.69 Approximate graph of the hyperbolic sine function.

Graph of hyperbolic cosine function

Figure 2.70 is an approximate graph of the function $y = \cosh x$. Its domain extends over the entire set of real numbers, and its range encompasses the set of real numbers y such that $y \geq 1$.

Graph of hyperbolic tangent function

Figure 2.71 is an approximate graph of the function $y = \tanh x$; its domain encompasses the entire set of real numbers.

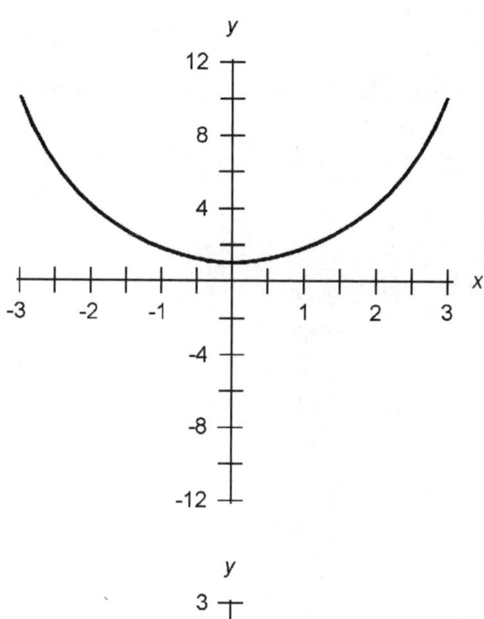

Figure 2.70 Approximate graph of the hyperbolic cosine function.

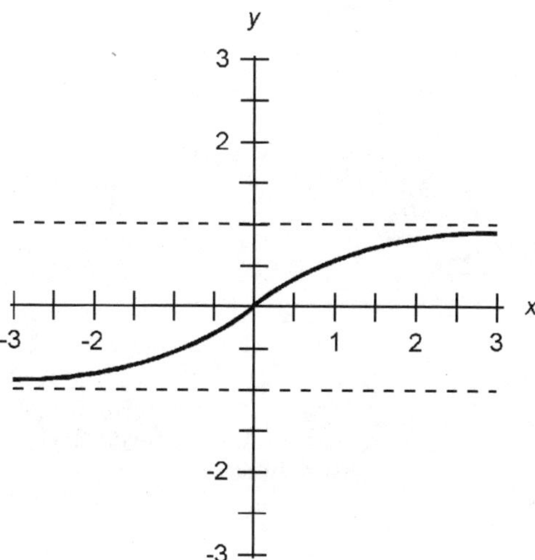

Figure 2.71 Approximate graph of the hyperbolic tangent function.

The range of the hyperbolic tangent function is limited to the set of real numbers y such that $-1 < y < 1$.

Graph of hyperbolic cosecant function

Figure 2.72 is an approximate graph of the function $y = \operatorname{csch} x$; its domain encompasses the set of real numbers x such that $x \neq 0$. The range of the hyperbolic cotangent function encompasses the set of real numbers y such that $y \neq 0$.

Graph of hyperbolic secant function

Figure 2.73 is an approximate graph of the function $y = \operatorname{sech} x$. Its domain encompasses the entire set of real numbers. Its range is limited to the set of real numbers y such that $0 < y \leq 1$.

Graph of hyperbolic cotangent function

Figure 2.74 is an approximate graph of the function $y = \coth x$; its domain encompasses the entire set of real numbers x such that $x \neq 0$. The range of the hyperbolic cotangent function encompasses the set of real numbers y such that $y < -1$ or $y > 1$.

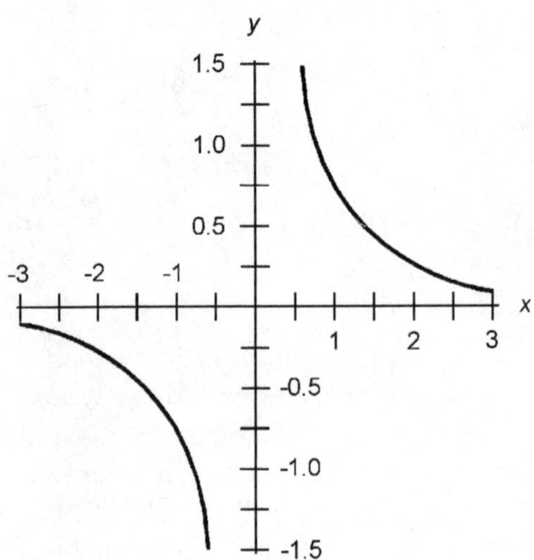

Figure 2.72 Approximate graph of the hyperbolic cosecant function.

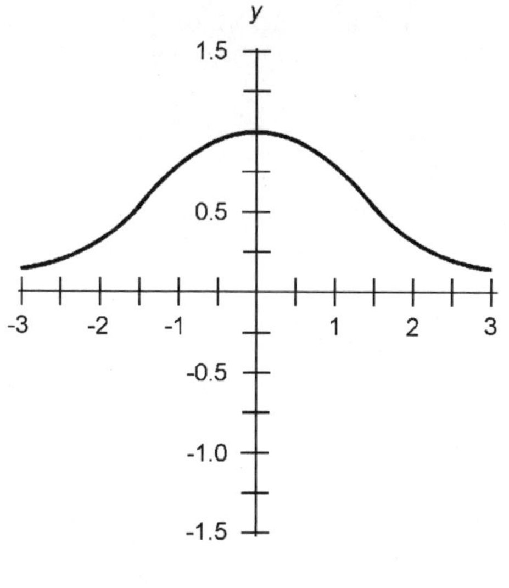

Figure 2.73 Approximate graph of the hyperbolic secant function.

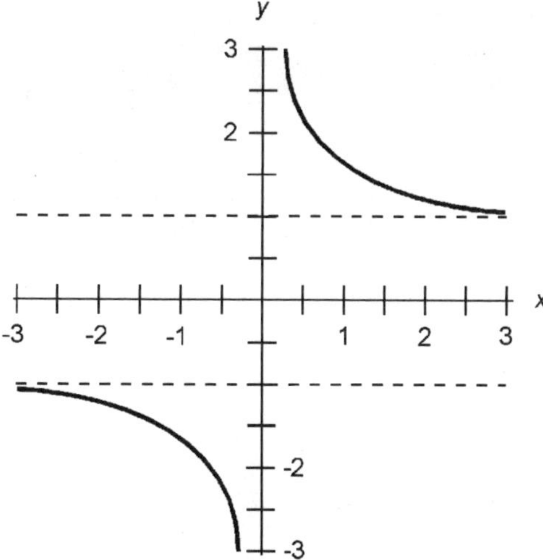

Figure 2.74 Approximate graph of the hyperbolic cotangent function.

Inverse hyperbolic functions

Each of the six hyperbolic functions has an inverse. These inverse relations are known as *hyperbolic arcsine, hyperbolic arccosine, hyperbolic arctangent, hyperbolic arccosecant, hyperbolic arcsecant*, and *hyperbolic arccotangent*. In formulas and equations, they are abbreviated arcsinh or $\sinh^{-1}$, arccosh or $\cosh^{-1}$, arctanh or $\tanh^{-1}$, arccsch or csch^{-1}, arcsech or sech^{-1}, and arccoth or $\coth^{-1}$ respectively. They are functions when their domains are restricted, as shown in the graphs of Figs. 2.75 through 2.80.

Graph of hyperbolic arcsine function

Figure 2.75 is an approximate graph of the function $y = \sinh^{-1} x$; both its domain and range encompass the entire set of real numbers.

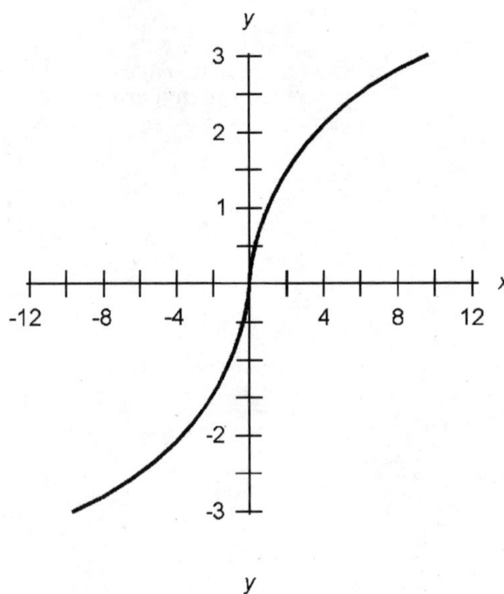

Figure 2.75 Approximate graph of the hyperbolic arcsine function.

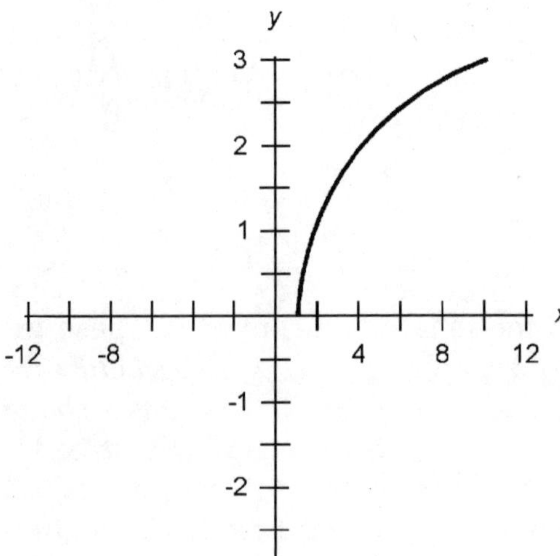

Figure 2.76 Approximate graph of the hyperbolic arccosine function.

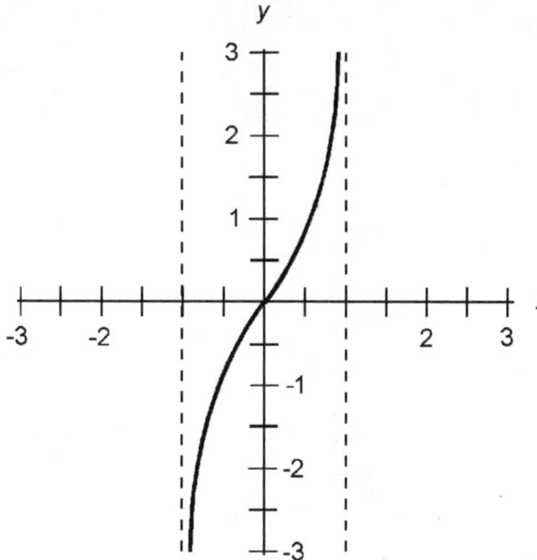

Figure 2.77 Approximate graph of the hyperbolic arctangent function.

Graph of hyperbolic arccosine function

Figure 2.76 is an approximate graph of the function $y = \cosh^{-1} x$. The domain includes real numbers x such that $x \geq 1$. The range of the hyperbolic arccosine function is limited to the non-negative reals, that is, to real numbers y such that $y \geq 0$.

Graph of hyperbolic arctangent function

Figure 2.77 is an approximate graph of the function $y = \tanh^{-1} x$. The domain is limited to real numbers x such that $-1 < x < 1$. The range of the hyperbolic arctangent function spans the entire set of real numbers.

Graph of hyperbolic arccosecant function

Figure 2.78 is an approximate graph of the function $y = \operatorname{csch}^{-1} x$. Both the domain and the range of the hyperbolic arccosecant function include all real numbers except zero.

Graph of hyperbolic arcsecant function

Figure 2.79 is an approximate graph of the function $y = \operatorname{sech}^{-1} x$. The domain of this function is limited to real numbers

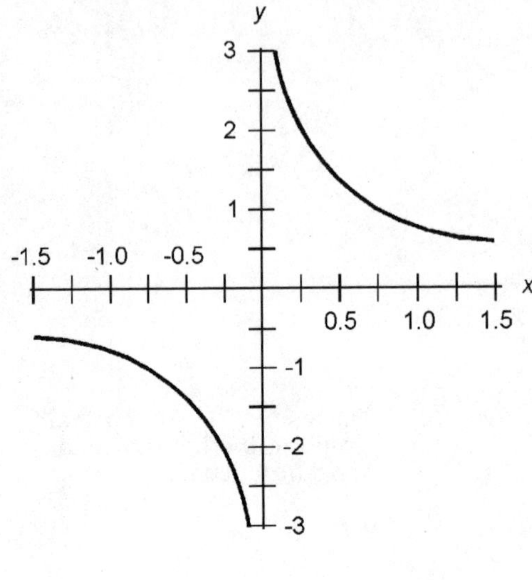

Figure 2.78 Approximate graph of the hyperbolic arccosecant function.

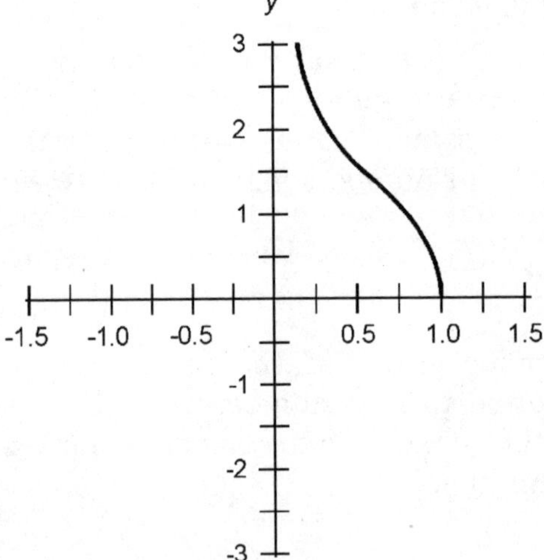

Figure 2.79 Approximate graph of the hyperbolic arcsecant function.

x such that $0 < x \leq 1$. The range of the hyperbolic arcsecant function is limited to the non-negative reals, that is, to real numbers y such that $y \geq 0$.

Graph of hyperbolic arccotangent function

Figure 2.80 is an approximate graph of the function $y = \coth^{-1} x$. The domain of this function includes all real numbers x such that $x < -1$ or $x > 1$. The range of the hyperbolic arccotangent function includes all real numbers except zero.

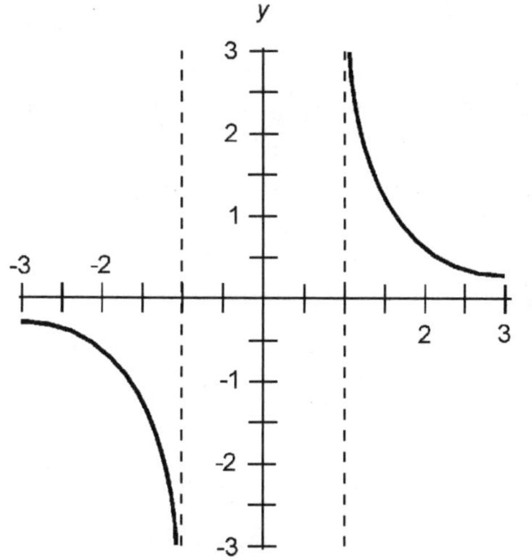

Figure 2.80 Approximate graph of the hyperbolic arccotangent function.

Inverse hyperbolic functions as natural logarithms

Let x be a real number. The values of the inverse hyperbolic functions of x can be defined in logarithmic terms, where ln represents the natural (base-e) logarithm function, and the domains (values of x) are restricted as defined in the preceding paragraphs and in Figs. 2.75 through 2.80. The following equations hold:

$$\sinh^{-1} x = \ln (x + (x^2 + 1)^{1/2})$$

$$\cosh^{-1} x = \ln (x + (x^2 - 1)^{1/2})$$

$$\tanh^{-1} x = 0.5 \ln ((1 + x)/(1 - x))$$

$$\text{csch}^{-1} x = \ln (x^{-1} + (x^{-2} + 1)^{1/2})$$

$$\text{sech}^{-1} x = \ln (x^{-1} + (x^{-2} - 1)^{1/2})$$

$$\coth^{-1} x = 0.5 \ln ((x + 1)/(x - 1))$$

Hyperbolic Identities

The following paragraphs depict common identities for hyperbolic functions. Unless otherwise specified, values of variables can span the real-number domains of the hyperbolic functions as defined above.

Pythagorean theorem for hyperbolic sine and cosine

The difference between the squares of the hyperbolic sine and hyperbolic cosine of a variable is always equal to either 1 or −1. The following formulas hold for all nonzero real numbers x:

$$\sinh^2 x - \cosh^2 x = -1$$

$$\cosh^2 x - \sinh^2 x = 1$$

Pythagorean theorem for hyperbolic cotangent and cosecant

The difference between the squares of the hyperbolic cotangent and hyperbolic cosecant of a variable is always equal to either 1 or −1. The following formulas hold for all nonzero real numbers x:

$$\operatorname{csch}^2 x - \coth^2 x = -1$$

$$\coth^2 x - \operatorname{csch}^2 x = 1$$

Pythagorean theorem for hyperbolic secant and tangent

The sum of the squares of the hyperbolic secant and hyperbolic tangent of a variable is always equal to 1. The following formula holds for all real numbers x:

$$\operatorname{sech}^2 x + \tanh^2 x = 1$$

Hyperbolic sine of negative variable

The hyperbolic sine of the negative of a variable is equal to the negative (additive inverse) of the hyperbolic sine of the variable. The following formula holds for all real numbers x:

$$\sinh -x = -\sinh x$$

Hyperbolic cosine of negative variable

The hyperbolic cosine of the negative of a variable is equal to the hyperbolic cosine of the variable. The following formula holds for all real numbers x:

$$\cosh -x = \cosh x$$

Hyperbolic tangent of negative variable

The hyperbolic tangent of the negative of a variable is equal to the negative (additive inverse) of the hyperbolic tangent of the variable. The following formula holds for all real numbers x:

$$\tanh -x = -\tanh x$$

Hyperbolic cosecant of negative variable

The hyperbolic cosecant of the negative of a variable is equal to the negative (additive inverse) of the hyperbolic cosecant of the variable. The following formula holds for all nonzero real numbers x:

$$\operatorname{csch} -x = -\operatorname{csch} x$$

Hyperbolic secant of negative variable

The hyperbolic secant of the negative of a variable is equal to the hyperbolic secant of the variable. The following formula holds for all real numbers x:

$$\operatorname{sech} -x = \operatorname{sech} x$$

Hyperbolic cotangent of negative variable

The hyperbolic cotangent of the negative of a variable is equal to the negative (additive inverse) of the hyperbolic cotangent of the variable. The following formula holds for all nonzero real numbers x:

$$\coth -x = -\coth x$$

Hyperbolic sine of double value

The hyperbolic sine of twice any given variable is equal to twice the hyperbolic sine of the original variable times the hyperbolic cosine of the original variable. The following formula holds for all real numbers x:

$$\sinh 2x = 2 \sinh x \cosh x$$

Hyperbolic cosine of double value

The hyperbolic cosine of twice any given variable can be found according to any of the following three formulas for all real numbers x:

$$\cosh 2x = \cosh^2 x + \sinh^2 x$$

$$\cosh 2x = 1 + 2 \sinh^2 x$$

$$\cosh 2x = 2 \cosh^2 x - 1$$

Hyperbolic tangent of double value

The hyperbolic tangent of twice any given variable can be found according to the following formula for all real numbers x:

$$\tanh 2x = (2 \tanh x) / (1 + \tanh^2 x)$$

Hyperbolic sine of half value

The hyperbolic sine of half any given variable can be found according the following formula for all non-negative real numbers x:

$$\sinh (x/2) = ((1 - \cosh x)/2)^{1/2}$$

For negative real numbers x, the formula is:

$$\sinh (x/2) = -((1 - \cosh x)/2)^{1/2}$$

Hyperbolic cosine of half value

The hyperbolic cosine of half any given variable can be found according to the following formula for all real numbers x:

$$\cos (x/2) = ((1 + \cos x)/2)^{1/2}$$

Hyperbolic tangent of half value

The hyperbolic tangent of half any given variable can be found according to the following formula for all non-negative real numbers x:

$$\tanh(x/2) = ((\cosh x - 1)/(\cosh x + 1))^{1/2}$$

For negative real numbers x, the formula is:

$$\tanh(x/2) = -((\cosh x - 1) / (\cosh x + 1))^{1/2}$$

The following formula applies for all real numbers x:

$$\tanh(x/2) = \sinh x/(\cosh x + 1)$$

The following formula applies for all nonzero real numbers x:

$$\tanh(x/2) = (\cosh x - 1)/\sinh x$$

Hyperbolic sine of sum

The hyperbolic sine of the sum of two variables x and y can be found according to the following formula:

$$\sinh(x + y) = \sinh x \cosh y + \cosh x \sinh y$$

Hyperbolic cosine of sum

The hyperbolic cosine of the sum of two variables x and y can be found according to the following formula:

$$\cosh(x + y) = \cosh x \cosh y + \sinh x \sinh y$$

Hyperbolic tangent of sum

The hyperbolic tangent of the sum of two variables x and y can be found according to the following formula, provided $\tanh x \tanh y \neq -1$:

$$\tanh(x + y) = (\tanh x + \tanh y)/(1 + \tanh x \tanh y)$$

Hyperbolic sine of difference

The hyperbolic sine of the difference between two variables x and y can be found according to the following formula:

$$\sinh(x - y) = \sinh x \cosh y - \cosh x \sinh y$$

Hyperbolic cosine of difference

The hyperbolic cosine of the difference between two variables x and y can be found according to the following formula:

$$\cosh (x - y) = \cosh x \cosh y - \sinh x \sinh y$$

Hyperbolic tangent of difference

The hyperbolic tangent of the difference between two variables x and y can be found according to the following formula, provided $\tanh x \tanh y \neq 1$:

$$\tanh (x - y) = (\tanh x - \tanh y)/(1 - \tanh x \tanh y)$$

Logarithms

A *logarithm* (sometimes called a *log*) is an exponent to which a constant is raised to obtain a given number. Suppose the following relationship exists among three real numbers a, and x, and y:

$$a^y = x$$

Then y is the *base-a logarithm* of x. This expression is written

$$y = \log_a x$$

The two most common logarithm bases are $a = 10$ and $a = e \approx 2.71828$.

Common logarithms

Base-10 logarithms are also known as *common logarithms* or *common logs*. In equations, common logarithms are denoted by writing "log" without a subscript. For example:

$$\log 10 = 1.000$$

Figure 2.81 is an approximate linear-coordinate graph of the function $y = \log x$. Figure 2.82 is the same graph in semilog coordinates. The domain is limited to the positive reals, that is, to real numbers x such that $x > 0$. The range of the common log function encompasses the entire set of real numbers.

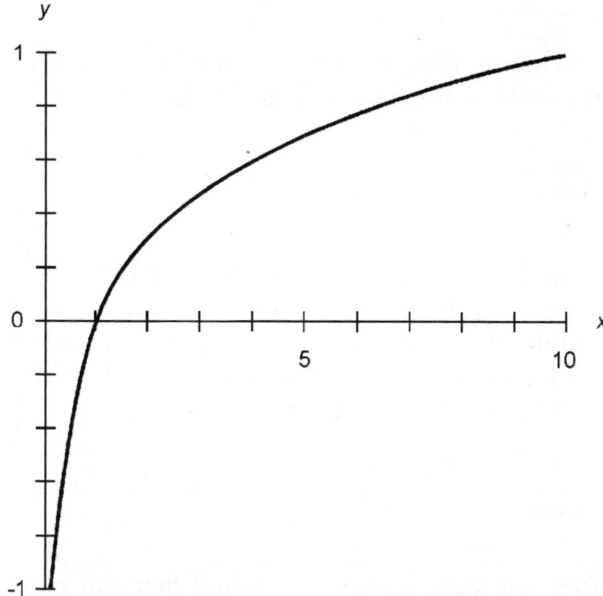

Figure 2.81 Approximate linear-coordinate graph of the common logarithm function.

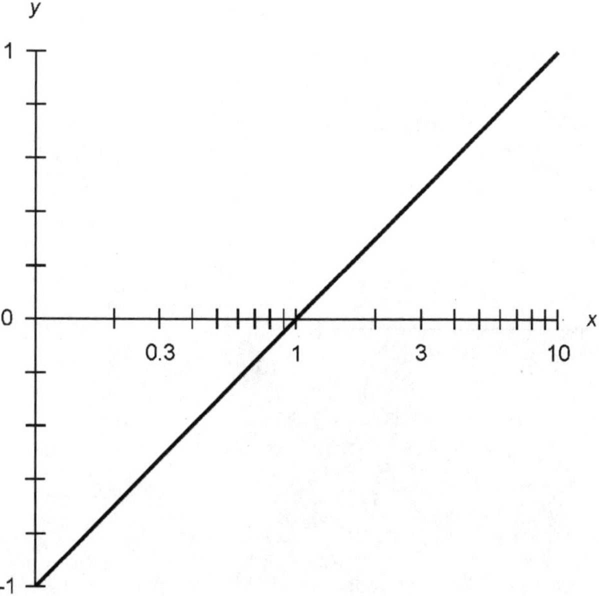

Figure 2.82 Approximate semilog-coordinate graph of the common logarithm function.

Natural logarithms

Base-e logarithms are also called *natural logs* or *Napierian logs.* In equations, the natural-log function is usually denoted "ln" or "$\log_e$." For example:

$$\ln 2.71828 = \log_e 2.71828 \approx 1.00000$$

Figure 2.83 is an approximate linear-coordinate graph of the function $y = \ln x$. Figure 2.84 is the same graph in semilog coordinates. The domain is limited to the positive reals, that is, to real numbers x such that $x > 0$. The range of the natural log function encompasses the entire set of real numbers.

Common logarithm in terms of natural logarithm

Let x be a positive real number. The common logarithm of x can be expressed in terms of the natural logarithms of x and 10:

$$\log x = \ln x/\ln 10 \approx 0.434 \ln x$$

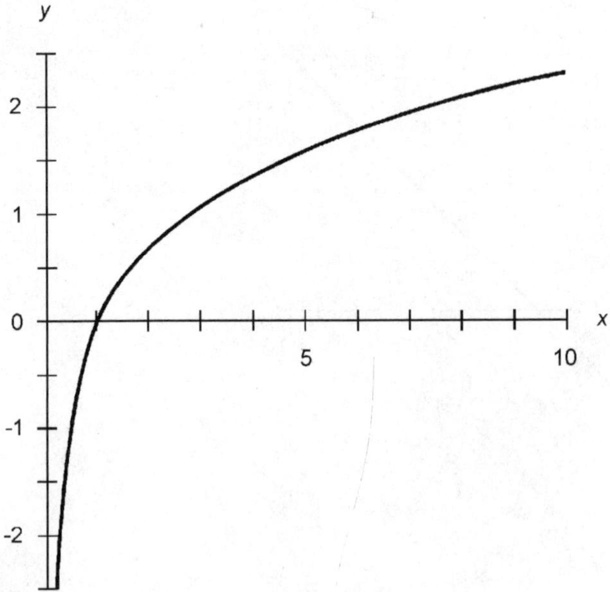

Figure 2.83 Approximate linear-coordinate graph of the natural logarithm function.

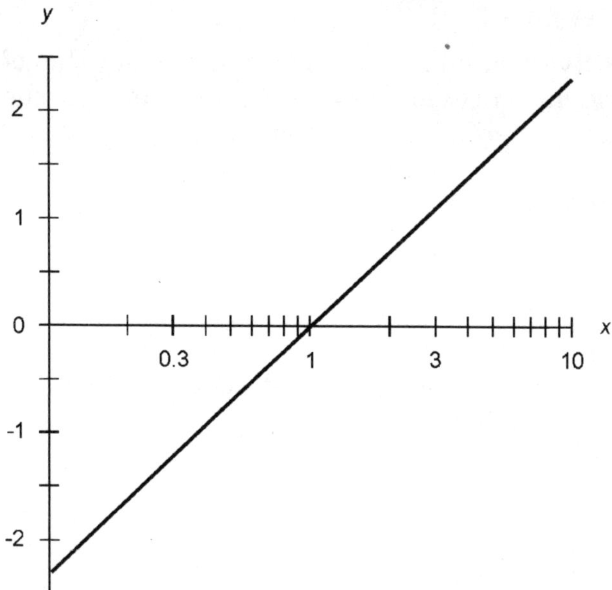

Figure 2.84 Approximate semilog-coordinate graph of the natural logarithm function.

Natural logarithm in terms of common logarithm

Let x be a positive real number. The natural logarithm of x can be expressed in terms of the common logarithms of x and e:

$$\ln x = \log x / \log e \approx 2.303 \log x$$

Common logarithm of product

Let x and y be positive real numbers. The common logarithm of the product is equal to the sum of the common logarithms of the individual numbers:

$$\log xy = \log x + \log y$$

Natural logarithm of product

Let x and y be positive real numbers. The natural logarithm of the product is equal to the sum of the natural logarithms of the individual numbers:

$$\ln xy = \ln x + \ln y$$

Common logarithm of ratio

Let x and y be positive real numbers. The common logarithm of their ratio, or quotient, is equal to the difference between the common logarithms of the individual numbers:

$$\log (x/y) = -\log (y/x) = \log x - \log y$$

Natural logarithm of ratio

Let x and y be positive real numbers. The natural logarithm of their ratio, or quotient, is equal to the difference between the natural logarithms of the individual numbers:

$$\ln (x/y) = -\ln (y/x) = \ln x - \ln y$$

Common logarithm of power

Let x be a positive real number; let y be any real number. The common logarithm of x raised to the power y can be reduced to a product, as follows:

$$\log x^y = y \log x$$

Natural logarithm of power

Let x be a positive real number; let y be any real number. The natural logarithm of x raised to the power y can be reduced to a product, as follows:

$$\ln x^y = y \ln x$$

Common logarithm of reciprocal

Let x be a positive real number. The common logarithm of the reciprocal (multiplicative inverse) of x is equal to the additive inverse of the common logarithm of x, as follows:

$$\log (1/x) = -\log x$$

Natural logarithm of reciprocal

Let x be a positive real number. The natural logarithm of the reciprocal (multiplicative inverse) of x is equal to the additive inverse of the natural logarithm of x, as follows:

$$\ln (1/x) = -\ln x$$

Common logarithm of root

Let x be a positive real number; let y be any real number except zero. The common logarithm of the yth root of x (also denoted as x to the $1/y$ power) is given by:

$$\log (x^{1/y}) = (\log x)/y$$

Natural logarithm of root

Let x be a positive real number; let y be any real number except zero. The natural logarithm of the yth root of x (also denoted as x to the $1/y$th power) is given by:

$$\ln (x^{1/y}) = (\ln x)/y$$

Common logarithm of power of 10

The common logarithm of 10 to any real-number power is always equal to that real number:

$$\log (10^x) = x$$

Natural logarithm of power of *e*

The natural logarithm of e to any real-number power is always equal to that real number:

$$\ln (e^x) = x$$

Natural logarithm of complex number

Let c be a complex number in polar form:

$$c = r \cos \theta + j(r \sin \theta)$$

where r represents the length of the complex vector in the complex-number plane, and θ represents the angle in radians counterclockwise from the abscissa (the positive real-number axis). Let k be an integer. The natural logarithm of c is periodic and can be depicted by the following formula:

$$\ln c = \ln r + j(\theta + 2k\pi)$$

Exponential Functions

An *exponential* is a number that results from the raising of a constant to a given power. Suppose the following relationship exists among three real numbers a, x, and y:

$$a^x = y$$

Then y is the *base-a exponential* of x. The two most common exponential-function bases are $a = 10$ and $a = e \approx 2.71828$.

Common exponentials

Base-10 exponentials are also known as *common exponentials*. For example:

$$10^{-3.000} = 0.001$$

Figure 2.85 is an approximate linear-coordinate graph of the function $y = 10^x$. Figure 2.86 is the same graph in semilog coordinates. The domain encompasses the entire set of real numbers. The range is limited to the positive reals, that is, to real numbers y such that $y > 0$.

Natural exponentials

Base-e exponentials are also known as *natural exponentials*. For example:

$$e^{-3.000} \approx 2.71828^{-3.000} \approx 0.04979$$

Figure 2.87 is an approximate linear-coordinate graph of the function $y = e^x$. Figure 2.88 is the same graph in semilog coordinates. The domain encompasses the entire set of real numbers. The range is limited to the positive reals, that is, to real numbers y such that $y > 0$.

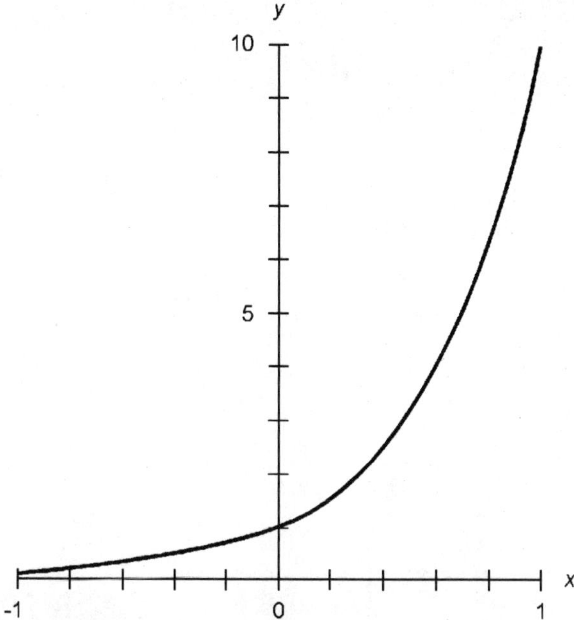

Figure 2.85 Approximate linear-coordinate graph of the common exponential function.

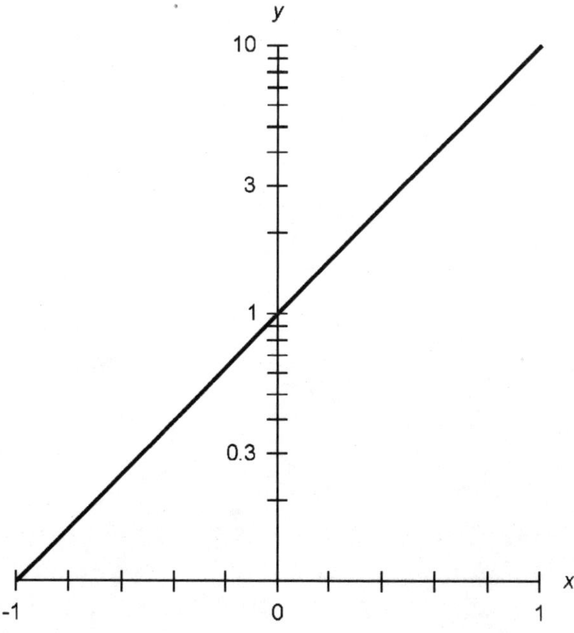

Figure 2.86 Approximate semilog graph of the common exponential function.

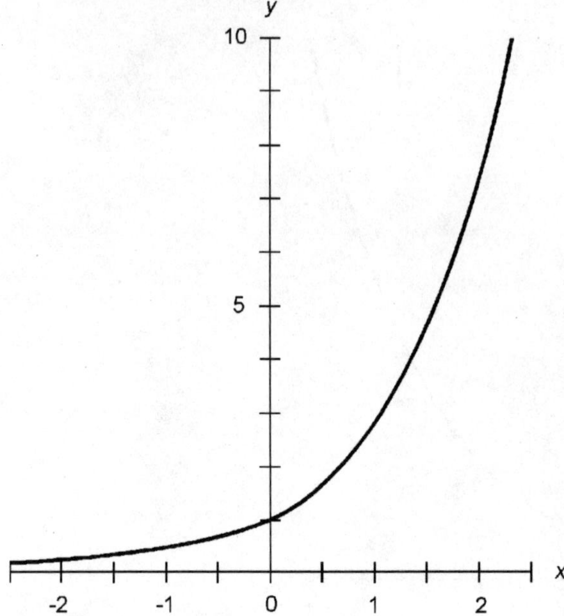

Figure 2.87 Approximate linear-coordinate graph of the natural exponential function.

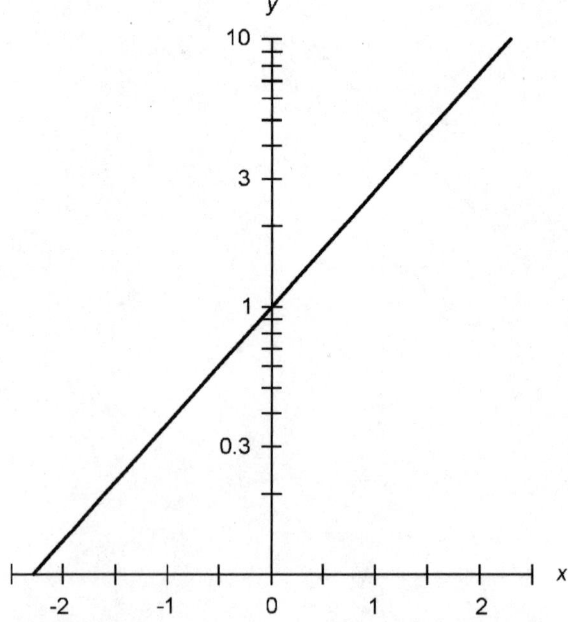

Figure 2.88 Approximate semilog graph of the natural exponential function.

Reciprocal of common exponential

Let x be a real number. The reciprocal of the common exponential of x is equal to the common exponential of the additive inverse of x:

$$1/(10^x) = 10^{-x}$$

Reciprocal of natural exponential

Let x be a real number. The reciprocal of the natural exponential of x is equal to the natural exponential of the additive inverse of x:

$$1/(e^x) = e^{-x}$$

Product of common exponentials

Let x and y be real numbers. The product of the common exponentials of x and y is equal to the common exponential of the sum of x and y:

$$(10^x)(10^y) = 10^{(x+y)}$$

Product of natural exponentials

Let x and y be real numbers. The product of the natural exponentials of x and y is equal to the natural exponential of the sum of x and y:

$$(e^x)(e^y) = e^{(x+y)}$$

Ratio of common exponentials

Let x and y be real numbers. The ratio (quotient) of the common exponentials of x and y is equal to the common exponential of the difference between x and y:

$$10^x/10^y = 10^{(x-y)}$$

Ratio of natural exponentials

Let x and y be real numbers. The ratio (quotient) of the natural exponentials of x and y is equal to the natural exponential of the difference between x and y:

$$e^x/e^y = e^{(x-y)}$$

Exponential of common exponential

Let x and y be real numbers. The yth power of the quantity 10^x is equal to the common exponential of the product xy:

$$(10^x)^y = 10^{(xy)}$$

Exponential of natural exponential

Let x and y be real numbers. The yth power of the quantity e^x is equal to the natural exponential of the product xy:

$$(e^x)^y = e^{(xy)}$$

Product of common and natural exponentials

Let x be a real number. The product of the common and natural exponentials of x is equal to the exponential of x to the base $10e$:

$$(10^x)(e^x) = (10e)^x \approx (27.1828)^x$$

Let x be a nonzero real number. The product of the common and natural exponentials of $1/x$ is equal to the exponential of $1/x$ to the base $10e$:

$$(10^{1/x})(e^{1/x}) = (10e)^{1/x} \approx (27.1828)^{1/x}$$

Ratio of common to natural exponential

Let x be a real number. The ratio (quotient) of the common exponential of x to the natural exponential of x is equal to the exponential of x to the base $10/e$:

$$10^x/e^x = (10/e)^x \approx (3.6788)^x$$

Let x be a nonzero real number. The ratio (quotient) of the common exponential of $1/x$ to the natural exponential of $1/x$ is equal to the exponential of $1/x$ to the base $10/e$:

$$(10^{1/x})/(e^{1/x}) = (10/e)^{1/x} \approx (3.6788)^{1/x}$$

Ratio of natural to common exponential

Let x be a real number. The ratio (quotient) of the natural exponential of x to the common exponential of x is equal to the exponential of x to the base $e/10$:

$$e^x/10^x = (e/10)^x \approx (0.271828)^x$$

Let x be a nonzero real number. The ratio (quotient) of the natural exponential of $1/x$ to the common exponential of $1/x$ is equal to the exponential of $1/x$ to the base $e/10$:

$$(e^{1/x})/(10^{1/x}) = (e/10)^{1/x} \approx (0.271828)^{1/x}$$

Common exponential of ratio

Let x and y be real numbers, with the restriction $y \neq 0$. The common exponential of the ratio (quotient) of x to y is equal to the exponential of $1/y$ to the base 10^x:

$$10^{x/y} = (10^x)^{1/y}$$

Natural exponential of ratio

Let x and y be real numbers, with the restriction $y \neq 0$. The natural exponential of the ratio (quotient) of x to y is equal to the exponential of $1/y$ to the base e^x:

$$e^{x/y} = (e^x)^{1/y}$$

Natural exponential of imaginary number

Let jx be an imaginary number, where x is a non-negative real number expressed as an angle in radians, with the restriction that the angle be reduced to its simplest form, that is, $0 \leq x < 2\pi$. (If $x \geq 2\pi$, a natural-number multiple of 2π can be subtracted to obtain an equivalent value of x such that $0 \leq x < 2\pi$. If $x < 0$, a natural-number multiple of 2π can be added to obtain an equivalent value of x such that $0 \leq x < 2\pi$.) The following equations hold:

$$e^{jx} = \cos x + j \sin x$$

$$e^{-jx} = \cos x - j \sin x$$

Complex number in exponential form

Let c be a complex number $a + jb$, where a and b are real numbers. Let r represent the length of the complex vector in the complex-number plane, and x represent the angle of the vector in radians counterclockwise from the abscissa (the positive real-number axis). Then the following equations hold:

$$r = (a^2 + b^2)^{1/2}$$

$$x = \tan^{-1}(b/a)$$

$$c = r \cos x + j(r \sin x)$$

(If $x \geq 2\pi$, a natural-number multiple of 2π can be subtracted to obtain an equivalent value of x such that $0 \leq x < 2\pi$. If $x < 0$, a natural-number multiple of 2π can be added to obtain an equivalent value of x such that $0 \leq x < 2\pi$.) The value of c can be depicted as the natural exponential of an imaginary number:

$$c = re^{jx}$$

Chapter

3

Applied Mathematics, Calculus, and Differential Equations

This chapter outlines principles and formulas in scientific notation, logic, sequences and series, differentiation, integration, differential equations, and probability.

Scientific Notation

The term *scientific notation* refers to various methods of expressing numbers and variables, including the use of subscripts, superscripts, and powers of 10.

Subscripts

Subscripts modify the meanings of units, constants, and variables. A subscript is placed to the right of the main character (without spacing), is set in smaller type than the main character, and is set below the base line. Numeric subscripts are not italicized; alphabetic subscripts are sometimes italicized. Examples of subscripted quantities are:

Z_0	read "Z sub nought"; stands for characteristic impedance
R_{out}	read "R sub out"; stands for output resistance
y_n	read "y sub n"; represents a variable

Superscripts

Superscripts represent exponents (the raising of a base quantity to a power). Superscripts are usually numerals, but they are sometimes alphabetic characters. Italicized, lowercase English letters from the second half of the alphabet (*n* through *z*) are generally used to represent variable exponents. A superscript is placed to the right of the main character (without spacing), is set in smaller type than the main character, and is set above the base line. Examples of superscripted quantities are:

2^3	read "two cubed"; represents $2 \times 2 \times 2$
e^x	read "*e* to the *x*th"; represents the exponential function of *x*
$y^{1/2}$	read "*y* to the one-half"; represents the square root of *y*

Power-of-10 notation

Extreme numerical values can be represented by an exponential scheme known as *power-of-10 notation*. A numeral in this form is written as follows:

$$m.n \times 10^z$$

where *m* (to the left of the radix point) is a number from the set {1, 2, 3, 4, 5, 6, 7, 8, 9}, *n* (to the right of the radix point) is a non-negative integer, and *z* (the power of 10) can be any integer. Here are some examples of numbers written in this form:

$$2.56 \times 10^6$$

$$8.0773 \times 10^{-18}$$

$$1.000 \times 10^0$$

In some countries, power-of-10 notation requires that that $m = 0$. In this rarely-used form, the above numbers appear as:

$$0.256 \times 10^7$$

$$0.80773 \times 10^{-17}$$

$$0.1000 \times 10^1$$

The "times sign"

The multiplication sign in a power-of-10 expression can be denoted in various ways. Instead of the common cross symbol (×), an asterisk (*) can be used, so the above expressions become:

$$2.56 * 10^{6}$$

$$8.0773 * 10^{-18}$$

$$1.000 * 10^{0}$$

Another alternative is to use a small dot raised above the base line (·), so the expressions appear as:

$$2.56 \cdot 10^{6}$$

$$8.0773 \cdot 10^{-18}$$

$$1.000 \cdot 10^{0}$$

This small dot should not be confused with a radix point, as in the expression:

$$m.n \cdot 10^{z}$$

in which the dot between m and n is a radix point and lies along the base line, while the dot between n and 10^{z} is a multiplication symbol and lies above the base line.

Plain-text exponents

Sometimes it is necessary to express numbers in power-of-10 notation using plain text. This is the case, for example, when transmitting information within the body of an e-mail message (rather than as an attachment). Some electronic calculators and computers use this system. The uppercase letter E indicates that the quantity immediately following is an exponent. In this format, the above expressions are written as:

2.56E6

8.0773E−18

1.000E0

Sometimes the exponent is always written with two numerals

and always includes a plus sign or a minus sign, so the above expressions appear as:

2.56E+06

8.0773E−18

1.000E+00

Another alternative is to use an asterisk to indicate multiplication, and the symbol ^ to indicate a superscript, so the expressions appear as:

2.56 * 10^6

8.0773 * 10^−18

1.000 * 10^0

In all of these examples, the numerical values represented are identical. Respectively, if written out in full, they are:

2,560,000

0.0000000000000000080773

1.000

Rules for use

In printed literature, it is common practice to use power-of-10 notation only when z (the power of 10) is fairly large or small. If $-2 \leq z \leq 2$, numbers are written out in full as a rule, and the power of 10 is not shown. If $z = -3$ or $z = 3$, numbers are sometimes written out in full, and are sometimes depicted in power-of-10 notation. If $z \leq -4$ or $z \geq 4$, values are expressed in power-of-10 notation as a rule. Calculators set to display quantities in power-of-10 notation usually do so for all numbers, even those for which the power of 10 is zero.

Addition and subtraction of numbers is best done by writing numbers out in full, if possible. Thus, for example:

$$(3.045 \times 10^2) + (6.853 \times 10^3)$$

$$= 304.5 + 6853 = 7157.5$$

$$= 7.1575 \times 10^3$$

When numbers are multiplied or divided in power-of-10 notation, the decimal numbers (to the left of the multiplication symbol) are multiplied or divided by each other. Then the powers of 10 are added (for multiplication) or subtracted (for division). Finally, the product or quotient is reduced to standard form. An example is:

$$(3.045 \times 10^2)(6.853 \times 10^3)$$
$$= 20.867385 \times 10^5 = 2.0867385 \times 10^6$$

Truncation

The process of *truncation* deletes all the numerals to the right of a certain point in the decimal part of an expression. Some electronic calculators use this process to fit numbers within their displays. For example, the number 3.830175692803 can be shortened in steps as follows:

3.830175692803

3.83017569280

3.8301756928

3.830175692

3.83017569

3.8301756

3.830175

3.83017

3.8301

3.830

3.83

3.8

3

Rounding

Rounding is a more accurate, and preferred, method of rendering numbers in shortened form. In this process, when a given digit (call it r) is deleted at the right-hand extreme of an expression, the digit q to its left (which becomes the new r after the old r is deleted) is not changed if $0 \leq r \leq 4$. If $5 \leq r \leq 9$, then q is increased by 1 ("rounded up"). Some electronic calculators use rounding rather than truncation. If rounding is used, the number 3.830175692803 can be shortened in steps as follows:

3.830175692803

3.83017569280

3.8301756928

3.830175693

3.83017569

3.8301757

3.830176

3.83018

3.8302

3.830

3.83

3.8

4

Significant figures

When calculations are performed using power-of-10 notation, the number of significant figures in the result cannot be greater than the number of significant figures in the shortest expression in the calculation.

In the foregoing example showing addition, the sum, 7.1575×10^3, must be cut down to four significant figures because the addends have only four significant figures. If the resultant is truncated, it becomes 7.157×10^3. If rounded, it becomes 7.158×10^3.

In the foregoing example showing multiplication, the resultant, 2.0867385×10^6, must be cut down to four significant figures because the multiplicands have only four significant figures. If the resultant is truncated, it becomes 2.086×10^6. If rounded, it becomes 2.087×10^6.

"Downsizing" of resultants is best done at the termination of a calculation process, if that process involves more than one computation.

Boolean Algebra

Boolean algebra is a simple system of mathematical logic using the numbers 0 and 1. These logic states are also called *low* and *high,* respectively. There are three fundamental operations: negation, multiplication, and addition. Boolean operations behave differently than their counterparts in conventional real-number and complex-number algebra.

Negation

Let X be a logical quantity, also known as a *Boolean variable.* The negation of X is written as $-X$, X', or X with a line over it. In the following paragraphs, the dash ($-$) is used to represent Boolean negation. The following rules apply:

$$\text{If } X = 0, \text{ then } -X = 1$$

$$\text{If } X = 1, \text{ then } -X = 0$$

Multiplication

Let X and Y be Boolean variables. The Boolean product can be written in any of these ways:

$$XY$$

$$X \times Y$$

$$X \cdot Y$$

$$X * Y$$

In the following paragraphs, the $\times$ symbol is used. The following rules apply:

$$\text{If } X = 0 \text{ and } Y = 0, \text{ then } X \times Y = 0$$

$$\text{If } X = 0 \text{ and } Y = 1, \text{ then } X \times Y = 0$$

$$\text{If } X = 1 \text{ and } Y = 0, \text{ then } X \times Y = 0$$

$$\text{If } X = 1 \text{ and } Y = 1, \text{ then } X \times Y = 1$$

Addition

Let X and Y be Boolean variables. The Boolean sum is written X + Y. The following rules apply:

$$\text{If } X = 0 \text{ and } Y = 0, \text{ then } X + Y = 0$$

$$\text{If } X = 0 \text{ and } Y = 1, \text{ then } X + Y = 1$$

$$\text{If } X = 1 \text{ and } Y = 0, \text{ then } X + Y = 1$$

$$\text{If } X = 1 \text{ and } Y = 1, \text{ then } X + Y = 1$$

Even-multiple negation

The negation of any logical quantity X, carried out an even number of times, is always equal to X. The following equations hold:

$$-(-X) = X$$

$$-(-(-(-X))) = X$$

$$-(-(-(-(-(-X))))) = X$$

etc.

Odd-multiple negation

The negation of any logical quantity X, carried out an odd number of times, is always equal to −X. The following equations hold:

$$-(-(-X)) = -X$$

$$-(-(-(-(-X)))) = -X$$

$$-(-(-(-(-(-(-X)))))) = -X$$

etc.

Additive identity

The sum of 0 and any logical quantity X is always equal to X. Therefore, 0 is known as the *additive identity element*. The following Boolean equations hold:

$$X + 0 = X$$

$$0 + X = X$$

Addition of 1

The sum of 1 and any logical quantity X is always equal to 1. The following Boolean equations hold:

$$X + 1 = 1$$

$$1 + X = 1$$

Addition of identical quantities

When any logical quantity X is added to itself, the result is equal to X. The following Boolean equation holds:

$$X + X = X$$

Addition of opposites

The sum of any logical quantity X and its negative is always equal to 1. The following Boolean equations hold:

$$X + (-X) = 1$$

$$-X + X = 1$$

Commutativity of addition

The order in which a sum is performed does not matter. For any two logical quantities X and Y, the following Boolean equation holds:

$$X + Y = Y + X$$

Associativity of addition

The manner in which a sum is grouped does not matter. For any three logical quantities X, Y, and Z, the following Boolean equation holds:

$$(X + Y) + Z = X + (Y + Z)$$

Negation of a sum

The negation of the sum of two quantities is equal to the product of the negations of the individual quantities. This is also known as *DeMorgan's rule for addition.* For any two logical quantities X and Y, the following Boolean equation holds:

$$-(X + Y) = (-X) \times (-Y)$$

Multiplicative identity

The product of 1 and any logical quantity X is always equal to X. Therefore, 1 is known as the *Boolean multiplicative identity element.* The following Boolean equations hold:

$$X \times 1 = X$$

$$1 \times X = X$$

Multiplication by zero

The product of 0 and any logical quantity X is always equal to 0. The following Boolean equations hold:

$$X \times 0 = 0$$

$$0 \times X = 0$$

Multiplication of identical quantities

When any logical quantity X is multiplied by itself, the result is equal to X. The following Boolean equation holds:

$$X \times X = X$$

Multiplication of opposites

The product of any logical quantity X and its negative is always equal to 0. The following Boolean equations hold:

$$X \times (-X) = 0$$

$$-X \times X = 0$$

Commutativity of multiplication

The order in which a product is performed does not matter. For any two logical quantities X and Y, the following Boolean equation holds:

$$X \times Y = Y \times X$$

Associativity of multiplication

The manner in which a product is grouped does not matter. For any three logical quantities X, Y, and Z, the following Boolean equation holds:

$$(X \times Y) \times Z = X \times (Y \times Z)$$

Negation of a product

The negation of the product of two quantities is equal to the sum of the negations of the individual quantities. This is also known as *DeMorgan's rule for multiplication.* For any two logical quantities X and Y, the following Boolean equation holds:

$$-(X \times Y) = (-X) + (-Y)$$

Distributivity

Boolean multiplication distributes over Boolean addition. For any three logical quantities X, Y, and Z, the following equation holds:

$$X \times (Y + Z) = (X \times Y) + (X \times Z)$$

Product added to a quantity

For any two logical quantities X and Y, the following Boolean equation holds:

$$X + (X \times Y) = X$$

Quantity added to a product

For any two logical quantities X and Y, the following Boolean equation holds:

$$(X \times (-Y)) + Y = X + Y$$

Propositional Logic

Propositional logic, also known as *propositional calculus, sentential logic,* or *sentential calculus,* is similar to Boolean algebra but involves more operations. Propositions or sentences are symbolized by letters from the middle of the alphabet, such as P, Q, R, and S. Logic values are T (true) and F (false). Standard operations and relations are *negation, conjunction, disjunction, implication,* and *equivalence.*

Negation

Logical negation is also known as the *NOT operation.* Let P be a proposition. The negation of P is written as ¬P or ~P. In the following paragraphs, the tilde (~) is used. See Table 3.1 for logic values.

Conjunction

Logical conjunction is also known as the *AND operation.* Let P and Q be propositions. The conjunction of P and Q is written $P \wedge Q$. See Table 3.1 for logic values.

TABLE 3.1 Propositional Logic Values

P	Q	~P	$P \wedge Q$	$P \vee Q$	$P \rightarrow Q$	$P \leftrightarrow Q$
F	F	T	F	F	T	T
F	T	T	F	T	T	F
T	F	F	F	T	F	F
T	T	F	T	T	T	T

Disjunction

Logical disjunction is also known as the *OR operation*. Let P and Q be propositions. The disjunction of P and Q is written $P \vee Q$. See Table 3.1 for logic values.

Implication

Logical implication is also known as *if-then*. Let P and Q be propositions. The statement "If P, then Q," also read as "P implies Q," is written $P \rightarrow Q$. (The arrow symbol is not to be confused with the identical symbol denoting the fact that the value of a sequence or series approaches a constant.) See Table 3.1 for logic values.

Equivalence

Logical equivalence is also know as *if-and-only-if* or *iff*. Let P and Q be propositions. The statement "P if and only if Q" is written $P \leftrightarrow Q$. See Table 3.1 for logic values.

Even-multiple negation

The negation of any proposition P, carried out an even number of times, is logically equivalent to the original proposition P. The following statements are valid:

$$\sim(\sim P) \leftrightarrow P$$

$$\sim(\sim(\sim(\sim P))) \leftrightarrow P$$

$$\sim(\sim(\sim(\sim(\sim(\sim P))))) \leftrightarrow P$$

etc.

Odd-multiple negation

The negation of any proposition P, carried out an odd number of times, is logically equivalent to $\sim P$. The following statements are valid:

$$\sim(\sim(\sim P)) \leftrightarrow \sim P$$

$$\sim(\sim(\sim(\sim(\sim P)))) \leftrightarrow \sim P$$

$$\sim(\sim(\sim(\sim(\sim(\sim(\sim P)))))) \leftrightarrow \sim P$$

etc.

Idempotence of disjunction

The disjunction of any proposition P and itself is logically equivalent to P. The following statement is valid:

$$P \vee P \leftrightarrow P$$

Excluded middle

The disjunction of any proposition P and its negative is a logically valid statement. That is:

$$P \vee \sim P$$

Commutativity of disjunction

The order in which a disjunction is stated does not matter. For any two propositions P and Q, the following statement is valid:

$$P \vee Q \leftrightarrow Q \vee P$$

Associativity of disjunction

The manner in which a disjunction is grouped does not matter. For any three propositions P, Q, and R, the following statement is valid:

$$(P \vee Q) \vee R \leftrightarrow P \vee (Q \vee R)$$

DeMorgan's law for disjunction

The negation of the disjunction of two propositions is equal to the conjunction of the negations of the propositions. For any two propositions P and Q, the following statement is valid:

$$\sim(P \vee Q) \leftrightarrow (\sim P) \wedge (\sim Q)$$

Idempotence of conjunction

The conjunction of any proposition P and itself is logically equivalent to P. The following statement is valid:

$$P \wedge P \leftrightarrow P$$

Commutativity of conjunction

The order in which a conjunction is stated does not matter. For any two propositions P and Q, the following statement is valid:

$$P \wedge Q \leftrightarrow Q \wedge P$$

Associativity of conjunction

The manner in which a conjunction is grouped does not matter. For any three propositions P, Q, and R, the following statement is valid:

$$(P \wedge Q) \wedge R \leftrightarrow P \wedge (Q \wedge R)$$

DeMorgan's law for conjunction

The negation of the conjunction of two propositions is equal to the disjunction of the negations of the propositions. For any two propositions P and Q, the following statement is valid:

$$\sim(P \wedge Q) \leftrightarrow (\sim P) \vee (\sim Q)$$

Modus ponens

For any two propositions P and Q, if P is true and P implies Q, then Q is true. The following logical statement is valid:

$$(P \wedge (P \rightarrow Q)) \rightarrow Q$$

Modus tollens

For any two propositions P and Q, if Q is false and P implies Q, then P is false. The following logical statement is valid:

$$(\sim Q \wedge (P \rightarrow Q)) \rightarrow \sim P$$

Proof by cases

For any two propositions P and Q, if P implies Q and the negation of P implies Q, then Q is true. The following logical statement is valid:

$$((P \rightarrow Q) \wedge (\sim P \rightarrow Q)) \rightarrow Q$$

Elimination of cases

For any two propositions P and Q, the conjunction of ~P with the disjunction of P and Q implies Q. The following logical statement is valid:

$$(\sim P \wedge (P \vee Q)) \rightarrow Q$$

Contradiction

For any proposition P, if P and ~P are simultaneously true, then anything follows. The following statement is valid:

$$(P \wedge \sim P) \rightarrow Q$$

Contraposition

For any two propositions P and Q, if P implies Q, then ~Q implies ~P. Also, if ~Q implies ~P, then P implies Q. The following statement is valid:

$$(P \rightarrow Q) \leftrightarrow (\sim Q \rightarrow \sim P)$$

Transitivity of implication

For any three propositions P, Q, and R, if P implies Q and Q implies R, then P implies R. The following statement is valid:

$$((P \rightarrow Q) \wedge (Q \rightarrow R)) \rightarrow (P \rightarrow R)$$

Transitivity of equivalence

For any three propositions P, Q, and R, if P is logically equivalent to Q and Q is logically equivalent to R, then P is logically equivalent to R. The following statement is valid:

$$((P \leftrightarrow Q) \wedge (Q \leftrightarrow R)) \rightarrow (P \leftrightarrow R)$$

Sequences and Series

This section contains definitions and formulas for sequences and series commonly encountered in engineering and science.

Infinite sequence

An *infinite sequence* A is a function f whose domain is the set $\boldsymbol{N}$ of positive integers. The following formula applies:

$$A = a_1, a_2, a_3, \ldots, a_n, \ldots = f(1), f(2), f(3), \ldots$$

Occasionally the set of non-negative integers is specified for $\boldsymbol{N}$:

$$A = a_0, a_1, a_2, \ldots, a_n, \ldots = f(0), f(1), f(2), \ldots$$

Positive infinite sequence

A *positive infinite sequence* is an infinite sequence, all of whose terms are real numbers greater than zero.

Negative infinite sequence

A *negative infinite sequence* is an infinite sequence, all of whose terms are real numbers less than zero.

Alternating infinite sequence

An *alternating infinite sequence* is an infinite sequence of positive and negative terms such that, if a_n and a_{n+1} are successive terms, the following statements hold:

$$\text{If } a_n < 0, \text{ then } a_{n+1} > 0$$

$$\text{If } a_n > 0, \text{ then } a_{n+1} < 0$$

Bounded infinite sequence

A *bounded infinite sequence* is a sequence such that, for every $n \in \boldsymbol{N}$, the defining function f has the following property:

$$p \le f(n) \le q$$

for some $p \in \boldsymbol{N}$ and some $q \in \boldsymbol{N}$.

Nondecreasing infinite sequence

A *nondecreasing infinite sequence* is a sequence such that, for every $n \in \boldsymbol{N}$, the defining function f has the following property:

$$f(n + 1) \ge f(n)$$

Nonincreasing infinite sequence

A *nonincreasing infinite sequence* is a sequence such that, for every $n \in \mathbf{N}$, the defining function f has the following property:

$$f(n + 1) \leq f(n)$$

Convergent infinite sequence

A *convergent infinite sequence* is a sequence in which the value of the defining function f approaches a specific real number s as n increases without bound:

$$f(n) \rightarrow s \text{ as } n \rightarrow \infty$$

(The arrow symbol is not to be confused with the identical symbol denoting the fact that one logical proposition implies another.) Example of convergent infinite sequences are:

$$A = f(n) = 1/(2^n) = \tfrac{1}{2}, \tfrac{1}{4}, \tfrac{1}{8}, \tfrac{1}{16}, \tfrac{1}{32}, \ldots$$

$$B = g(n) = 1/(-2^n) = -\tfrac{1}{2}, \tfrac{1}{4}, -\tfrac{1}{8}, \tfrac{1}{16}, -\tfrac{1}{32}, \ldots$$

The values of both of these infinite sequences converge toward zero.

Uniqueness of limit

Suppose $A = f(n)$ is a convergent sequence such that the following statements are both valid:

$$f(n) \rightarrow s_1 \text{ as } n \rightarrow \infty$$

$$f(n) \rightarrow s_2 \text{ as } n \rightarrow \infty$$

Then $s_1 = s_2$. In other words, an infinite sequence can never converge to more than one value.

Divergent infinite sequence

A *divergent infinite sequence* is a sequence in which the value of the defining function f increases and/or decreases without bound as n increases without bound:

$$f(n) \rightarrow +\infty \text{ as } n \rightarrow \infty$$

and/or

$$f(n) \rightarrow -\infty \text{ as } n \rightarrow \infty$$

(The arrow symbol is not to be confused with the identical symbol denoting the fact that one logical proposition implies another.) Examples of a divergent infinite sequence are:

$$f(n) = 2^n = 2, 4, 8, 16, 32, \ldots$$

$$f(n) = (-2)^n = -2, 4, -8, 16, -32, \ldots$$

Term-by-term reciprocal

Let A be an infinite sequence, none of whose terms are equal to zero. Let B be the infinite sequence whose terms are reciprocals of the corresponding terms of A:

$$A = a_1, a_2, a_3, \ldots$$

$$B = 1/a_1, 1/a_2, 1/a_3, \ldots$$

Then if A diverges, B converges toward zero.

Boundedness and convergence

Let A be an infinite sequence $a_1, a_2, a_3, \ldots$. The following statements are always true: if A is bounded and nondecreasing, then A is convergent; if A is bounded and nonincreasing, then A is convergent.

Alteration of initial terms in sequence

Let A and B be infinite sequences that are identical except for the first k terms; the first k terms have different values a_n and b_n, as follows:

$$A = a_1, a_2, a_3, \ldots, a_k, a_{k+1}, a_{k+2}, a_{k+3}, \ldots$$

$$B = b_1, b_2, b_3, \ldots, b_k, a_{k+1}, a_{k+2}, a_{k+3}, \ldots$$

Then the following statements are always true: if A is convergent, then B is convergent; and if A is divergent, then B is divergent.

Infinite series

An *infinite series* consists of the elements of an infinite sequence arranged in a specific order, separated by addition symbols:

$$a_1 + a_2 + a_3 + \dots + a_n + \dots$$

or

$$a_0 + a_1 + a_2 + \dots + a_n + \dots$$

Positive infinite series

A *positive infinite series* is the sum of an infinite sequence, all of whose terms are real numbers greater than zero.

Negative infinite series

A *negative infinite series* is an the sum of an infinite sequence, all of whose terms are real numbers less than zero.

Alternating infinite series

An *alternating infinite series* is the sum of an infinite sequence of positive and negative terms such that, if a_n and a_{n+1} are successive terms, the following statements hold:

$$\text{If } a_n < 0, \text{ then } a_{n+1} > 0$$

$$\text{If } a_n > 0, \text{ then } a_{n+1} < 0$$

Partial sum

The *partial sum* S_n of an infinite series is the sum of its first n terms:

$$S_n = a_1 + a_2 + a_3 + \dots + a_n$$

Convergent infinite series

A *convergent infinite series* is an infinite series whose partial sum S_n approaches a specific finite number S as n increases without bound:

$$S_n \rightarrow S \text{ as } n \rightarrow \infty$$

$$\therefore$$

$$S = a_1 + a_2 + a_3 + \ldots$$

An example of a convergent infinite series is:

$$C = \tfrac{1}{2} + \tfrac{1}{4} + \tfrac{1}{8} + \ldots = 1$$

Uniqueness of sum

Suppose S is a convergent infinite series such that the following statements are both valid for the partial sums:

$$S_n \rightarrow T_1 \text{ as } n \rightarrow \infty$$

$$S_n \rightarrow T_2 \text{ as } n \rightarrow \infty$$

Then $T_1 = T_2$. In other words, the partial sum of an infinite series can never converge to more than one value.

Divergent infinite series

A *divergent infinite series* is an infinite series that is not convergent; its partial sum S_n does not approach any specific finite number as n increases without bound. An example of a divergent infinite series is:

$$D = 1 + 2 + 3 + 4 + \ldots$$

Conditionally convergent infinite series

A *conditionally convergent infinite series* is an infinite series that is convergent for certain values of a parameter x, but is divergent for other values of x. An example of a conditionally convergent infinite series is:

$$C_{cc} = 1 - x + x^2 - x^3 + x^4 - x^5 + \ldots$$

This infinite series converges to $1/(1 + x)$ if $-1 < x < 1$, but diverges if $x \leq -1$ or $x \geq 1$.

Absolutely convergent infinite series

An *absolutely convergent infinite series* is an infinite series such that the sum of the absolute values of all its terms is convergent. Let

$$S = a_1 + a_2 + a_3 + \ldots$$

be an infinite series. Then S is absolutely convergent if and only if the following series converges:

$$S_{\text{abs}} = |a_1| + |a_2| + |a_3| + \ldots$$

Alteration of initial terms in series

Let S and T be infinite series that are identical except for the first k terms; the first k terms have different values a_n and b_n, as follows:

$$S = a_1 + a_2 + a_3 + \ldots + a_k + a_{k+1} + a_{k+2} + a_{k+3}, \ldots$$

$$T = b_1 + b_2 + b_3 + \ldots + b_k + a_{k+1} + a_{k+2} + a_{k+3}, \ldots$$

Then the following statements are always true: if S is convergent, then T is convergent; and if S is divergent, then T is divergent.

Terms in convergent infinite series

If S is a convergent infinite series, then the corresponding sequence A converges toward zero. Let S, a partial sum S_n, and A be denoted as follows:

$$S = a_1 + a_2 + a_3 + \ldots$$

$$S_n = a_1 + a_2 + a_3 + \ldots + a_n$$

$$A = a_1, a_2, a_3, \ldots$$

If $S_n \rightarrow S$ as $n \rightarrow \infty$, then $a_n \rightarrow 0$ as $n \rightarrow \infty$, where S is a specific, unique real number.

Terms in divergent infinite series

Let S be an infinite series. Suppose the corresponding sequence A does not converge toward zero. Then S is divergent. Let S, a partial sum S_n, and A be denoted as follows:

$$S = a_1 + a_2 + a_3 + \ldots$$

$$S_n = a_1 + a_2 + a_3 + \ldots + a_n$$

$$A = a_1, a_2, a_3, \ldots$$

If it is not the case that $a_n \to 0$ as $n \to \infty$, then there exists no real number such that $S_n \to S$ as $n \to \infty$.

Multiple of convergent infinite series

Let S be a convergent infinite series, and let k be a nonzero real number. If each term of S is muliplied by k, the resulting series kS converges. Let S, a partial sum S_n, kS, and a partial sum kS_n be denoted as follows:

$$S = a_1 + a_2 + a_3 + \ldots$$

$$S_n = a_1 + a_2 + a_3 + \ldots + a_n$$

$$kS = ka_1 + ka_2 + ka_3 + \ldots$$

$$kS_n = ka_1 + ka_2 + ka_3 + \ldots + ka_n$$

If $S_n \to S$ as $n \to \infty$, then $kS_n \to kS$ as $n \to \infty$.

Factorial

For a given positive integer n, the number n *factorial* (written $n!$) is the product of all the positive integers up to, and including, n:

$$n! = 1 \times 2 \times 3 \times 4 \times \ldots \times n$$

Factorials of n become large rapidly as n increases. Factorials of the first several positive integers are:

$$1! = 1$$

$$2! = 2$$

$$3! = 6$$

$$4! = 24$$

$$5! = 120$$

$$6! = 720$$

$$7! = 5{,}040$$

$$8! = 40{,}320$$

$$9! = 362{,}880$$

$$10! = 3{,}628{,}800$$

$$11! = 39{,}916{,}800$$

$$12! = 479{,}001{,}600$$

Arithmetic series

An *arithmetic series* is a series A such that:

$$f(a_{n+1}) = a_n + d$$

$$\therefore$$

$$A = a_1 + (a_1 + d) + (a_1 + 2d) + (a_1 + 3d) + \ldots$$

where d is a constant called the *difference*. For example, if $a_1 = 5$ and $d = 2$, then:

$$A = 5 + 7 + 9 + 11 + 13 \ldots$$

Geometric series

A *geometric series* is a series G such that:

$$f(a_{n+1}) = a_n/r$$

$$\therefore$$

$$G = a_1 + (a_1/r) + (a_1/r^2) + (a_1/r^3) + \ldots$$

where r is a constant called the *ratio*. For example, if $a_1 = 3$ and $r = 2$, then:

$$G = 3 + 3/2 + 3/4 + 3/8 + 3/16 + \ldots$$

Harmonic series

A *harmonic series* is a series $H = a_1 + a_2 + a_3 + \ldots + a_n + \ldots$ such that the series consisting of the reciprocal of each term is an arithmetic series A:

$$1/f(a_{n+1}) = 1/(a_n + d)$$

$$\therefore$$

$$H = 1/b_1 + 1/b_2 + 1/b_3 + \ldots + 1/b_n + \ldots$$

Where d is a constant. For example, if $a_1 = 1$ and $d = 3$, then:

$$H = 1, 1/4, 1/7, 1/10, 1/13, \ldots$$

Power series

A *power series* is a series P such that the following equation holds for coefficients a_i (where i is a non-negative integer subscript) and a variable x:

$$P = a_0 + a_1x + a_2x^2 + a_3x^3 + \ldots + a_nx^n + \ldots$$

where $a_1, a_2, a_3, \ldots + a_n, \ldots$ is a sequence. For example, if the sequence of coefficients is 2, 4, 6, 8, ... then:

$$P = 2 + 4x + 6x^2 + 8x^3 + \ldots$$

Arithmetic-geometric series

An *arithmetic-geometric series* is a series C such that, for constants a and b and a variable x:

$$C = a + (a + b)x + (a + 2b)x^2 + (a + 3b)x^3 + \ldots$$

$$+ (a + (n - 1)b)x^{(n-1)}$$

Taylor series

A *Taylor series*, also known as a *Taylor expansion*, is a series T such that the following equation holds for some function f, its successive derivatives f', f'', f''', ..., some constant a, and some variable x:

$$T =$$

$$f(a) + ((x - a)(f'(a))$$

$$+$$

$$((x - a)^2(f''(a))/2$$

$$+$$

$$((x - a)^3(f'''(a))/3!$$

$$+$$

$$\ldots$$

Maclaurin series

A *Maclaurin series* M is a Taylor series with $a = 0$. The following holds for some function f, its successive derivatives f', f'', f''', ..., and some variable x:

$$M =$$

$$f(0)$$

$$+$$

$$(x(f'(0))$$

$$+$$

$$(x^2(f''(0))/2$$

$$+$$

$$(x^3(f'''(0))/3!$$

$$+$$

$$\ldots$$

Binomial series

Let x be a variable, let a be a real number, and let n be a positive integer. Then the value of $(x + a)^n$ can be found by summing a finite series known as the *Binomial series*:

$$(x + a)^n =$$

$$a^n$$

$$+$$

$$na^{n-1}x$$

$$+$$

$$(n(n-1)/2!)a^{n-2}x^2$$

$$+$$

$$(n(n - 1)(n - 2)/3!)a^{n-3}x^3$$

$$+$$

$$(n(n - 1)(n - 2)(n - 3)/4!)a^{n-4}x^4$$

$$+$$

$$\cdots$$

$$+$$

$$x^n$$

General Fourier series

A *Fourier series* represents a periodic function F having period $2L$, such that the following equation holds for some variable x, some sequence $a_1, a_2, a_3, \ldots + a_n, \ldots$, and some sequence $b_1, b_2, b_3, \ldots + b_n, \ldots$:

$$F =$$

$$a_0/2 + a_1 \cos(\pi x/L) + b_1 \sin(\pi x/L)$$

$$+$$

$$a_2 \cos(2\pi x/L) + b_2 \sin(2\pi x/L)$$

$$+$$

$$a_3 \cos(3\pi x/L) + b_3 \sin(3\pi x/L) + \ldots$$

$$+$$

$$a_n \cos(n\pi x/L) + b_n \sin(n\pi x/L)$$

$$+$$

$$\ldots$$

Fourier series for square wave

Figure 3.1 is a graph of a *square wave* with peak values of $\pi/4$ and $-\pi/4$, and with period 2π. This function can be represented by the following infinite series:

$$f(x) = \sin x + (\sin 3x)/3 + (\sin 5x)/5 + (\sin 7x)/7 + (\sin 9x)/9 + \ldots$$

Fourier series for ramp wave

Figure 3.2 is a graph of a *ramp wave* with peak values of $\pi/2$ and $-\pi/2$, and with period 2π. This function can be represented by the following infinite series:

$$f(x) = \sin x - (\sin 2x)/2 + (\sin 3x)/3 - (\sin 4x)/4 + (\sin 5x)/5 - \ldots$$

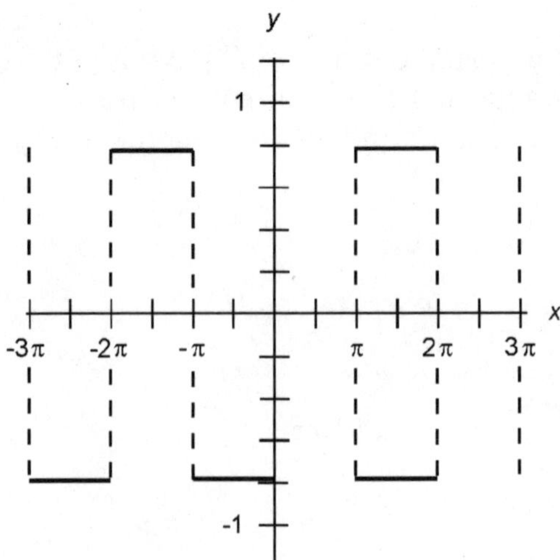

Figure 3.1 Square wave.

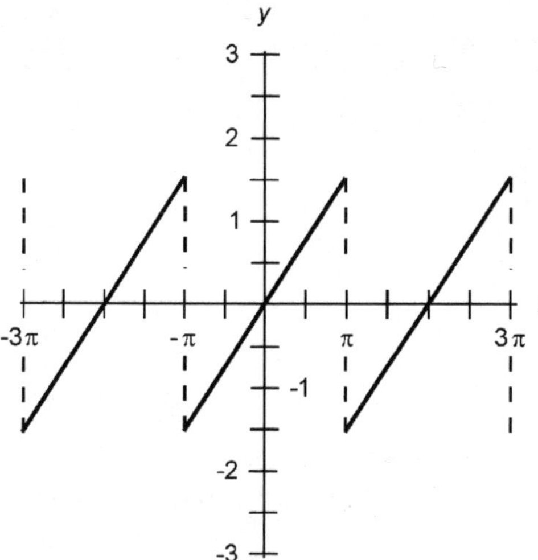

Figure 3.2 Ramp wave.

Fourier series for triangular wave

Figure 3.3 is a graph of a *triangular wave* with peak values of $\pi^2/8$ and $-\pi^2/8$, and with period 2π. This function can be represented by the following infinite series:

$$f(x) = \cos x + (\cos 3x)/3^2 + (\cos 5x)/5^2$$
$$+ (\cos 7x)/7^2 + (\cos 9x)/9^2 + \ldots$$

Fourier series for half-rectified sine wave

Figure 3.4 is a graph of a *half-rectified sine wave* with peak values of $\pi/2$ and 0, and with period 2π. This function can be represented by the following infinite series:

$$f(x) = \tfrac{1}{2} + (\pi/4)\sin x - (\cos 2x)/3 - (\cos 4x)/(3 \times 5)$$
$$- (\cos 6x)/(5 \times 7) - (\cos 8x)/(7 \times 9) - \ldots$$

Fourier series for full-rectified sine wave

Figure 3.5 is a graph of a *full-rectified sine wave* with peak values of $\pi/4$ and 0, and with period π. This function can be represented by the following infinite series:

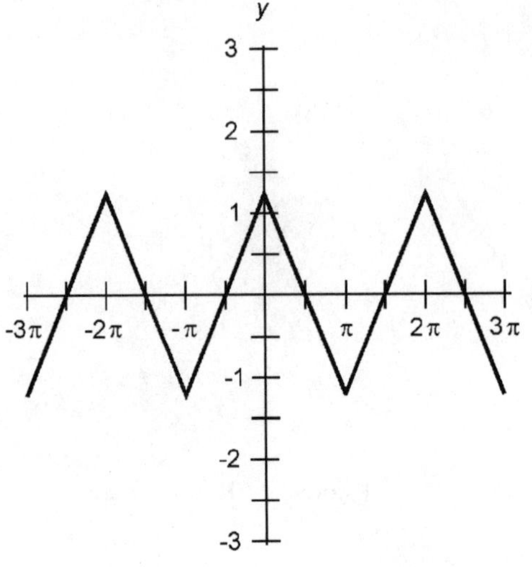

Figure 3.3 Triangular wave.

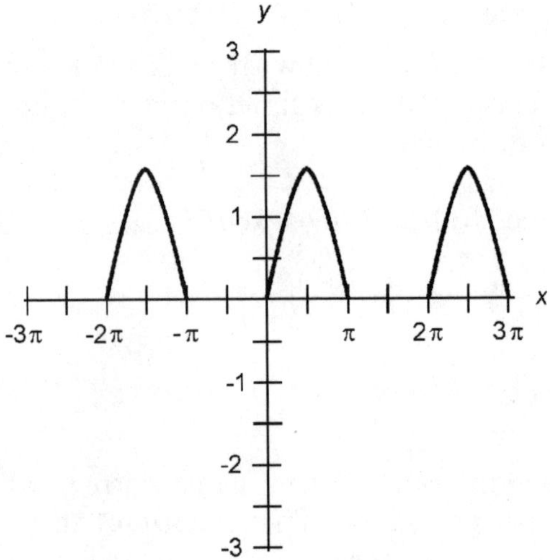

Figure 3.4 Half-rectified sine wave.

$$f(x) = \tfrac{1}{2} - (\cos 2x)/3 - (\cos 4x)/(3 \times 5)$$
$$- (\cos 6x)/(5 \times 7) - (\cos 8x)/(7 \times 9) - \ldots$$

Scalar Differentiation

Let f be a real-number function, let x_0 be an element of the domain of f, and let y_0 be an element of the range of f such that $y_0 = f(x_0)$. Suppose that f is continuous in the vicinity of (x_0,y_0)

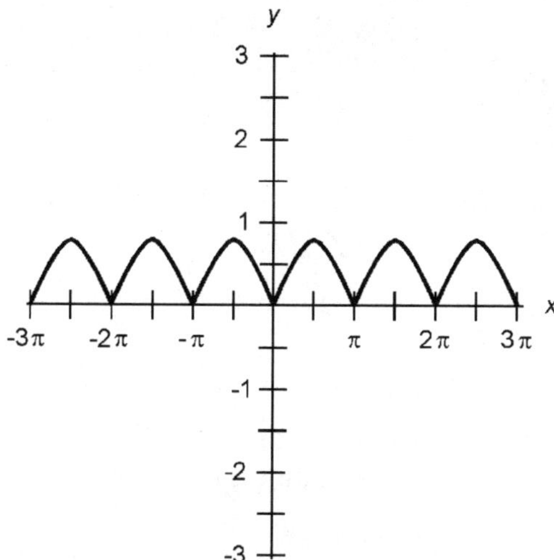

Figure 3.5 Full-rectified sine wave.

as shown in Fig. 3.6. Let Δx represent a small change in x, and let Δy represent the change in $y = f(x)$ that occurs as a result of Δx. Then the *derivative* of f at (x_0, y_0) is defined as:

$$f'(x_0) = \mathrm{Lim}_{\Delta x \to 0} (\Delta y / \Delta x)$$

If f is differentiable at all points in its domain, then the *derivative of f* is defined and can be denoted in several ways:

$$f'(x) = d / dx\ (f) = df/dx = dy/dx$$

In Fig. 3.6, the slope of line L approaches $f'(x_0)$ as $\Delta x \to 0$.

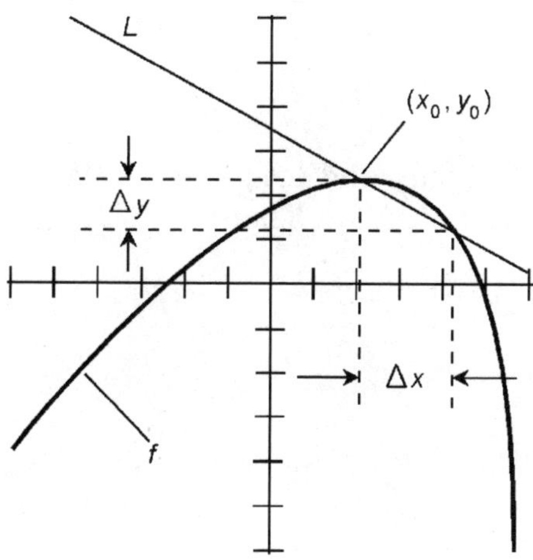

Figure 3.6 Derivative represented by the slope of a curve at a point.

For this reason, the derivative $f'(x_0)$ is graphically described as the slope of a line tangent to the curve of f at the point (x_0,y_0).

Second derivative

The *second derivative* of a function f is the derivative of its derivative. This can be denoted in various ways:

$$f''(x) = d^2/dx^2\ (f) = d^2f/dx^2 = d^2y/dx^2$$

Higher-order derivatives

The *nth derivative* of a function f is the derivative taken in succession n times, where n is a positive integer. This can be denoted as follows:

$$f^{(n)}(x) = d^n/dx^n\ (f) = d^nf/dx^n = d^ny/dx^n$$

Derivative of constant

The derivative of a constant is always equal to zero. Let f be a function of x such that $f(x) = c$, where c is a real number. Then:

$$d(c)/dx = 0$$

Derivative of sum of two functions

Let f and g be two different functions, and let $f + g = f(x) + g\ (x)$ for all x in the domains of both f and g. Then:

$$d(f + g)/dx = df/dx + dg/dx$$

Derivative of difference of two functions

Let f and g be two different functions, and let $f - g = f(x) - g(x)$ for all x in the domains of both f and g. Then:

$$d(f - g)/dx = df/dx - dg/dx$$

Derivative of function multiplied by a constant

Let f be a function, let x be an element of the domain of f, and let c be a constant. Then:

$$d(cf)/dx = c(df/dx)$$

Derivative of product of two functions

Let f and g be two different functions. Define the product of f and g as follows:

$$f \times g = f(x) \times g(x)$$

for all x in the domains of both f and g. Then:

$$d(f \times g)/dx = f \times (dg/dx) + g \times (df/dx)$$

Derivative of product of three functions

Let f, g, and h be three different functions. Define the product of f, g, and h as follows:

$$f \times g \times h = f(x) \times g(x) \times h(x)$$

for all x in the domains of f, g, and h. Then:

$$\begin{aligned} d(f \times g \times h)/dx &= f \times g \times (dh/dx) \\ &\quad + f \times h \times (dg/dx) + g \times h \times (df/dx) \end{aligned}$$

Derivative of quotient of two functions

Let f and g be two different functions, and define $f/g = f(x)/g(x)$ for all x in the domains of both f and g. Then:

$$d(f/g)/dx = (g \times (df/dx) - f \times (dg/dx))/g^2$$

where $g^2 = g(x) \times g(x)$, not to be confused with d^2g/dg^2.

Reciprocal derivatives

Let f be a function, and let x and y be variables such that $y = f(x)$. The following formulas hold:

$$dy/dx = 1/(dx/dy)$$

$$dx/dy = 1/(dy/dx)$$

Derivative of function raised to a power

Let f be a function, let x be an element of the domain of f, and let n be a positive integer. Then:

$$d(f^n)/dx = n(f^{n-1}) \times df/dx$$

where f^n denotes f multiplied by itself n times, not to be confused with the nth derivative.

Chain rule

Let f and g be two different functions of a variable x. The derivative of the composite function $f(g(x))$ is given by the following formula:

$$(f(g(x)))' = f'(g(x)) \times g'(x)$$

Partial derivative of two-variable function

Let f be a real-number function of two variables x and y. Let (x_0,y_0) be an element of the domain of f. Suppose that f is continuous in the vicinity of (x_0,y_0). Let Δx represent a small change in x. The *partial derivative of f with respect to x* at the point (x_0,y_0) is obtained by treating y as a constant:

$$\partial f/\partial x = \text{Lim}_{\Delta x \to 0}\ (f(x_0+\Delta x,y_0) - f(x_0,y_0))/\Delta x$$

Let Δy represent a small change in y. The *partial derivative of f with respect to y* at the point (x_0,y_0) is obtained by treating x as a constant:

$$\partial f/\partial y = \text{Lim}_{\Delta y \to 0}\ (f(x_0,y_0+\Delta y) - f(x_0,y_0))/\Delta y$$

Partial derivative of multivariable function

Let f be a real-number function of n variables $x_1, x_2, ..., x_n$. Let $(x_{10},x_{20},...,x_{n0})$ be an element of the domain of f. Suppose that f is continuous in the vicinity of $(x_{10},x_{20},...,x_{n0})$. Let Δx_i represent a small change in x_i, where x_i is one of the variables of the domain of f. The *partial derivative of f with respect to x_i* at the point $(x_{10},x_{20},...,x_{n0})$ is defined as the following limit as Δx_i approaches zero:

$$\partial f/\partial x_i = \text{Lim}\ (f(x_{10},x_{20},\ldots,x_{i0}+\Delta x_i,\ldots x_{n0}) - f(x_{10},x_{20},\ldots,x_{n0}))/\Delta x_i$$

That is, all the variables except x_i are treated as constants.

Tangent to curve at point (x_0,y_0)

Let f be a function such that $y = f(x)$. Let (x_0,y_0) be a point on the graph of f, and suppose f is continuous at (x_0,y_0). Let L be a line tangent to the graph of f, and suppose L passes through (x_0,y_0) as shown in Fig. 3.7. Suppose the derivative of f at (x_0, y_0) is equal to some real number m. Then the equation of line L is given by the following:

$$y - y_0 = m(x - x_0)$$

If the derivative of f at (x_0,y_0) is zero, then the equation of the line L tangent to f at that point is given by:

$$y = y_0$$

The derivative of f at (x_0,y_0) is undefined when the equation of the line L tangent to f at that point is given by:

$$x = x_0$$

Normal to curve at point (x_0,y_0)

Let f be a function such that $y = f(x)$. Let (x_0,y_0) be a point on the graph of f, and suppose f is continuous at (x_0,y_0). Let L be a line normal (perpendicular) to the graph of f, and suppose L

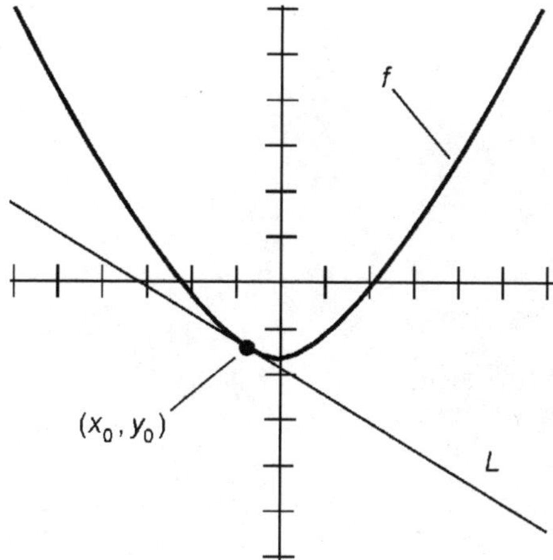

Figure 3.7 Line tangent to function at a point.

passes through (x_0,y_0) as shown in Fig. 3.8. Suppose the derivative of f at (x_0,y_0) is equal to some nonzero real number m. Then the equation of line L is given by the following:

$$y - y_0 = (-x + x_0)/m$$

If the derivative of f at (x_0,y_0) is zero, then the equation of the line L normal to f at that point is given by:

$$x = x_0$$

The derivative of f at (x_0,y_0) is undefined when the equation of the line L normal to f at that point is given by:

$$y = y_0$$

Angle of intersection between curves

Let f and g be functions. Let (x_0,y_0) be a point at which the graphs of f and g intersect, and suppose f and g are both continuous at (x_0,y_0), as shown in Fig. 3.9. Suppose the derivative of f at (x_0,y_0) is equal to some nonzero real number m, and the derivative of g at (x_0,y_0) is equal to some nonzero real number n. Then the acute angle θ at which the graphs intersect is given by the following:

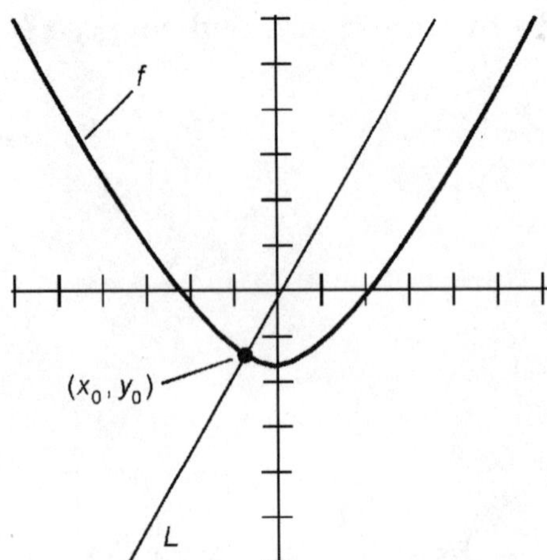

Figure 3.8 Line normal to function at a point.

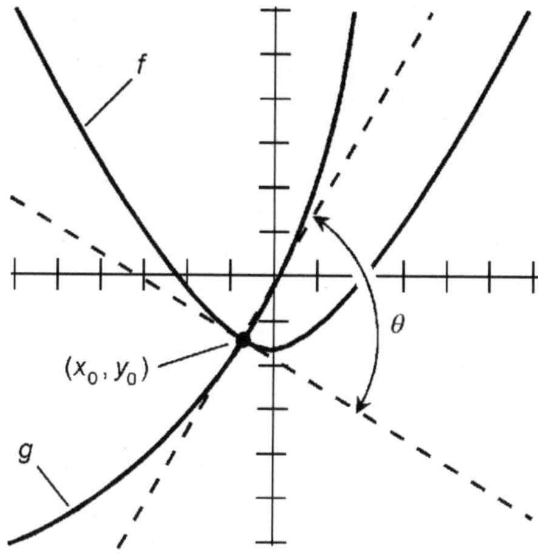

Figure 3.9 Angle of intersection between two curves at a point.

$$\theta = \tan^{-1}((m - n)/(mn + 1))$$

if $m > n$, and

$$\theta = 180° - \tan^{-1}((m - n)/(mn + 1))$$

if $m < n$, for angle measures in degrees. If $m = n$, then $\theta = 180°$, and the two curves are tangent to each other at the point (x_0,y_0). For angle measures in radians, substitute π for 180°.

Local maximum

Let f be a function such that $y = f(x)$. Let (x_0,y_0) be a point on the graph of f, and suppose f is twice differentiable at (x_0,y_0). Suppose the following are both true:

$$f'(x_0) = 0$$

$$f''(x_0) < 0$$

Then (x_0,y_0) is a *local maximum* in the graph of f. An example is shown in Fig. 3.10.

Local minimum

Let f be a function such that $y = f(x)$. Let (x_0,y_0) be a point on the graph of f, and suppose f is twice differentiable at (x_0,y_0). Suppose the following are both true:

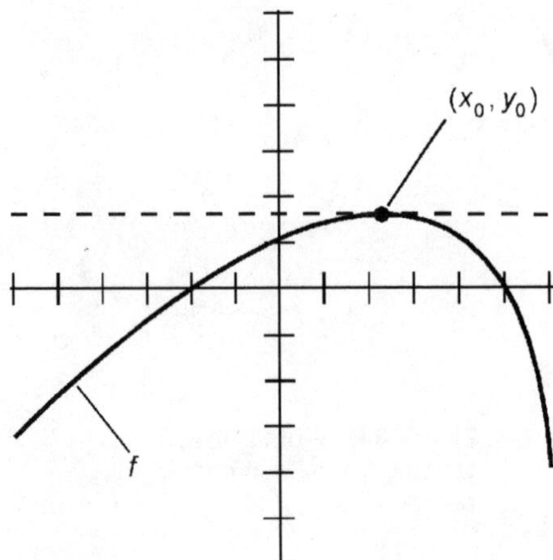

Figure 3.10 Local maximum value of a function.

$$f'(x_0) = 0$$
$$f''(x_0) > 0$$

Then (x_0,y_0) is a *local minimum* in the graph of f. An example is shown in Fig. 3.11.

Inflection point

Let f be a function such that $y = f(x)$. Let (x_0,y_0) be a point on the graph of f, and suppose f is twice differentiable at (x_0,y_0). Suppose the following are both true:

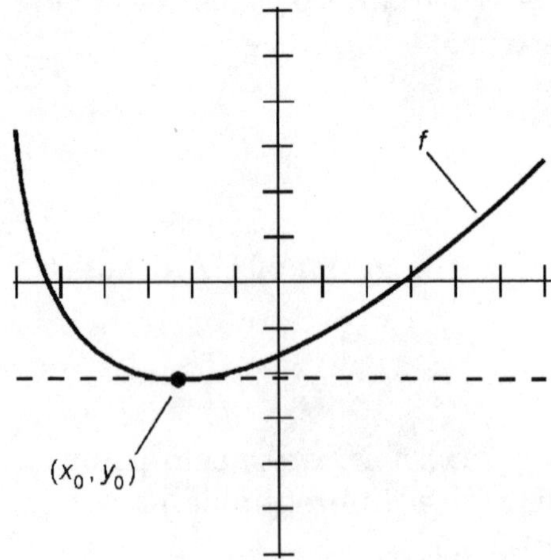

Figure 3.11 Local minimum value of a function.

$$f'(x_0) = 0$$

$$f''(x_0) = 0$$

Then (x_0,y_0) is an *inflection point* in the graph of f. Examples are shown in Fig. 3.12. At A, the curve is concave upward-to-downward:

$$f''(x) > 0 \text{ for } x < x_0$$

$$f''(x) < 0 \text{ for } x > x_0$$

At B, the curve is concave downward-to-upward:

$$f''(x) < 0 \text{ for } x < x_0$$
$$f''(x) > 0 \text{ for } x > x_0$$

Derivative of sine wave

The derivative of a sine wave is a cosine wave. This is the equivalent of a 90-degree phase shift (Fig. 3.13). The amplitude of the resultant wave depends on the amplitude and frequency of the sine wave.

Derivative of up-ramp wave

The derivative of an up-ramp wave is a positive constant (Fig. 3.14). The magnitude of the resultant depends on the amplitude and frequency of the up-ramp wave.

Derivative of down-ramp wave

The derivative of a down-ramp wave is a negative constant (Fig. 3.15). The magnitude of the resultant depends on the amplitude and frequency of the down-ramp wave.

Derivative of triangular wave

The derivative of a triangular wave is a square wave (Fig. 3.16). The amplitude of the resultant depends on the amplitude and frequency of the triangular wave.

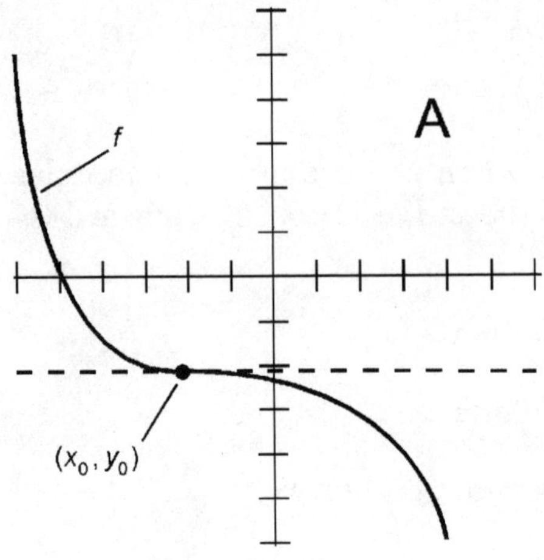

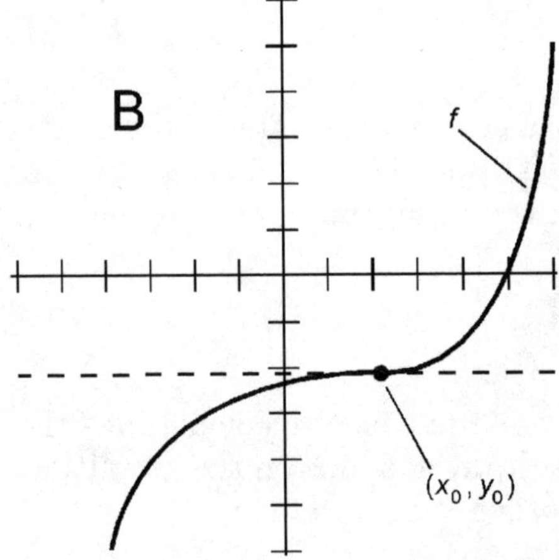

Figure 3.12 Inflection point in the graph of a function. At A, function decreasing in the region; at B, function increasing in the region.

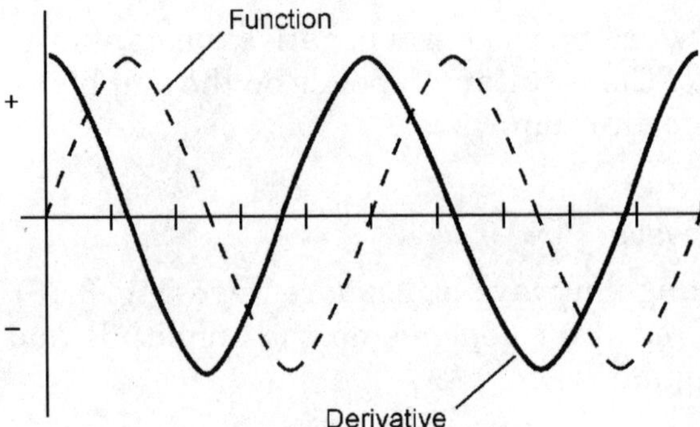

Figure 3.13 Derivative of a sine wave.

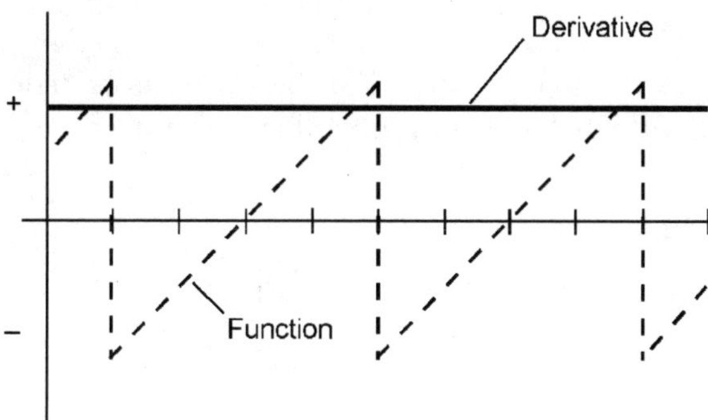

Figure 3.14 Derivative of an up-ramp wave.

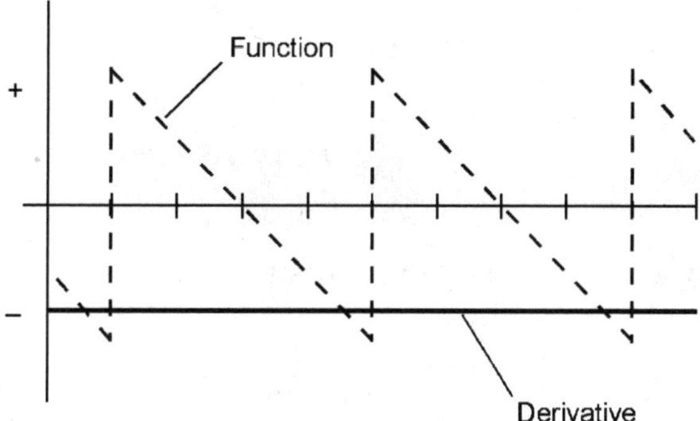

Figure 3.15 Derivative of a down-ramp wave.

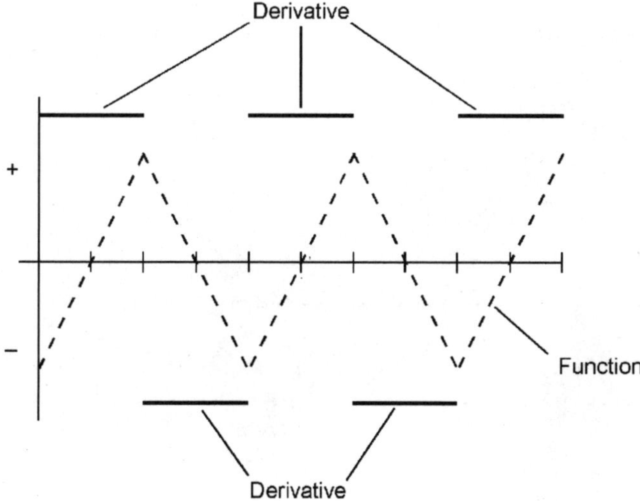

Figure 3.16 Derivative of a triangular wave.

Derivative of square wave

The derivative of a square wave is zero (Fig. 3.17). This is true regardless of the amplitude or the frequency of the square wave.

Derivatives of common functions

Chapter 6 contains a table of derivatives of common mathematical functions.

Vector Differentiation

Let **i**, **j**, and **k** represent unit vectors in Cartesian *xyz*-space, as follows:

$$\mathbf{i} = (1,0,0)$$

$$\mathbf{j} = (0,1,0)$$

$$\mathbf{k} = (0,0,1)$$

A vector function **G** of a scalar variable v consists of the sum of three scalar functions G_1, G_2, and G_3:

$$\mathbf{G}(v) = G_1(v)\mathbf{i} + G_2(v)\mathbf{j} + G_3(v)\mathbf{k}$$

The derivative of $\mathbf{G}(v)$ with respect to v is defined as:

$$d\mathbf{G}(v)/dv = \text{Lim}_{\Delta v \to 0} (\mathbf{G}(v + \Delta v) - \mathbf{G}(v))/\Delta v$$

$$= (dG_1(v)/dv)\mathbf{i} + (dG_2(v)/dv)\mathbf{j} + (dG_3(v)/dv)\mathbf{k}$$

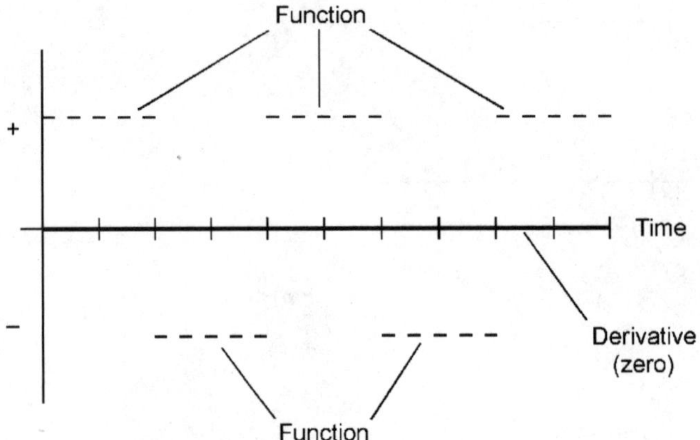

Figure 3.17 Derivative of a square wave.

If $\mathbf{G}(x,y,z)$ is a vector function of three variables x, y, and z, the partial derivatives $\partial G_1(x,y,z)/\partial x$, $\partial G_2(x,y,z)/\partial y$, and $\partial G_3(x,y,z)/\partial z$ are determined for each variable by holding the other two variables constant.

Derivative of dot product of two vector functions

Let $\mathbf{F}(v)$ and $\mathbf{G}(v)$ be vector functions of a scalar variable v. The derivative of the dot product $\mathbf{F}(v) \bullet \mathbf{G}(v)$ with respect to v is given by the following formula:

$$d(\mathbf{F}(v) \bullet \mathbf{G}(v))/dv = \mathbf{F}(v) \bullet d\mathbf{G}(v)/dv + \mathbf{G}(v) \bullet d\mathbf{F}(v)/dv$$

Derivative of cross product of two vector functions

Let $\mathbf{F}(v)$ and $\mathbf{G}(v)$ be vector functions of a scalar variable v. The derivative of the cross product $\mathbf{F}(v) \times \mathbf{G}(v)$ with respect to v is given by the following formula:

$$d(\mathbf{F}(v) \times \mathbf{G}(v))/dv = \mathbf{F}(v) \times d\mathbf{G}(v)/dv + d\mathbf{F}(v)/dv \times \mathbf{G}(v)$$

Del operator

The *del operator*, also called *nabla* and symbolized ∇, is defined as follows:

$$\nabla = \mathbf{i}(\partial/\partial x) + \mathbf{j}(\partial/\partial y) + \mathbf{k}(\partial/\partial z)$$

Distributivity of del involving scalar functions

Let $F(x,y,z)$ and $G(x,y,z)$ be scalar functions of three variables x, y, and z. Then the del operator distributes over addition of the functions. The following formula applies:

$$\nabla\ (F(x,y,z) + G(x,y,z)) = \nabla\ F(x,y,z) + \nabla\ G(x,y,z)$$

Distributivity of del involving dot product

Let $\mathbf{F}(v)$ and $\mathbf{G}(v)$ be vector functions of a scalar variable v. Then the dot product of the del operator and the sum of the functions is equal to the sum of the dot products of the del operator and the individual functions. The following formula applies:

$$\nabla \cdot (\mathbf{F}(v) + \mathbf{G}(v)) = \nabla \cdot \mathbf{F}(v) + \nabla \cdot \mathbf{G}(v)$$

Distributivity of del involving cross product

Let $\mathbf{F}(v)$ and $\mathbf{G}(v)$ be vector functions of a scalar variable v. Then the cross product of the del operator and the sum of the functions is equal to the sum of the cross products of the del operator and the individual functions, taken in the same order. The following formula applies:

$$\nabla \times (\mathbf{F}(v) + \mathbf{G}(v)) = \nabla \times \mathbf{F}(v) + \nabla \times \mathbf{G}(v)$$

Del of the dot product

Let $\mathbf{F}(v)$ and $\mathbf{G}(v)$ be vector functions of a scalar variable v. The following formula holds:

$$\begin{aligned} &\nabla\, (\mathbf{F}(v) \cdot \mathbf{G}(v)) \\ &= (\mathbf{G}(v) \cdot \nabla)\mathbf{F}(v) + (\mathbf{F}(v) \cdot \nabla)\mathbf{G}(v) \\ &+ \mathbf{G}(v) \times (\nabla \times \mathbf{F}(v)) + \mathbf{F}(v) \times (\nabla \times \mathbf{G}(v)) \end{aligned}$$

Del dot the cross product

Let $\mathbf{F}(v)$ and $\mathbf{G}(v)$ be vector functions of a scalar variable v. The following formula holds:

$$\begin{aligned} &\nabla \cdot (\mathbf{F}(v) \times \mathbf{G}(v)) \\ &= \mathbf{G}(v) \cdot (\nabla \times \mathbf{F}(v)) - \mathbf{F}(v) \cdot (\nabla \times \mathbf{G}(v)) \end{aligned}$$

Del cross the cross product

Let $\mathbf{F}(v)$ and $\mathbf{G}(v)$ be vector functions of a scalar variable v. The following formula holds:

$$\begin{aligned} &\nabla \times (\mathbf{F}(v) \times \mathbf{G}(v)) \\ &= (\mathbf{G}(v) \cdot \nabla)\mathbf{F}(v) - \mathbf{G}(v)(\nabla \cdot \mathbf{F}(v)) \\ &- (\mathbf{F}(v) \cdot \nabla)\mathbf{G}(v) + \mathbf{F}(v)(\nabla \cdot \mathbf{G}(v)) \end{aligned}$$

Divergence

Let $\mathbf{G}(x,y,z)$ be a vector function of scalar variables x, y, and z as defined above. The *divergence* of $\mathbf{G}(x,y,z)$, written div $\mathbf{G}$, is defined as the dot product of the del operator and $\mathbf{G}(x,y,z)$, as follows:

$$\text{div }\mathbf{G} = \nabla \bullet \mathbf{G}(x,y,z) = \partial G_1(x,y,z)/\partial x + \partial G_2(x,y,z)/\partial y + \partial G_3(x,y,z)/\partial z$$

Curl

Let $\mathbf{G}(x,y,z)$ be a vector function of scalar variables x, y, and z as defined above. The *curl* of $\mathbf{G}(x,y,z)$, written curl $\mathbf{G}$, is defined as the cross product of the del operator and $\mathbf{G}(x,y,z)$, as follows:

$$\text{curl }\mathbf{G} = (\partial G_3(x,y,z)/\partial y - \partial G_2(x,y,z)/\partial x)\mathbf{i} + (\partial G_1(x,y,z)/\partial z - \partial G_3(x,y,z)/\partial x)\mathbf{j} + (\partial G_2(x,y,z)/\partial x - \partial G_1(x,y,z)/\partial y)\mathbf{k}$$

Laplacian

Let $\mathbf{G}(x,y,z)$ be a vector function of scalar variables x, y, and z as defined above. The *Laplacian* of $\mathbf{G}(x,y,z)$, written Laplacian $\mathbf{G}$, is defined as follows:

$$\text{Laplacian }\mathbf{G} = \nabla^2\,\mathbf{G} = \partial^2\mathbf{G}/\partial x^2 + \partial^2\mathbf{G}/\partial y^2 + \partial^2\mathbf{G}/\partial z^2$$

For a scalar function $F(x,y,z)$, the Laplacian is defined as follows:

$$\text{Laplacian } F = \nabla^2\,F = \partial^2 F/\partial x^2 + \partial^2 F/\partial y^2 + \partial^2 F/\partial z^2$$

Gradient

Let $F(x,y,z)$ be a scalar function of scalar variables x, y, and z as defined above. The *gradient* of $F(x,y,z)$, written grad F, is defined as follows:

$$\text{grad } F = \nabla\,F = (\partial F/\partial x)\mathbf{i} + (\partial F/\partial y)\mathbf{j} + (\partial F/\partial z)\mathbf{k}$$

Scalar Integration

In general, *integration* is the opposite of differentiation. *Integral calculus* is used to find areas, volumes, and accumulated quantities.

Indefinite integral

Let f be a defined and continuous real-number function of a variable x. The *antiderivative* or *indefinite integral* of f is a function F such that $dF/dx = f$. This is written as follows:

$$\int f(x)\, dx = F(x) + c$$

where c is a real number and dx is the differential of x (customarily annotated with all indefinite integrals).

Definite integral

Let f be a defined and continuous real-number function of a variable x. Let a and b be values in the domain of f such that $a < b$. Let F be the antiderivative of f as defined above. The *definite integral* of f from a to b is defined as follows:

$$\int_a^b f(x)\, dx = F(b) - F(a)$$

The constant of integration subtracts from itself, thereby canceling additively. The definite integral can be depicted as the area under the curve of f in rectangular coordinates (Fig. 3.18). Regions above the x axis are considered to have positive area; regions below the x axis are considered to have negative area.

Constant of integration

There exist an infinite number of antiderivatives for any given function, all of which differ by real-number values. If the function $F_a(x)$ is an antiderivative of $f(x)$, then $F_b(x) = F_a(x) + c$ is also an antiderivative of $f(x)$, and c is known as the *constant of integration*.

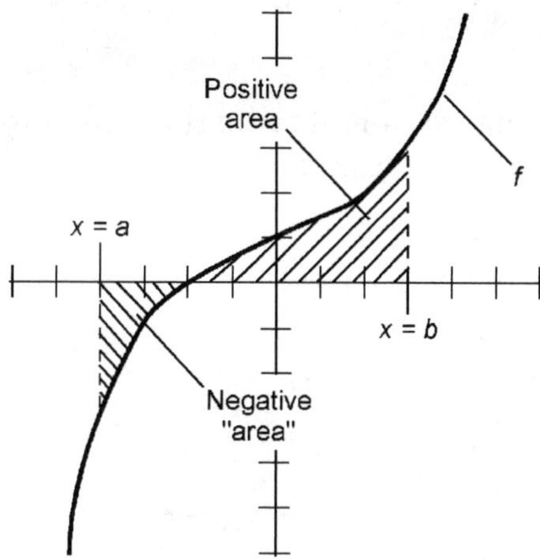

Figure 3.18 Definite integral represented as the area under a curve between two points.

Linearity

Let f and g be defined, continuous functions of x. Let a and b be real-number constants. Then:

$$\int (a \times f(x) + b \times g(x))\, dx = a \times \int f(x)\, dx + b \times \int g(x)\, dx$$

Integration by parts

Let f and g be defined, continuous functions of x, and let F be an antiderivative of f. Then:

$$\int (f(x) \times g(x))\, dx = F(x) \times g(x) - \int (F(x) \times dg/dx)\, dx$$

Indefinite integral of constant

Let k be a constant; let c be the constant of integration. The following formula applies:

$$\int k\, dx = kx + c$$

Indefinite integral of variable

Let x be a variable. Let c be the constant of integration. The following formula applies:

$$\int x\, dx = x^2/2 + c$$

Indefinite integral of variable multiplied by constant

Let x be a variable. Let k be a constant, and let c be the constant of integration. The following formula applies:

$$\int kx\, dx = kx^2/2 + c$$

Indefinite integral of function multiplied by constant

Let $f(x)$ be a function of a variable x. Let k be a constant. The following formula applies:

$$\int kf(x)\, dx = k \int f(x)\, dx$$

Indefinite integral of reciprocal

Let x be a variable. Let c be the constant of integration. The following formula applies:

$$\int (1/x)\, dx = \ln |x| + c$$

Indefinite integral of reciprocal multiplied by constant

Let x be a variable. Let k be a constant, and let c be the constant of integration. The following formula applies:

$$\int (k/x)\, dx = k \ln |x| + c$$

Indefinite integral of sum of functions

Let $f_1(x)$, $f_2(x)$, $f_3(x)$, ..., and $f_n(x)$ be functions of a variable x. The following formula applies:

$$\int (f_1(x) + f_2(x) + f_3(x) + \ldots + f_n(x))\, dx$$
$$= \int f_1(x)\, dx + \int f_2(x)\, dx + \int f_3(x)\, dx + \ldots + \int f_n(x)\, dx$$

Indefinite integral of variable raised to integer power

Let x be a variable. Let k be an integer with the restriction that $k \neq -1$, and let c be the constant of integration. The following formula applies:

$$\int x^k\, dx = (x^{k+1}/(k + 1)) + c$$

Indefinite integral of constant raised to variable power

Let x be a variable. Let k be a constant such that $k > 0$ and $k \neq 1$. Let c be the constant of integration. The following formula applies:

$$\int k^x\, dx = (k^x/(\ln k)) + c$$

Indefinite integral of sine wave

The antiderivative (indefinite integral) of a sine wave is shown in Fig. 3.19. The amplitude and displacement of the resultant wave depend on the amplitude and frequency of the sine wave.

Indefinite integral of up-ramp wave

The antiderivative (indefinite integral) of an up-ramp wave is shown in Fig. 3.20. The magnitude of the resultant depends on the amplitude and frequency of the up-ramp wave.

Indefinite integral of down-ramp wave

The antiderivative (indefinite integral) of a down-ramp wave is shown in Fig. 3.21. The magnitude of the resultant depends on the amplitude and frequency of the down-ramp wave.

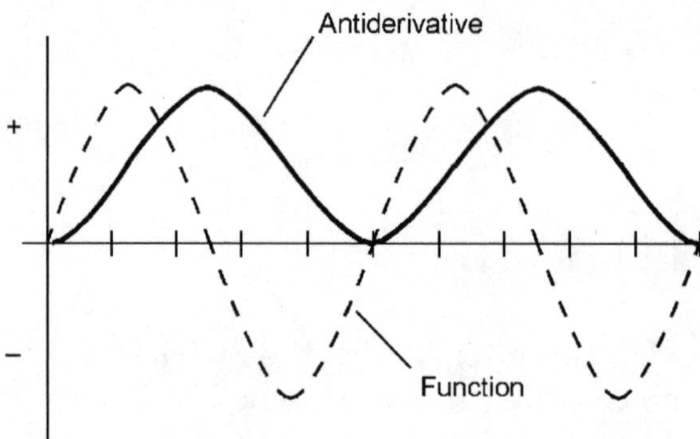

Figure 3.19 Antiderivative (indefinite integral) of a sine wave.

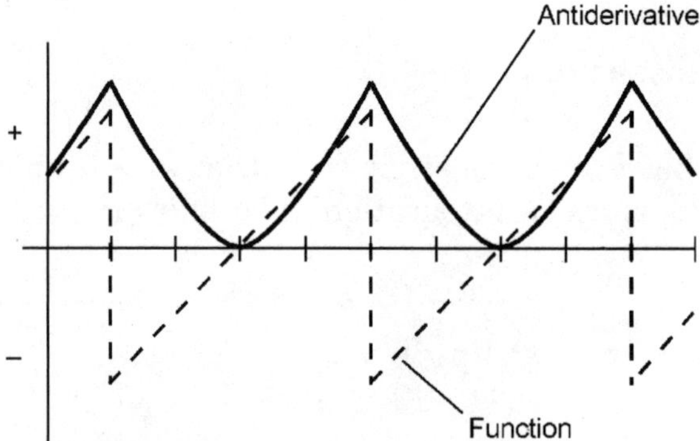

Figure 3.20 Antiderivative (indefinite integral) of an up-ramp wave.

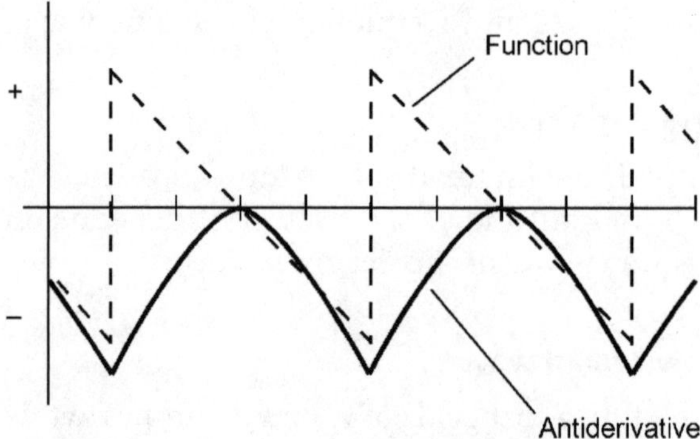

Figure 3.21 Antiderivative (indefinite integral) of a down-ramp wave.

Indefinite integral of triangular wave

The antiderivative (indefinite integral) of a triangular wave is shown in Fig. 3.22. The amplitude of the resultant depends on the amplitude and frequency of the triangular wave.

Indefinite integral of square wave

The antiderivative (indefinite integral) of a square wave is shown in Fig. 3.23. The amplitude of the resultant depends on the amplitude of the square wave.

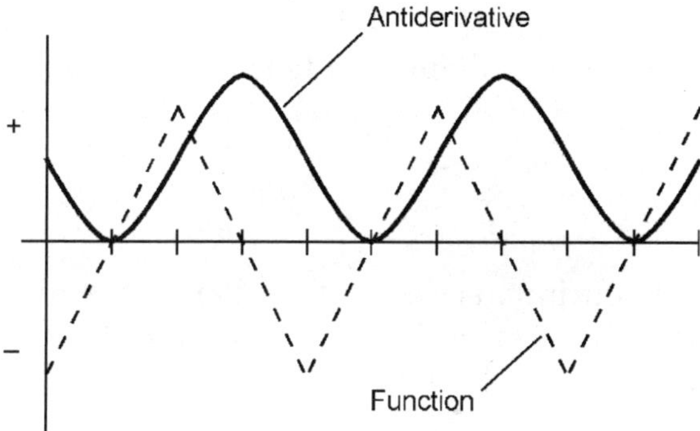

Figure 3.22 Antiderivative (indefinite integral) of a triangular wave.

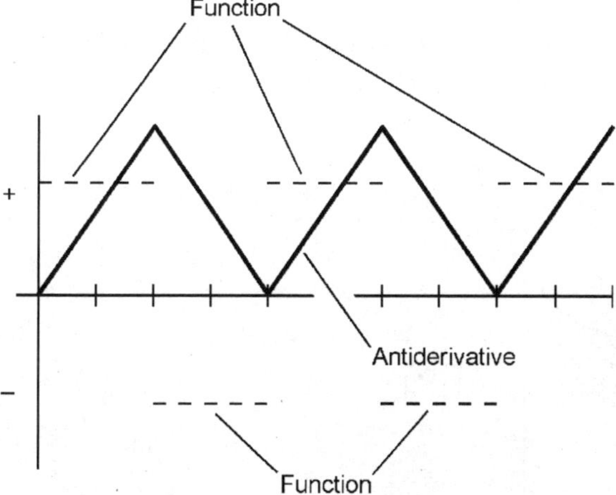

Figure 3.23 Antiderivative (indefinite integral) of a square wave.

Average value of function over interval

Let $f(x)$ be a function that is continuous over the domain from $x = a$ to $x = b$, where a and b are real numbers and $a < b$. Let $F(x)$ be the antiderivative of $f(x)$. Then the average value, A, of $f(x)$ over the open, half-open, or closed interval bounded by a and b is given by the following formula (Fig. 3.24):

$$A = (F(b) - F(a))/(b - a)$$

Indefinite integrals of common functions

Chapter 6 contains a table of indefinite integrals of common mathematical functions.

Vector Integration

Let **F** be a defined and continuous vector function of a scalar variable v. The *antiderivative* or *indefinite integral* of **F** is a vector function **G** such that $d\mathbf{G}/dv = \mathbf{F}$. This is written as follows:

$$\int \mathbf{F}(v)\, dv = \mathbf{G}(v) + \mathbf{c}$$

where **c** is a constant vector and dv is the differential of v (customarily annotated with all indefinite integrals).

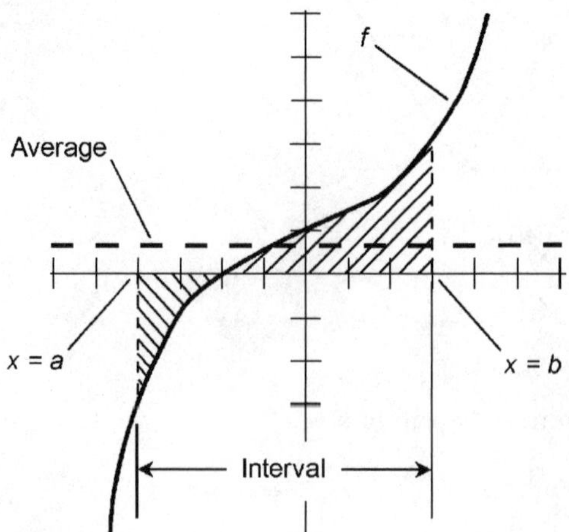

Figure 3.24 Determination of the average value of a function over an interval.

Definite integral

Let **F** be a defined and continuous vector function of a variable v. Let a and b be values in the domain of **F** such that $a < b$. Let **G** be the antiderivative of **F** as defined above. The *definite integral* of **F** from a to b is defined as follows:

$$\int_a^b \mathbf{F}(v)\, dv = \mathbf{G}(b) - \mathbf{G}(a)$$

The constant vector of integration subtracts from itself, thereby vanishing.

Line integral

Let P and Q be two points in Cartesian xyz-space, connected by a curve C. Define the following:

$$d\mathbf{r} = dx\mathbf{i} + dy\mathbf{j} + dz\mathbf{k}$$

where

$$\mathbf{i} = (1,0,0)$$

$$\mathbf{j} = (0,1,0)$$

$$\mathbf{k} = (0,0,1)$$

Let **G** be a vector function that follows C from P and Q, consisting of three scalar functions G_1, G_2, and G_3 such that:

$$\mathbf{G} = G_1\mathbf{i} + G_2\mathbf{j} + G_3\mathbf{k}$$

The *line integral* of **G** over the curve C is defined as follows:

$$\int_C \mathbf{G} \cdot d\mathbf{r} = \int G_1\, dx + \int G_2\, dy + \int G_3\, dz$$

Direction of line integral

The line integral along a curve in a given direction is equal to the negative of the line integral along the same curve in the opposite direction. The following formula holds for a vector function **G** as defined above:

$$\int_P^Q \mathbf{G} \cdot d\mathbf{r} = -\int_Q^P \mathbf{G} \cdot d\mathbf{r}$$

Separation of paths

Let C be a curve connecting points P and Q; let R be some intermediate point on the curve between P and Q. The following formula holds for a vector function **G** as defined above:

$$\int_P^Q \mathbf{G} \cdot d\mathbf{r} = \int_P^R \mathbf{G} \cdot d\mathbf{r} + \int_R^Q \mathbf{G} \cdot d\mathbf{r}$$

Alternatively, let C be a curve connecting points P and R, let D be a curve connecting points R and Q, and let $C + D$ denote the composite curve connecting points P and Q. The following formula holds:

$$\int_{C+D} \mathbf{G} \cdot d\mathbf{r} = \int_C \mathbf{G} \cdot d\mathbf{r} + \int_D \mathbf{G} \cdot d\mathbf{r}$$

Integration around a closed curve

The line integral of a conservative vector field around a closed curve is always equal to zero. A conservative vector field is one that can be written as the del of a function, for example, $\mathbf{G} = \nabla f(x,y,z)$. The line integral along a closed curve in the counterclockwise direction is generally symboled as follows:

$$\oint_C \mathbf{G} \cdot d\mathbf{r}$$

Surface integral

Let S be a surface in xyz-space; let R be the projection of S onto the xy-plane. Let **G** be a vector function; let **N** be a unit vector normal to S in a region dS (Fig. 3.25). The surface integral of **G** over S is given by the following formula:

$$\int_S \mathbf{G} \cdot \mathbf{N}\, dS = \iint_D \mathbf{G} \cdot (\mathbf{r}_u \times \mathbf{r}_v)\, dR$$

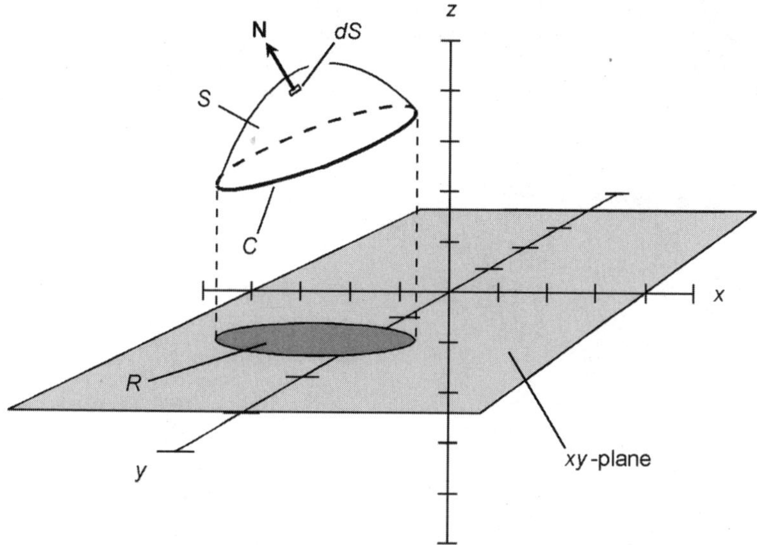

Figure 3.25 Integral over surface S, expressed as double integral over projection R in xy-plane; Stokes' theorem.

Divergence Theorem

Let S be a surface in Cartesian xyz-space that encloses a solid having volume V. Let $\mathbf{G}$ be a vector function; let $\mathbf{N}$ be a vector normal to S in an arbitrarily small region dS as shown in Fig. 3.26. The following formula, known as the *Divergence Theorem* or *Gauss' theorem,* states that:

$$\int_V \nabla \cdot \mathbf{G}\, dV = \int_S \mathbf{G} \cdot dS$$

Stokes' theorem

Let S be a surface in Cartesian xyz-space that is bounded by a closed curve C, as shown in Fig. 3.25. Let $\mathbf{G}$ be a vector function; let $\mathbf{N}$ be a vector normal to S in an arbitrarily small region dS. Let $d\mathbf{r}$ be defined as follows:

$$d\mathbf{r} = dx\mathbf{i} + dy\mathbf{j} + dz\mathbf{k}$$

where

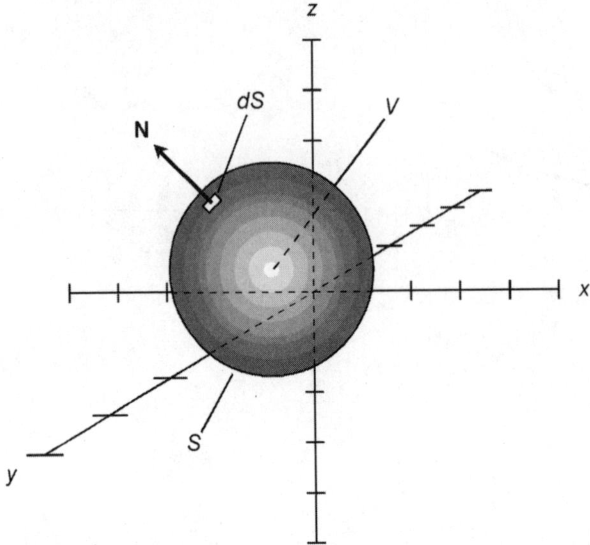

Figure 3.26 Divergence Theorem.

$$\mathbf{i} = (1,0,0)$$

$$\mathbf{j} = (0,1,0)$$

$$\mathbf{k} = (0,0,1)$$

The following formula, known as *Stokes' theorem*, states that:

$$\oint_C \mathbf{G} \cdot d\mathbf{r} = \int_S (\nabla \times \mathbf{G}) \cdot dS$$

Differential Equations

A *differential equation* contains one or more derivatives or differentials as variables. The derivatives can be of any order (first, second, third, etc.). There are numerous forms. Some of the most common forms, and their solutions, are given in this section.

Linear differential equation of first order

Let $f(x)$ and $g(x)$ be distinct functions. Let x and y be variables. Let c be the constant of integration. A *linear differential equation of first order* takes the following form:

$$dy/dx + y\,f(x) = g(x)$$

The solution is given by:

$$y\,e^{\int f(x)\,dx} = \int g(x)\,e^{\int f(x)\,dx}\,dx + c$$

Homogeneous differential equation

Let f be a function; let x and y be variables, with the restriction that $x \neq 0$. Let $v = y/x$. Let c be the constant of integration. A *homogeneous differential equation* takes the following form:

$$dy/dx = f(v)$$

The solution is given by:

$$\ln |x| = \int 1/((f(v) - v)\,dv + c$$

Separation of variables

Let F, G, M, and N be functions; let x and y be variables, with the restriction that $M(x) \neq 0$ and $G(y) \neq 0$. Let k be a constant. Suppose we are given a differential equation of the following form:

$$F(x)\,G(y)\,dx + M(x)\,N(y)\,dy = 0$$

The solution, according to *separation of variables,* is as follows:

$$\int (F(x)/M(x))\,dx + \int (N(y)/G(y))\,dy = k$$

Linear, homogeneous, second-order differential equation

Let x, y, and z be variables; let a, b, m, n, p, and q be real-number constants. Let j represent the unit imaginary number (the positive square root of -1). Let s and t be the roots of the following quadratic equation:

$$z^2 + az + b = 0$$

A *linear, homogeneous, second-order differential equation* takes the following form:

$$d^2y/dx^2 + a(dy/dx) + by = 0$$

The solution(s) of the differential equation can occur in any of three different ways, depending on the nature of the solutions s, t to the quadratic equation.

1. If s and t are both real numbers and $s \neq t$, then the general solution to the differential equation is given by:

$$y = me^{sx} + ne^{tx}$$

2. If s and t are both real numbers and $s = t$ (call them both s in this situation), then the general solution to the differential equation is given by:

$$y = me^{sx} + nxe^{sx}$$

3. If s and t are complex conjugates such that $s = p + jq$ and $t = p - jq$, then the general solution to the differential equation is given by:

$$y = e^{px} (m \cos qx + n \sin qx)$$

In this instance, p and q can be derived from a and b via the quadratic formula:

$$p = -a/2$$

$$q = (b - a^2/4)^{1/2}$$

Linear, non-homogeneous, second-order differential equation

Let x, y, and z be variables; let a, b, m, n, p, and q be real-number constants. Let j represent the unit imaginary number (the positive square root of -1). Let s and t be the roots of the following quadratic equation:

$$z^2 + az + b = 0$$

A *linear, non-homogeneous, second-order differential equation* takes the following form:

$$d^2y/dx^2 + a(dy/dx) + by = F(x)$$

The solution(s) of the differential equation can occur in any of three different ways, depending on the nature of the solutions to the quadratic equation.

1. If s and t are both real numbers and $s \neq t$, then the solution to the differential equation is given by:

$$y = me^{sx} + ne^{tx}$$
$$+ (e^{sx}/(s - t)) \int e^{-sx} F(x)\, dx$$
$$+ (e^{tx}/(s - t)) \int e^{-tx} F(x)\, dx$$

2. If s and t are both real numbers and $s = t$ (call them both s in this situation), then the solution to the differential equation is given by:

$$y = me^{sx} + nxe^{sx}$$
$$+ xe^{sx} \int e^{-sx} F(x)\, dx$$
$$- e^{sx} \int xe^{-sx} F(x)\, dx$$

3. If s and t are complex conjugates such that $s = p + jq$ and $t = p - jq$, then the solution to the differential equation is given by:

$$y = e^{px} (m \cos qx + n \sin qx)$$
$$+ (e^{px} \sin qx/q) \int e^{-px} F(x) \cos qx\, dx$$
$$- (e^{px} \cos qx/q) \int e^{-px} F(x) \sin qx\, dx$$

In this instance, p and q can be derived from a and b via the quadratic formula:

$$p = -a/2$$

$$q = (b - a^2/4)^{1/2}$$

Euler differential equation

Let x and y be variables; let t be a parameter such that $x = e^t$. Let a and b be constants, and let $F(x)$ be a function of x. A *Euler differential equation* takes the following form:

$$x^2(d^2y/dx^2) + ax(dy/dx) + by = F(x)$$

This equation can be put into the following form:

$$d^2y/dt^2 + (a - 1)(dy/dt) + by = F(e^t)$$

The Euler equation can then solved as a linear, second-order, non-homogeneous differential equation.

Bernoulli differential equation

Let $F(x)$ and $G(x)$ be distinct functions. Let x and y be variables. Let n be a natural number, and let c be the constant of integration. A *Bernoulli differential equation* takes the following form:

$$dy/dx + yF(x) = y^n G(x)$$

The solution is given by:

$$y^{(1-n)} e^{(1-n) \int F(x)\, dx} = (1 - n) \int G(x)\, e^{(1-n) \int F(x)\, dx}\, dx + c$$

Exact differential equation

Let $f(x,y)$ and $g(x,y)$ be distinct functions of two variables x and y. Let c be a constant. An *exact differential equation* takes the following form:

$$f(x,y)\, dx + g(x,y)\, dy = 0$$

such that

$$\partial f(x,y)/\partial y = \partial g(x,y)/\partial x$$

The solution is given by the following equation, where ∂x denotes integration with respect to x, treating y as a constant:

$$\int f(x,y)\ \partial x + (g(x,y) - \partial/\partial y \int f(x,y)\ \partial x)\ dy = c$$

Probability

Probability can be expressed as a ratio between 0 and 1, or as a percentage between 0 and 100 inclusive. The probability of an event H taking place in a given situation is written $p(H)$. This section outlines important principles in probability theory.

Probability with discrete random variable

Let X be a discrete random variable, that is, one that can attain n possible values, all equally likely. Suppose an event H results from exactly m different values of X, where $m \leq n$. Then the probability $p(H)$ that the event H will result from any given value of X, expressed as a ratio, is given by the following formula:

$$p(H) = m/n$$

Expressed as a percentage, the probability $p_{\%}(H)$ is:

$$p_{\%}(H) = 100m/n$$

Probability with continuous random variable

Let X be a continuous random variable, that is, one that can attain all possible real values. Let $F(X)$ be the distribution function that defines the probability associated with defined regions of the domain, represented by open, half-open, or closed intervals bounded by X_1 and X_2, where $X_1 < X_2$ as shown in Fig. 3.27. Let H be the event that results from situations defined as follows:

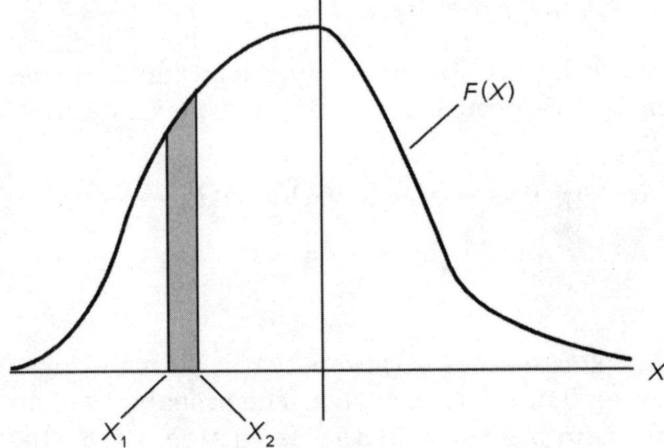

Figure 3.27 Probability defined with continuous random variable.

$$X_1 < X < X_2 \text{ for } (X_1,X_2)$$

$$X_1 < X \leq X_2 \text{ for } (X_1,X_2]$$

$$X_1 \leq X < X_2 \text{ for } [X_1,X_2)$$

$$X_1 \leq X \leq X_2 \text{ for } [X_1,X_2]$$

Then the probability $p(H)$ that H will result from any given value of X, expressed as a ratio, is given by the following formula:

$$p(H) = \frac{\int_{X_1}^{X_2} F(X)\,dX}{\int_{-\infty}^{\infty} F(X)\,dX}$$

Expressed as a percentage, the probability $p_{\%}(\mathrm{H})$ is:

$$p_{\%}(H) = 100\,\frac{\int_{X_1}^{X_2} F(X)\,dX}{\int_{-\infty}^{\infty} F(X)\,dX}$$

Independent events

Two events H_1 and H_2 are said to be *independent* if and only if the probability, expressed as a ratio, of their occurring simultaneously is equal to the product of the probabilities, expressed as ratios, of their occurring separately. The following equation holds:

$$p(H_1 \cap H_2) = p(H_1)\, p(H_2)$$

Mutually exclusive events

Let H_1 and H_2 be two events that are *mutually exclusive;* that is, they have no elements in common:

$$H_1 \cap H_2 = \varnothing$$

The probability, expressed as a ratio, of either event occurring is equal to the sum of the probabilities, expressed as ratios, of their occurring separately. The following equation holds:

$$p(H_1 \cup H_2) = p(H_1) + p(H_2)$$

Complementary events

If two events H_1 and H_2 are *complementary,* then the probability, expressed as a ratio, of either event occurring is equal to 1 minus the probability, expressed as a ratio, of the other event occurring. The following equations hold:

$$p(H_2) = 1 - p(H_1)$$

$$p(H_1) = 1 - p(H_2)$$

Nondisjoint events

Let H_1 and H_2 be two events that are *nondisjoint;* that is, they have at least one element in common:

$$H_1 \cap H_2 \neq \varnothing$$

The probability, expressed as a ratio, of either event occurring is equal to the sum of the probabilities, expressed as ratios, of their occurring separately, minus the probability, expressed as a ratio, of both occurring simultaneously. The following equation holds:

$$p(H_1 \cup H_2) = p(H_1) + p(H_2) - p(H_1 \cap H_2)$$

Conditional probability

Let H_1 and H_2 be two events. Suppose that H_1 has occurred. The probability that H_2 will occur is called the *conditional probability* of H_2 given H_1. This is determined mathematically as follows:

$$p(H_2|H_1) = (p(H_1 \cap H_2))/p(H_2)$$

Tree diagrams

Suppose there are n_1 ways in which an event H_1 can occur, and n_2 ways in which an event H_2 can occur. The total number of ways, n, in which the two events, taken together, can occur, is equal to the product of the ways in which either event can occur:

$$n = n_1 n_2$$

This can be illustrated by means of a branched figure called a *tree diagram*. The diagram shows all the various combinations geometrically. If $n_1 \neq n_2$, there are two different ways in which a given situation can be depicted using tree diagrams. An example is shown in Fig. 3.28 for $n_1 = 3$ and $n_2 = 4$.

Permutations

Suppose that, out of q objects, r objects are taken at a time in a specific order. Also suppose that $q \geq r$, where both q and r are positive integers. The possible number of choices is symbolized $_qP_r$ and can be calculated as follows:

$$_qP_r = q!/(q - r)!$$

where ! denotes *factorial*, the product of all positive integers less than or equal to a specified integer.

Combinations

Suppose that, out of q objects, r objects are taken at a time in no particular order. Also suppose that $q \geq r$, where both q and

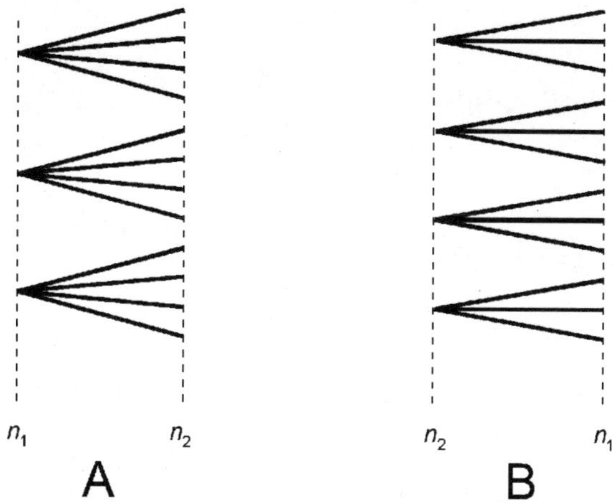

Figure 3.28 Tree diagrams for events where $n_1 = 3$ and $n_2 = 4$. At A, branch points are from n_1; at B, branch points are from n_2.

r are positive integers. The possible number of such *combinations* is symbolized ${}_qC_r$ and can be calculated as follows:

$${}_qC_r = {}_nP_r/r! = q!/(r!(q - r)!)$$

Expectation

Let X be a discrete random variable. The *expectation* of X, also called the *mean* and symbolized μ, is equal to the sum of the products of all possible values of X and their respective probabilities. Suppose there are n possible values of X, such that:

$$x_i \in \{x_1, x_2, \ldots, x_n\}$$

Then the following formula applies:

$$\mu = \sum x_i\, p(x_i)$$
$$= x_1\, p(x_1) + x_2\, p(x_2) + \ldots + x_n\, p(x_n)$$

Variance

Let X be a discrete random variable. The *variance* of X, symbolized σ^2, is equal to the sum of the squares of the differences of the expectation μ and all possible values of X, multiplied by the probability of X. Suppose there are n possible values of X, such that:

$$x_i \in \{x_1, x_2, ..., x_n\}$$

Then the following formula applies:

$$\sigma^2 = \sum (x_i - \mu)^2\, p(x_i)$$

$$= (x_1 - \mu)^2\, p(x_1) + (x_2 - \mu)^2\, p(x_2) + ... + (x_n - \mu)^2\, p(x_n)$$

Binomial random variable formula

Let X be a *binomial random variable,* that is, a discrete random variable in an experiment having these characteristics:

- A specific number n of trials, all of which are identical
- Two possible outcomes, say 0 and 1, for each trial
- Equal probability for either outcome in any given trial

Let q represent the probability of the outcome 1 in a given trial; let r represent the probability of the outcome 0 in a given trial. Let X be the number of outcomes equal to 1 in n trials, then the following formula holds:

$$p(X) = {}_nC_X q^X\, r^{n-X}$$

Expectation of binomial distribution

Let n be the number of trials in a binomial distribution. Let q represent the probability of the outcome 1 in any given trial. The *expectation,* μ, in this case is given by:

$$\mu = nq$$

Variance of binomial distribution

Let n be the number of trials in a binomial distribution. Let q represent the probability of the outcome 1 in any given trial; let r represent the probability of the outcome 0 in a given trial. The *variance,* σ^2, in this case is given by:

$$\sigma^2 = nqr = \mu r$$

Standard deviation of binomial distribution

Let n be the number of trials in a binomial distribution. Let q represent the probability of the outcome 1 in any given trial; let r represent the probability of the outcome 0 in a given trial. The *standard deviation,* denoted σ, is given by:

$$\sigma = (nqr)^{1/2} = (\mu r)^{1/2}$$

Normal distribution

Let X be a continuous random variable. Let x be any given data point in X. Let $E(z)$ represent the exponential function for a real-number variable z:

$$E(z) = e^z$$

where e is the natural logarithm base and $e \approx 2.71828$. In a *normal distribution,* X has a density function of the form:

$$g(x) = (E(-(x - \mu)^2/2\sigma^2))/\sigma(2\pi)^{1/2}$$

The normal distribution has a well-known symmetry, and is sometimes called the *bell-shaped curve* for this reason (Fig. 3.29). It has these characteristics:

- Approximately 68% of the data points are within the range $\pm\sigma$ of μ.

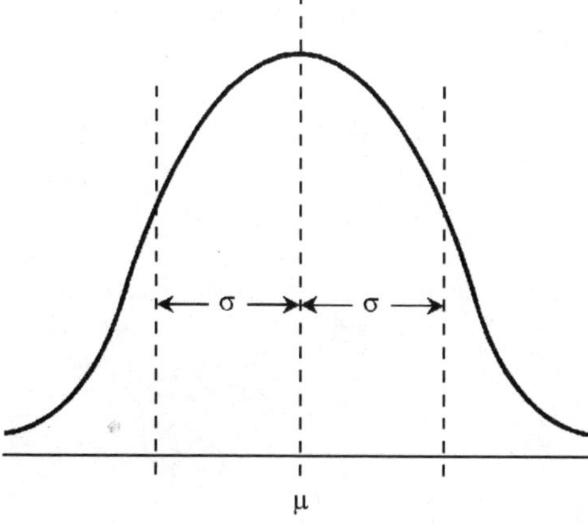

Figure 3.29 The normal distribution (bell-shaped curve) has a characteristic symmetry.

- Approximately 95% of the data points are within the range $\pm 2\sigma$ of μ.
- More than 99.5% of the data points are within the range $\pm 3\sigma$ of μ.

Uniform distribution

Let X be a continuous random variable. Let x be any given data point in X. Let a and b be real numbers. In a *uniform distribution, X* has a density function of the form:

$$g(x) = (b - a)^{-1}$$

The following constraint applies:

$$a < x < b$$

The mean, variance, and standard deviation are given by:

$$\mu = (a + b)/2$$

$$\sigma^2 = (b - a)^2/2$$

$$\sigma = ((b - a)^2/2)^{1/2}$$

Beta distribution

Define the *gamma function,* $\Gamma(z)$, of a complex variable z and a parameter t as follows:

$$\Gamma(z) = \int_0^\infty t^{z-1} e^{-t} \, dt$$

Let X be a continuous random variable. Let x be any given data point in X. Let a and b be real numbers. In a *beta distribution, X* has a density function of the form:

$$g(x) = (\Gamma(a + b + 2) \, x^a \, (1 - x)^b)/(\Gamma(a + 1)(\Gamma(b + 1))$$

The following constraints apply:

$$0 < x < 1$$

$$a > -1$$

$$b > -1$$

The mean, variance, and standard deviation are given by:

$$\mu = (a + 1)/(a + b + 2)$$

$$\sigma^2 = (a + 1)(b + 1)/((a + b + 2)^2(a + b + 3))$$

$$\sigma = (a + 1)^{1/2}(b + 1)^{1/2}/((a + b + 2)\ (a + b + 3)^{1/2})$$

Gamma distribution

Let X be a continuous random variable. Let x be any given data point in X. Let a and b be real numbers. In a *gamma distribution*, X has a density function of the form:

$$g(x) = x^a\ e^{-x/b}/(\Gamma(1 + a)b^{1+a})$$

The following constraints apply:

$$x > 0$$

$$a > -1$$

$$b > 0$$

The mean, variance, and standard deviation are given by:

$$\mu = ab + b$$

$$\sigma^2 = ab^2 + b^2$$

$$\sigma = (a + 1)^{1/2}b$$

Chi-square distribution

Let X be a continuous random variable. Let x be any given data point in X. Let n be a natural number. In a *chi-square distribution with n degrees of freedom,* X has a density function of the form:

$$g(x) = (x^{(n-2)/2}\ e^{-x/2})/(2^{n/2}\ \Gamma(n/2))$$

The following constraint applies:

$$x > 0$$

The mean, variance, and standard deviation are given by:

$$\mu = n$$

$$\sigma^2 = 2n$$

$$\sigma = (2n)^{1/2}$$

Exponential distribution

Let X be a continuous random variable. Let x be any given data point in X. Let a be a real number. In an *exponential distribution*, X has a density function of the form:

$$g(x) = e^{-x/a}/a$$

The following constraints apply:

$$x > 0$$

$$a > 0$$

The mean, variance, and standard deviation are given by:

$$\mu = a$$

$$\sigma^2 = a^2$$

$$\sigma = a$$

Linear pseudorandom number generation

A string of pseudorandom numbers can be obtained by repeatedly performing a linear function on integers, thereby obtaining an infinite sequence of digits. Choose an initial modulus (radix) number k. Common choices are prime numbers less than 10, that is, $k = 3$, $k = 5$, and $k = 7$, and also $k = 10$. This yields the following digit sets:

$$k = 3: \{0, 1, 2\}$$

$$k = 5: \{0, 1, 2, 3, 4\}$$

$$k = 7: \{0, 1, 2, 3, 4, 5, 6\}$$

$$k = 10: \{0, 1, 2, 3, 4, 5, 6, 7, 8, 9\}$$

Let n_1 be the starting number, taken from one of the sets above. Let

$$n_2 = pn_1 + q$$

calculated in the radix k, where p and q are integer constants taken from one of the sets above. In general, let

$$n_i + 1 = pn_i + q$$

where i designates the position of a given number in the sequence. The digits are simply put down end-to-end. Suppose $k = 10$, $p = 2$, $q = 3$, and $n_1 = 5$. Then:

$$n_1 = 5$$

$$n_2 = 2 \times 5 + 3 = 13$$

$$n_3 = 2 \times 13 + 3 = 29$$

$$n_4 = 2 \times 29 + 3 = 61$$

$$n_5 = 2 \times 61 + 3 = 125$$

The resulting sequence S of pseudorandom numbers starts out as follows:

$$S = 5, 1, 3, 2, 9, 6, 1, \ldots$$

Measurement error

Let x_a represent the actual value of a quantity to be measured. Let x_m represent the measured value of the quantity, in the same units as x_a. Then the *absolute error*, D_a (in the same units as x_a), is given by:

$$D_a = x_m - x_a$$

The *proportional error*, D_p, is given by:

$$D_p = (x_m - x_a)/x_a$$

The *percentage error,* $D_{\%}$, is given by:

$$D_{\%} = 100(x_m - x_a)/x_a$$

Error values and percentages are positive if $x_m > x_a$, and negative if $x_m < x_a$.

Chapter

4

Electricity, Electronics, and Communications

This chapter contains information relevant to direct current, alternating current, magnetism, transformers, digital circuits, filters, semiconductors, electron tubes, antennas, and measurement apparatus.

Direct Current

This section contains formulas involving direct-current (DC) charge quantity, amperage, voltage, resistance, power, and energy.

The coulomb

The standard unit of electrical charge quantity, symbolized by Q, is the *coulomb,* equivalent to the charge contained in approximately 6.24×10^{18} electrons.

Charge vs current and time

Let Q represent charge quantity in coulombs, let I represent direct current in amperes, and let t represent time in seconds. Then:

$$Q = \int I \, dt$$

This principle is illustrated in Fig. 4.1. If the current remains constant over time, then the above formula can be simplified to:

$$Q = It$$

Coulomb's Law

Let F represent force in newtons, let Q_X and Q_Y represent the charges on two distinct objects X and Y, and let d represent the distance between the charge centers of X and Y. Then:

$$F = (Q_X Q_Y)/d^2$$

If the charges are alike in polarity (+/+ or −/−), then F is positive (repulsive). If the charges are opposite in polarity (−/+ or +/−), then F is negative (attractive).

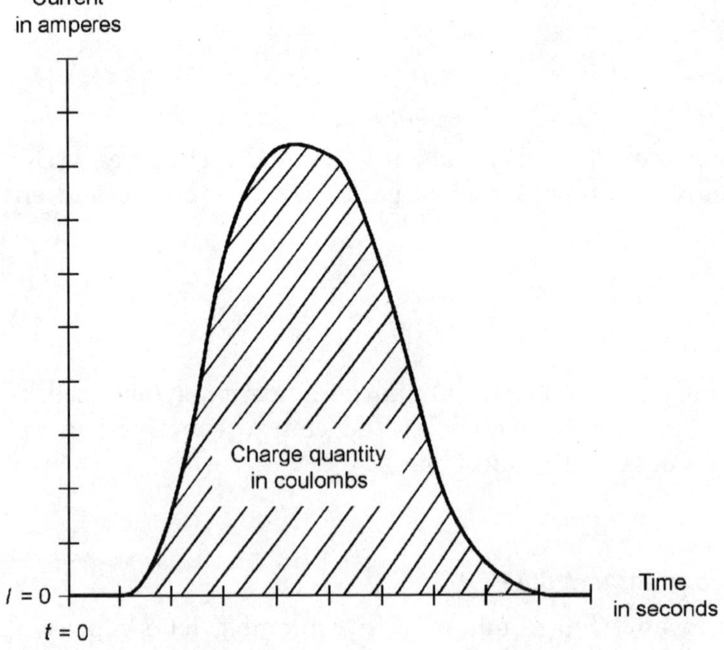

Figure 4.1 Electrical charge as a function of current and time.

The ampere

The standard unit of direct current, also called *DC amperage* and symbolized by I, is the *ampere,* equivalent to one coulomb of charge moving past a point in one second, always in the same direction.

Charging and discharging

Let $I_{c/d}$ represent instantaneous charging or discharging current in amperes. Let t represent time in hours. Then the accumulated charge or discharge quantity Q_{Ah}, in ampere-hours, is:

$$Q_{Ah} = \int I_{c/d}\, dt$$

If the rate of charging or discharging is constant, then:

$$Q_{Ah} = I_{c/d}t$$

Current vs charge and time

Let Q represent charge quantity in coulombs. Let t represent time in seconds. Then the instantaneous charging or discharging current $I_{c/d}$, in amperes, is:

$$I_{c/d} = dQ/dt$$

If the rate of charge or discharge is constant over an interval beginning at time t_1 and ending at time t_2, then:

$$I_{c/d} = (Q_2 - Q_1)/(t_2 - t_1)$$

where Q_1 is the charge at time t_1, and Q_2 is the charge at time t_2. In these formulas, positive values of $I_{c/d}$ represent a charging condition; negative values of $I_{c/d}$ represent a discharging condition.

Ohm's Law for DC amperage

Let V represent the voltage (in volts) across a component or device; let R represent the resistance (in ohms) of the component or device. Then the current I (in amperes) through the component or device is:

$$I = V/R$$

Current vs voltage and power

Let V represent the potential difference (in volts) across a component or device; let P represent the power (in watts) dissipated, radiated, or supplied by the component or device. Then the current I (in amperes) through the component or device is:

$$I = P/V$$

Current vs voltage, energy, and time

Let V represent the potential difference (in volts) across a component or device; let E represent the energy (in joules) dissipated, radiated, or supplied by the component or device over a period of time t (in seconds). Then the current I (in amperes) through the component or device is:

$$I = E/(Vt)$$

Current vs resistance and power

Let R represent the resistance (in ohms) of a component or device; let P represent the power (in watts) radiated, dissipated, or supplied by the component or device. Then the current I (in amperes) through the component or device is:

$$I = (P/R)^{1/2}$$

Current vs resistance, energy, and time

Let R represent the resistance (in ohms) of a component or device; let E represent the energy (in joules) dissipated, radiated, or supplied by the component or device over a period of time t (in seconds). Then the current I (in amperes) through the component or device is:

$$I = (E/(Rt))^{1/2}$$

Kirchhoff's Law for DC amperage

The current going into any point in a DC circuit is the same as the current going out. An example is shown at Fig. 4.2. If I_{in}

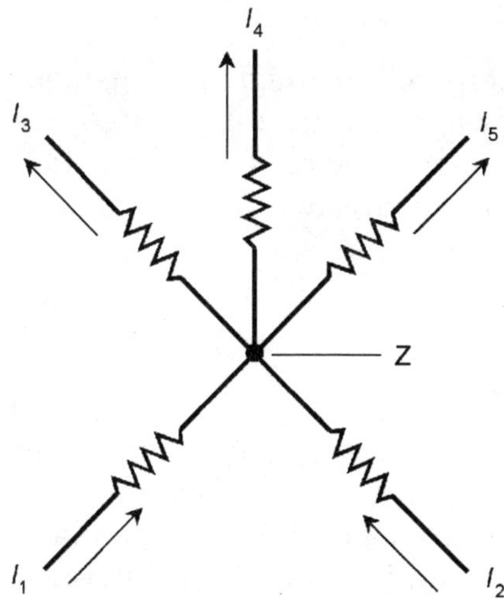

Figure 4.2 Kirchhoff's current law.

represents the total current entering the branch point Z and I_{out} represents the total current emerging from point Z, then:

$$I_{in} = I_{out}$$

$$I_{in} = I_1 + I_2$$

$$I_{out} = I_3 + I_4 + I_5$$

$$\therefore$$

$$I_1 + I_2 = I_3 + I_4 + I_5$$

The volt

The standard unit of DC voltage, also called *potential difference* or *electromotive force* (EMF), is the *volt*. Voltage is symbolized by V in this chapter; alternatively it can be symbolized by E unless doing so would confuse it with energy. One volt is the EMF required to drive one ampere of current through a resistance of one ohm.

Ohm's Law for DC voltage

Let I be the current (in amperes) through a component, and let R be the resistance (in ohms) of that component. Then the potential difference V (in volts) across the component is:

$$V = IR$$

Voltage vs current and power

Let I represent the current (in amperes) through a component or device; let P represent the power (in watts) dissipated, radiated, or supplied by the component or device. Then the potential difference V (in volts) across the component or device is:

$$V = P/I$$

Voltage vs current, energy, and time

Let I represent the current (in amperes) through a component or device; let E represent the energy (in joules) dissipated, radiated, or supplied by the component or device over a period of time t (in seconds). Then the potential difference V (in volts) across the component or device is:

$$V = E/(It)$$

Voltage vs resistance and power

Let R represent the resistance (in ohms) of a component or device; let P represent the power (in watts) dissipated, radiated, or supplied by the component or device. Then the potential difference V (in volts) across the component or device is:

$$V = (PR)^{1/2}$$

Voltage vs resistance, energy, and time

Let R represent the resistance (in ohms) of a component or device; let E represent the energy (in joules) dissipated, radiated, or supplied by the component or device over a period of time t (in seconds). Then the potential difference V (in volts) across the component or device is:

$$V = (ER/t)^{1/2}$$

Kirchhoff's Law for DC voltage

In a closed DC network, the sum of the voltages across all the components in any given loop, taking polarity into account, is zero. An example is shown at Fig. 4.3. The EMF of the battery is V; there are four components across which the potential differences are V_1, V_2, V_3, and V_4. The following equations hold in this case:

$$V + V_1 + V_2 + V_3 + V_4 = 0$$

$$V_1 + V_2 + V_3 + V_4 = -V$$

$$V = -(V_1 + V_2 + V_3 + V_4)$$

The ohm

The standard unit of DC resistance, symbolized by R, is the *ohm*. A component has a resistance of one ohm when an applied EMF of one volt across it results in a current of one ampere through it, or when a current of one ampere through the component produces a potential difference of one volt across it.

Ohm's Law for DC resistance

Let I be the current (in amperes) through a component, and let V be the potential difference (in volts) across it. Then the resistance R (in ohms) of the component is:

$$R = V/I$$

Resistance vs current and power

Let I be the current (in amperes) through a component or device, and let P be the power (in watts) radiated or supplied. Then the resistance R (in ohms) of the component or device is:

$$R = P/I^2$$

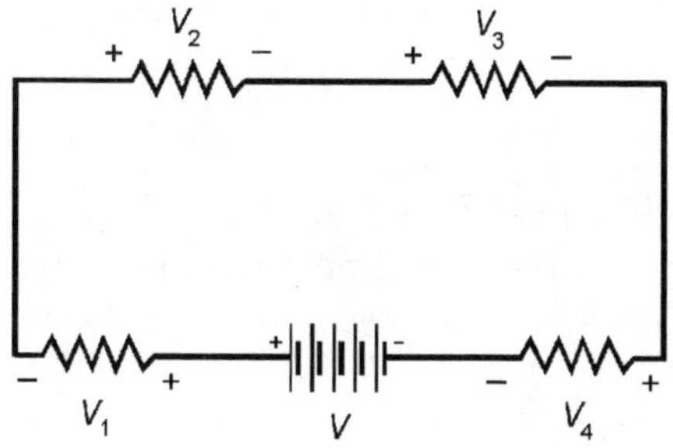

Figure 4.3 Kirchhoff's voltage law.

Resistance vs current, energy, and time

Let I be the current (in amperes) through a component or device, and let E be the energy (in joules) dissipated, radiated, or supplied over a period of time t (in seconds). Then the resistance R (in ohms) of the component or device is:

$$R = E/(I^2t)$$

Resistance vs voltage and power

Let V be the potential difference (in volts) across a component or device, and let P be the power (in watts) dissipated, radiated, or supplied. Then the resistance R (in ohms) of the component or device is:

$$R = V^2/P$$

Resistance vs voltage, energy, and time

Let V be the potential difference (in volts) across a component or device, and let E be the energy (in joules) dissipated, radiated, or supplied over a period of time t (in seconds). Then the resistance R (in ohms) of the component or device is:

$$R = V^2t/E$$

The watt

The standard unit of DC power, symbolized by P, is the *watt*. A component dissipates, radiates, or supplies one watt when it carries or provides a current of one ampere, and when the potential difference across it is one volt.

Power vs energy and time

Let E be the energy (in joules) dissipated, radiated, or supplied by a component or device over a period of time t (in seconds). Then the dissipated, radiated, or supplied power P (in watts) is:

$$P = E/t$$

Power vs current and voltage

Let I be the current (in amperes) through a component or device, and let V be the potential difference (in volts) across it. Then the dissipated, radiated, or supplied power P (in watts) is:

$$P = VI$$

Power vs current and resistance

Let I be the current (in amperes) through a component or device, and let R be its resistance (in ohms). Then the dissipated, radiated, or supplied power P (in watts) is:

$$P = I^2R$$

Power vs voltage and resistance

Let V be the potential difference (in volts) across a component or device, and let R be its resistance (in ohms). Then the dissipated, radiated, or supplied power P (in watts) is:

$$P = V^2/R$$

The joule

The standard unit of DC energy, symbolized by E, is the *joule*. A component dissipates, radiates, or supplies one joule when it dissipates, radiates, or supplies an average power of one watt over a time interval of one second.

Energy vs power and time

Let P be the power (in watts) dissipated, radiated, or supplied by a component or device over a period of time t (in seconds). Then the energy E (in joules) is:

$$E = \int P \, dt$$

If the power remains constant over the entire time interval, then

$$E = Pt$$

Energy vs current, voltage, and time

Let I be the current (in amperes) through a component or device, and let V be the potential difference (in volts) across it. Then the dissipated, radiated, or supplied energy E (in joules) over a period of time t (in seconds) is:

$$E = \int VI \, dt$$

If the current and voltage remain constant over the entire time interval, then

$$E = VIt$$

Energy vs current, resistance, and time

Let I be the current (in amperes) through a component or device, and let R be its resistance (in ohms). Then the dissipated or radiated energy E (in joules) over a period of time t (in seconds) is:

$$E = \int I^2R \, dt$$

If the current and resistance remain constant over the entire time interval, then

$$E = I^2Rt$$

Energy vs voltage, resistance, and time

Let V be the potential difference (in volts) across a component or device, and let R be its resistance (in ohms). Then the dissipated or radiated energy E (in joules) over a period of time t (in seconds) is:

$$E = \int (V^2/R) \, dt$$

If the current and resistance remain constant over the entire time interval, then

$$E = V^2t/R$$

Alternating Current

This section contains formulas involving alternating-current (AC) frequency, phase, amperage, voltage, impedance, power, and energy.

Frequency and phase

Frequency is usually symbolized by the letter f, *period* by the letter T, and *phase* angle by the Greek letter ϕ.

Frequency vs period

Let f be the frequency of an AC wave (in hertz), and let T be the period (in seconds). Then the following relations hold:

$$f = 1/T$$

$$T = 1/f$$

These relations are also valid for T in milliseconds (ms) and f in kilohertz (kHz); for T in microseconds (μs) and f in megahertz (MHz); for T in nanoseconds (ns) and f in gigahertz (GHz); and for T in picoseconds (ps) and f in terahertz (THz).

Phase angle vs time and frequency

Let f be the frequency of an AC wave (in hertz), and let t be the time (in seconds) following the instant t_0 at which the wave amplitude is zero and positive-going (Fig. 4.4). Then the phase angle ϕ, in degrees, is:

$$\phi = 360ft$$

If ϕ is expressed in radians, then:

$$\phi = 2\pi ft$$

These formulas are also valid for t in milliseconds (ms) and f in kilohertz (kHz); for t in microseconds (μs) and f in megahertz (MHz); for t in nanoseconds (ns) and f in gigahertz (GHz); and for t in picoseconds (ps) and f in terahertz (THz).

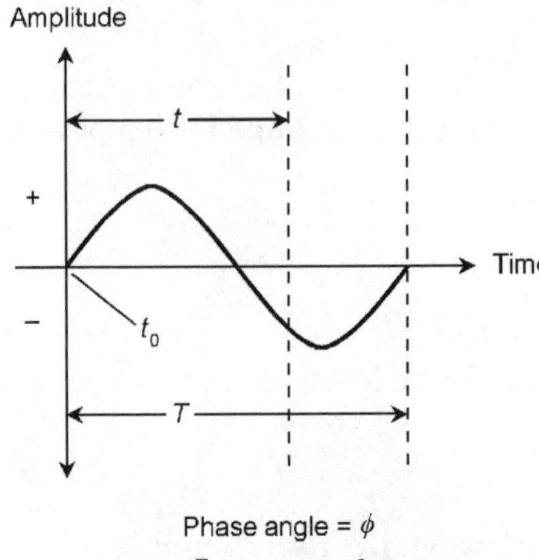

Figure 4.4 Relations among frequency (f), period (T), phase (ϕ), and time (t) for an AC sine wave cycle beginning at $t = t_0$.

Phase angle vs time and period

Let T be the period of an AC wave (in seconds) and let t be the time (in seconds) following the instant t_0 at which the wave amplitude is zero and positive-going. Then the phase angle ϕ, in degrees, is:

$$\phi = 360t/T$$

If ϕ is expressed in radians, then:

$$\phi = 2\pi t/T$$

These formulas are also valid for t and T in milliseconds (ms), microseconds (μs), nanoseconds (ns), and picoseconds (ps).

AC amplitude

The *amplitude* of an AC wave can be expressed in several ways. The definitions and formulas in the next several paragraphs apply to sinusoidal waveforms, and are expressed in terms of voltage (V). However, these definitions and formulas can also be used for current (I).

Instantaneous amplitude

The *instantaneous amplitude* (V_{inst}) of an AC sine wave constantly varies. In the example at Fig. 4.5, instantaneous amplitudes are shown as vertical displacement on the waveform.

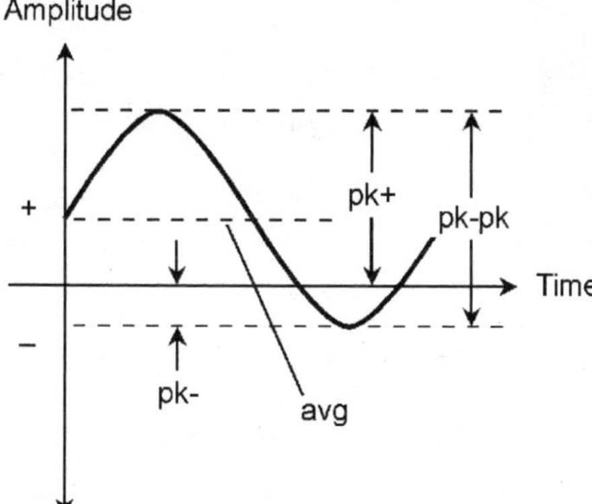

Figure 4.5 Positive peak (pk+), negative peak (pk−), peak-to-peak (pk-pk), and average (avg) values for an AC sine wave. See text for definition of root-mean-square (rms).

Positive peak amplitude

The *positive peak amplitude* (V_{pk+}) of an AC sine wave is the maximum deviation of V_{inst} in the positive direction. See Fig. 4.5.

Negative peak amplitude

The *negative peak amplitude* (V_{pk-}) of an AC sine wave is the maximum deviation of V_{inst} in the negative direction. See Fig. 4.5.

DC component

The *DC component* (V_{DC}) of an AC sine wave is the arithmetic mean of the positive and negative peak amplitudes:

$$V_{DC} = (V_{pk+} + V_{pk-})/2$$

Average amplitude

The *average amplitude* (V_{avg}) of an AC sine wave is the same as the DC component.

Peak amplitude when V_{DC} = 0

If $V_{DC} = 0$, then the positive and negative peak amplitudes are equal and opposite. This can be generalized as the *peak amplitude* (V_{pk}):

$$V_{pk} = V_{pk+} = -V_{pk-}$$

Peak-to-peak amplitude

The *peak-to-peak amplitude* (V_{pk-pk}) of an AC wave is the difference between the positive and negative peak amplitudes:

$$V_{pk-pk} = V_{pk+} - V_{pk-}$$

If $V_{DC} = 0$, then:

$$V_{pk-pk} = 2V_{pk+} = -2V_{pk-}$$

Instantaneous amplitude vs phase angle

Let V_{pk+} represent the positive peak amplitude of a wave. Let V_{DC} represent the DC component, and let ϕ represent the phase angle as measured from the point in time at which the instantaneous amplitude $V_{inst} = V_{DC}$ and is increasing positively. Then:

$$V_{inst} = V_{DC} + V_{pk+} \sin \phi$$

Effective amplitude

The *effective amplitude* of an AC wave is also known as the *direct-current (DC) equivalent amplitude* or the *root-mean-square (rms) amplitude*. Let V_{DC} represent the DC component. Then the rms amplitude, V_{rms}, is given by:

$$V_{rms} = V_{DC} + 2^{-1/2}(V_{pk+} - V_{DC})$$

$$\approx V_{DC} + 0.707\ (V_{pk+} - V_{DC})$$

If there is no DC component, then:

$$V_{rms} = 2^{-1/2}\ V_{pk} \approx 0.707\ V_{pk}$$

Impedance

Impedance, symbolized by the letter Z, is the opposition that a component or circuit offers to AC. Impedance is a two-dimensional quantity, consisting of two independent components, *resistance* and *reactance.*

Inductive reactance

Inductive reactance is symbolized jX_L. Its real-number coefficient, X_L, is always positive or zero.

jX_L vs frequency

If the frequency of an AC source is given (in hertz) as f, and the inductance of a component is given (in henrys) as L, then the vector expression for inductive reactance (in imaginary-number ohms), jX_L, is:

$$jX_L = j(2\pi fL) \approx j(6.28fL)$$

This formula also applies for f in kilohertz (kHz) and L in millihenrys (mH); for f in megahertz (MHz) and L in microhenrys (μH); for f in gigahertz (GHz) and L in nanohenrys (nH); and for f in terahertz (THz) and L in picohenrys (pH).

RL phase angle

The phase angle ϕ_{RL} in a resistance-inductance (*RL*) circuit is the arctangent of the ratio of the real-number coefficient of the inductive reactance to the resistance:

$$\phi_{RL} = \tan^{-1}(X_L/R)$$

Capacitive reactance

Capacitive reactance is symbolized jX_C. Its real-number coefficient, X_C, is always negative or zero.

jX_C vs frequency

If the frequency of an AC source is given (in hertz) as f, and the value of a capacitor is given (in farads) as C, then the vector expression for capacitive reactance (in imaginary-number ohms), jX_C, is given by:

$$jX_C = -j(1/(2\pi fC)) \approx -j(1/(6.28fC))$$

This formula also applies for f in megahertz (MHz) and C in microfarads (μF), and for f in terahertz (THz) and C in picofarads (pF).

RC phase angle

The phase angle ϕ_{RC} in a resistance-capacitance (*RC*) circuit is the arctangent of the ratio of the real-number coefficient of the capacitive reactance to the resistance:

$$\phi_{RC} = \tan^{-1}(X_C/R)$$

Complex impedances in series

Given two complex impedances $Z_1 = R_1 + jX_1$ and $Z_2 = R_2 + jX_2$ connected in series, the resultant complex impedance Z is their vector sum, given by:

$$Z = (R_1 + R_2) + j(X_1 + X_2)$$

Admittance

Admittance, symbolized by the letter Y, is the ease with which a component or circuit conducts AC. Admittance is a two-dimensional quantity, consisting of two independent components, *conductance* and *susceptance.*

AC conductance

In an AC circuit, electrical conductance behaves the same way as in a DC circuit. Conductance is symbolized by the capital letter G. The relationship between conductance and resistance is:

$$G = 1/R$$

The unit of conductance is the *siemens,* sometimes called the *mho.*

Inductive susceptance

Inductive susceptance is symbolized jB_L. Its real-number coefficient, B_L, is always negative or zero, being the negative reciprocal of the real-number coefficient of inductive reactance:

$$B_L = -1/X_L$$

The vector expression of inductive susceptance requires the j operator, as does the vector expression of inductive reactance. The reciprocal of j is its negative, so when calculating vector quantities B_L in terms of vector quantities X_L, the sign changes.

jB_L vs frequency

If the frequency of an AC source is given (in hertz) as f, and the value of an inductor is given (in henrys) as L, then the vector expression for inductive susceptance (in imaginary-number siemens), jB_L, is given by:

$$jB_L = -j(1/(2\pi fL)) \approx -j(1/(6.28fL))$$

The above formula also applies for f in kilohertz (kHz) and L in millihenrys (mH); for f in megahertz (MHz) and L in microhenrys (μH); for f in gigahertz (GHz) and L in nanohenrys (nH); and for f in terahertz (THz) and L in picohenrys (pH).

Capacitive susceptance

Capacitive susceptance is symbolized jB_C. Its real-number coefficient, B_C, is always positive or zero, being the negative reciprocal of the real-number coefficient of capacitive reactance:

$$B_C = -1/X_C$$

The expression of capacitive susceptance requires the j operator, as does the expression of capacitive reactance. The reciprocal of j is its negative, so when calculating B_C in terms of X_C, the sign changes.

jB_C vs frequency

If the frequency of an AC source is given (in hertz) as f, and the value of a capacitor is given (in farads) as C, then the vector expression for capacitive susceptance (in imaginary-number siemens), jB_C, is given by:

$$jB_C = j\,(2\pi fC) \approx j(6.28fC)$$

This formula also applies for f in megahertz (MHz) and C in microfarads (μF), and for f in terahertz (THz) and C in picofarads (pF).

Complex admittances in parallel

Given two complex admittances $Y_1 = G_1 + jB_1$ and $Y_2 = G_2 + jB_2$ connected in parallel, the resultant complex admittance Y is their vector sum, given by:

$$Y = (G_1 + G_2) + j\,(B_1 + B_2)$$

Complex impedances in parallel

To find the resultant complex impedance of two components in parallel, follow these steps in order:

1. Convert each real-number resistance to conductance: $G_n = 1/R_n$
2. Convert each imaginary-number reactance to susceptance, paying careful attention to the changes in sign of the real-number coefficients: $B_n = -1/X_n$
3. Sum the conductances and susceptances to get complex admittances
4. Use the above formula to find net complex admittance, consisting of a resultant conductance and a resultant susceptance
5. Convert the resultant real-number conductance back to resistance
6. Convert the resultant imaginary-number susceptance back to reactance, paying careful attention to the change in sign of the real-number coefficient: $X_n = -1/B_n$

The resulting expression $R + jX$ is the complex impedance of the two components in parallel.

AC amperage

The standard unit of alternating current, also called *AC amperage* and symbolized by I_{rms}, is the *ampere rms*.

Current vs voltage and reactance

Let V_{rms} be the AC voltage (in volts rms) across a component or device; let X be the real-number coefficient of the reactance (in ohms) of the component or device. Then the alternating current (in amperes rms), I_{rms}, is given by:

$$I_{rms} = |V_{rms}/X|$$

Current vs voltage, frequency, and inductance

Let V_{rms} be the AC voltage (in volts rms) across a component or device; let f be the AC frequency (in hertz); let L be the inductance (in henrys) of the component or device. Then the alternating current (in amperes rms), I_{rms}, is given by:

$$I_{rms} = V_{rms}/(2\pi fL) \approx V_{rms}/(6.28fL)$$

This formula also applies for f in kilohertz (kHz) and L in millihenrys (mH); for f in megahertz (MHz) and L in microhenrys (μH); for f in gigahertz (GHz) and L in nanohenrys (nH); and for f in terahertz (THz) and L in picohenrys (pH).

Current vs voltage, frequency, and capacitance

Let V_{rms} be the AC voltage (in volts rms) across a component or device; let f be the AC frequency (in hertz); let C be the capacitance (in farads) of the component or device. Then the alternating current (in amperes rms), I_{rms}, is given by:

$$I_{rms} = 2\pi V_{rms} fC \approx 6.28V_{rms} fC$$

This formula also applies for f in megahertz (MHz) and C in microfarads (μF), and for f in terahertz (THz) and C in picofarads (pF).

Current vs voltage and complex impedance

Let V_{rms} be the AC voltage (in volts rms) across a component or device; let the complex impedance of the component or device be $Z = R + jX$, where X is the real-number coefficient of the reactance (in ohms) of the component or device and R is the resistance (in ohms) of the component or device. Then the alternating current (in amperes rms), I_{rms}, is given by:

$$I_{rms} = V_{rms}/(R^2 + X^2)^{1/2}$$

AC voltage

The standard unit of AC voltage, also called *AC electromotive force* (*AC EMF*) and symbolized by V_{rms}, is the *volt rms*.

Voltage vs current and reactance

Let I_{rms} be the alternating current (in amperes rms) through a component or device; let X be the real-number coefficient of the reactance (in ohms) of the component or device. Then the AC voltage (in volts rms), V_{rms}, across the component or device is given by:

$$V_{rms} = |I_{rms}X|$$

Voltage vs current, frequency, and inductance

Let I_{rms} be the alternating current (in amperes rms) through a component or device; let f be the AC frequency (in hertz); let L be the inductance (in henrys) of the component or device. Then the AC voltage (in volts rms), V_{rms}, across the component or device is given by:

$$V_{rms} = 2\pi I_{rms} fL \approx 6.28 I_{rms} fL$$

This formula also applies for f in kilohertz (kHz) and L in millihenrys (mH); for f in megahertz (MHz) and L in microhenrys (μH); for f in gigahertz (GHz) and L in nanohenrys (nH); and for f in terahertz (THz) and L in picohenrys (pH).

Voltage vs current, frequency, and capacitance

Let I_{rms} be the alternating current (in amperes rms) through a component or device; let f be the AC frequency (in hertz); let C

be the capacitance (in farads) of the component or device. Then the AC voltage (in volts rms), V_{rms}, across the component or device is given by:

$$V_{rms} = I_{rms}/(2\pi fC) \approx I_{rms}/(6.28fC)$$

This formula also applies for f in megahertz (MHz) and C in microfarads (μF), and for f in terahertz (THz) and C in picofarads (pF).

Voltage vs current and complex impedance

Let I_{rms} be the alternating current (in amperes rms) through a component or device; let the complex impedance of the component or device be $Z = R + jX$, where X is the real-number coefficient of the reactance (in ohms) of the component or device and R is the resistance (in ohms) of the component or device. Then the AC voltage (in volts rms), V_{rms}, is given by:

$$V_{rms} = I_{rms} (R^2 + X^2)^{1/2}$$

AC power

There are three ways of expressing AC power: as *real power* (in watts rms), as *reactive power* (in watts reactive), or as *apparent power* (in volt-amperes).

Real power

Let V_{rms} be the AC voltage across a component or device (in volts rms); let I_{rms} be the alternating current through the component or device (in amperes rms); let ϕ be the phase angle between the voltage and current waves. Then the real power, P_R, dissipated or radiated by the component or device (in watts rms) is given by the following formula:

$$P_R = V_{rms} I_{rms} \cos \phi$$

Reactive power

Let V_{rms} be the AC voltage across a component or device (in volts rms); let I_{rms} be the alternating current through the component or device (in amperes rms); let ϕ be the phase angle between the voltage and current waves. Then the reactive power, P_X, manifested in the component or device (in watts reactive) is given by the following formula:

$$P_X = V_{rms} I_{rms} \sin \phi$$

Apparent power

Let V_{rms} be the AC voltage across a component or device (in volts rms); let I_{rms} be the alternating current through the component or device (in amperes rms); let P_R be the real power dissipated or radiated by the component or device (in watts rms); let P_X be the reactive power manifested in the component or device (in watts reactive). Then the apparent power, P_{VA}, dissipated or radiated by the component or device (in volt-amperes) is given by the following formulas:

$$P_{VA} = V_{rms} I_{rms}$$

$$P_{VA} = (P_R{}^2 + P_X{}^2)^{1/2}$$

AC energy

There are three ways of expressing AC energy: as *real energy* (in joules), as *reactive energy* (in joules reactive), or as *apparent energy* (in volt-ampere-seconds).

Real energy

Let V_{rms} be the AC voltage across a component or device (in volts rms); let I_{rms} be the alternating current through the component or device (in amperes rms); let ϕ be the phase angle between the voltage and current waves. Then the real energy, E_R, dissipated or radiated by the component or device (in joules) over a period of time t (in seconds) is given by the following formula:

$$E_R = V_{rms} I_{rms} t \cos \phi$$

Reactive energy

Let V_{rms} be the AC voltage across a component or device (in volts rms); let I_{rms} be the alternating current through the component or device (in amperes rms); let ϕ be the phase angle between the voltage and current waves. Then the reactive energy, E_X, manifested in the component or device (in joules reactive) over a period of time t (in seconds) is given by the following formula:

$$E_X = V_{rms} I_{rms} t \sin \phi$$

Apparent energy

Let V_{rms} be the AC voltage across a component or device (in volts rms); let I_{rms} be the alternating current through the component or device (in amperes rms); let E_R be the real energy dissipated or radiated by the component or device (in joules); let E_X be the reactive energy manifested in the component or device (in joules reactive). Then the apparent energy, E_{VAS}, dissipated or radiated by the component or device (in volt-ampere-seconds) over a period of time t (in seconds) is given by the following formulas:

$$E_{VAS} = V_{rms} I_{rms} t$$

$$E_{VAS} = (E_R^2 + E_X^2)^{1/2}$$

Magnetism, Inductors, and Transformers

This section contains formulas involving magnetic fields, magnetic circuits, and the behavior of inductor coils and transformers.

Reluctance

Reluctance is a measure of the opposition that a circuit offers to the establishment of a magnetic field. It is symbolized R and is measured in *ampere-turns per weber* in the SI system of units.

Reluctance of a magnetic core

Let s represent the length (in meters) of a path through a magnetic core; let μ represent the magnetic permeability of the core material (in tesla-meters per ampere); let A represent the cross-sectional area of the core (in square meters). Then the reluctance R (in ampere-turns per weber) is:

$$R = s/(A\mu)$$

The above formula also holds for s in centimeters, μ in gauss per oersted, and A in square centimeters.

Reluctances in series

Reluctances in series add like resistances in series. If R_1, R_2, R_3, ... R_n represent reluctances and R_S represents their series combination, then:

$$R_S = R_1 + R_2 + R_3 + \ldots + R_n$$

Reluctances in parallel

Reluctances in parallel add like resistances in parallel. If R_1, R_2, R_3, ... R_n represent reluctances and R_P represents their parallel combination, then:

$$R_P = 1/(1/R_1 + 1/R_2 + 1/R_3 + \ldots + 1/R_n)$$

For only two reluctances R_1 and R_2 in parallel, the composite parallel reluctance, R_P, is given by:

$$R_P = R_1R_2/(R_1 + R_2)$$

Magnetic flux density

Let Φ represent magnetic flux (in webers); let A represent the cross-sectional area of a region through which the flux lines pass at right angles (in square meters). Then the *magnetic flux density* (in teslas) is denoted B and is given by:

$$B = \Phi/A$$

The above formula also holds when B is specified in gauss, Φ is specified in maxwells, and A is specified in square centimeters.

Magnetic permeability

Let B represent magnetic flux density (in teslas); let H represent magnetic field intensity (in amperes per meter). Then the *magnetic permeability* (in tesla-meters per ampere) is denoted μ and is given by:

$$\mu = B/H$$

The above formula also holds when μ is expressed in gauss per oersted, B is expressed in gauss, and H is expressed in oersteds.

Magnetomotive force

Let N represent the number of turns in an air-core coil; let I represent the current through the coil (in amperes). Then the *magnetomotive force* (in ampere-turns) is denoted F and is given by:

$$F = NI$$

If F is specified in gilberts, then

$$F = 0.4\pi NI \approx 1.256NI$$

Magnetizing force

Let N represent the number of turns in an air-core coil; let I represent the current through the coil (in amperes); let s represent the length of a magnetic path through the coil (in meters). Then the *magnetizing force* (in ampere-turns per meter) is denoted H and is given by:

$$H = NI/s$$

If H is specified in oersteds and s is specified in centimeters, then

$$H = 0.4\pi NI/s \approx 1.256NI/s$$

Voltage induced by motion of conductor

Let B represent the intensity of a stationary, constant magnetic field (in webers per square meter); let s represent the length of a conductor (in meters); let v represent the velocity of the conductor (in meters per second) at right angles to the magnetic lines of flux. Then the *induced voltage* (in volts) between the ends of the conductor is denoted V and is given by:

$$V = Bsv$$

Voltage induced by variable magnetic flux

Let N represent the number of turns in a coil; let $d\Phi/dt$ represent the change in magnetic flux (in webers per second). Then the induced voltage (in volts), V, between the ends of the coil is given by:

$$V = N(d\Phi/dt)$$

Transformer efficiency

Let I_{pri} represent the current (in amperes) in the primary winding of a transformer; let I_{sec} represent the current (in amperes) in the secondary; let V_{pri} represent the rms sine-wave AC voltage across the primary; let V_{sec} represent the rms sine-wave AC voltage across the secondary. Then the *transformer efficiency* (as a ratio) is denoted *Eff* and is given by:

$$Eff = V_{sec}I_{sec}/(V_{pri}I_{pri})$$

Expressed as a percentage and denoted $Eff_{\%}$, the efficiency of the transformer is:

$$Eff_{\%} = 100\ V_{sec}I_{sec}/(V_{pri}I_{pri})$$

PS turns ratio

Let N_{pri} represent the number of turns in the primary winding of a transformer; let N_{sec} represent the number of turns in the secondary winding. Then the *primary-to-secondary turns ratio* is denoted *P:S* and is given by:

$$P{:}S = N_{pri}/N_{sec}$$

SP turns ratio

The *secondary-to-primary turns ratio* of a transformer is denoted *S:P* and is given by:

$$S{:}P = N_{sec}/N_{pri} = 1/(P{:}S)$$

Voltage transformation

Let V_{pri} represent the rms sine-wave AC voltage across the primary winding of a transformer (in volts). Then the rms sine-wave AC voltage across the secondary, V_{sec} (in volts), neglecting transformer losses, is given by either of the following formulas:

$$V_{sec} = (S{:}P)\ V_{pri}$$

$$V_{sec} = V_{pri}/(P{:}S)$$

Impedance transformation

Let *S:P* represent the secondary-to-primary turns ratio of a transformer; let $Z_{in} = R_{in} + j0$ represent a purely resistive (zero-reactance) impedance at the input (across the primary winding). Then the impedance at the output (across the secondary winding), Z_{out}, is also purely resistive, and is given by:

$$Z_{out} = (S{:}P)^2 Z_{in} = (S{:}P)^2 R_{in} + j0$$

Let $P{:}S$ represent the primary-to-secondary turns ratio of a transformer; let $Z_{sec} = R_{sec} + j0$ represent a purely resistive (zero-reactance) impedance connected across the secondary winding. Then the reflected impedance across the primary winding, Z_{pri}, is also purely resistive, and is given by:

$$Z_{pri} = (P{:}S)^2 Z_{sec} = (P{:}S)^2 R_{sec} + j0$$

Current demand

Let I_{load} represent the rms sine-wave alternating current drawn by a load connected to the secondary winding of a transformer (in amperes). Then the rms sine-wave alternating current demanded from a power source connected to the primary, I_{src} (in amperes), neglecting transformer losses, is given by:

$$I_{src} = (S{:}P)\, I_{load}$$

Ohmic power loss

Let I_{rms} represent the alternating current (in amperes rms) through an inductor or transformer winding; let V_{rms} represent the AC voltage (in volts rms) across the inductor or winding; let R represent the resistive component of the complex impedance of the inductor or winding (in ohms). Then the *ohmic power loss*, denoted P_Ω and expressed in watts, is given by either of the following two formulas:

$$P_\Omega = I_{rms}^{\,2} R$$

$$P_\Omega = V_{rms}^{\,2}/R$$

Eddy-current power loss

Let B represent the maximum flux density in an inductor or transformer core (in gauss); let s represent the thickness of the core material (in centimeters); let U represent the volume of the core material (in cubic centimeters); let f represent the frequency of the applied alternating current (in hertz); let k represent the core constant as specified by the manufacturer. Then the *eddy-current power loss* (in watts) is denoted P_I and is given by:

$$P_I = kUBs^2f^2$$

At 60 Hz, the utility line frequency commonly used in the United States:

$$P_I = 3600kUBs^2$$

In circuits where the AC line frequency is 50 Hz:

$$P_I = 2500kUBs^2$$

For silicon steel, a core material used in AC power-supply transformers, the core constant k is typically in the neighborhood of 4×10^{-12}. However, the value of k can differ substantially from this in the case of powerdered-iron cores.

Hysteresis power loss

Let $A_{B\text{-}H}$ represent the area of the measured hysteresis curve (B-H curve) for a core material at a specific frequency, where B is the flux density in gauss and H is the magnetizing force in oersteds. Then the *hysteresis power loss* (in watts) is denoted P_H and is given by:

$$P_H = (0.796 \times 10^{-8})\, A_{B\text{-}H}$$

Total power loss in transformer

Let V_{pri} represent the AC voltage (in volts rms) across the primary winding of a transformer operating with a specific constant load. Let V_{sec} represent the AC voltage (in volts rms) across the secondary; let I_{pri} represent the alternating current (in amperes rms) through the primary; let I_{sec} represent the alternating current (in amperes rms) through the secondary. Then, assuming zero reactance in the source or the load, the *total power loss* in the transformer, P_{loss} (in watts) is given by:

$$P_{loss} = V_{pri}\, I_{pri} - V_{sec}\, I_{sec}$$

Total power loss in inductor or winding

Let P_Ω represent the ohmic power loss (in watts) in an inductor or transformer winding; let P_I represent the eddy-current power loss (in watts); let P_H represent the hysteresis power loss (in watts); let P_Φ represent flux-leakage power loss (in watts). Then the total power loss P_{loss} (in watts) is given by:

$$P_{loss} = P_\Omega + P_I + P_H + P_\Phi$$

Resonance, Filters, and Noise

This section contains formulas relevant to resonance, filter design, and noise characteristics.

LC resonant frequency

Let L be the inductance (in henrys) and C be the capacitance (in farads) in an inductance-capacitance (LC) resonant circuit. Then the *LC resonant frequency* (in hertz) is denoted f_0 and is given by:

$$f_0 = 1/(2\pi L^{1/2} C^{1/2})$$

This formula also holds for f_0 in megahertz, L in microhenrys, and C in microfarads.

Quarter-wave cavity resonant frequency

Let s be the end-to-end length (in inches) of an air cavity. Then the fundamental *quarter-wave cavity resonant frequency* (in megahertz) is denoted f_0 and is given by:

$$f_0 = 2950/s$$

If s is in centimeters, then:

$$f_0 = 7500/s$$

Harmonic quarter-wave resonances occur at odd integral multiples of this frequency.

Half-wave cavity resonant frequency

Let s be the end-to-end length (in inches) of an air cavity. Then the fundamental *half-wave cavity resonant frequency* (in megahertz) is denoted f_0 and is given by:

$$f_0 = 4900/s$$

If s is in centimeters, then:

$$f_0 = 15{,}000/s$$

Harmonic half-wave resonances occur at all integral multiples of this frequency.

Quarter-wave transmission-line resonant frequency

Let s be the end-to-end length (in inches) of a section of transmission line whose velocity factor (as a ratio between 0 and 1) is v. Then the fundamental *quarter-wave transmission-line resonant frequency* (in megahertz) is denoted f_0 and is given by:

$$f_0 = 2950v/s$$

If s is in centimeters, then:

$$f_0 = 7500v/s$$

If s is in feet, then:

$$f_0 = 246v/s$$

Harmonic quarter-wave resonances occur at odd integral multiples of this frequency.

Half-wave transmission-line resonant frequency

Let s be the end-to-end length (in inches) of a section of transmission line whose velocity factor (as a ratio between 0 and 1) is v. Then the fundamental *half-wave transmission-line resonant frequency* (in megahertz) is denoted f_0 and is given by:

$$f_0 = 4900v/s$$

If s is in centimeters, then:

$$f_0 = 15{,}000v/s$$

If s is in feet, then:

$$f_0 = 492v/s$$

Harmonic half-wave resonances occur at all integral multiples of this frequency.

Lowpass filter

A *lowpass filter* offers little or no attenuation of signals at frequencies less than the *cutoff,* and significant attenuation of signals at frequencies greater than the cutoff.

Constant-*k* lowpass filter

Let f be the cutoff frequency (in hertz) for a constant-k lowpass LC filter as shown in Fig. 4.6. Let R be the load resistance (in ohms). Then the optimum inductance L (in henrys) is given by:

$$L = R/(\pi f)$$

The optimum capacitance C (in farads) for the filter shown in Fig. 4.6 is given by:

$$C = 1/(\pi f R)$$

Series *m*-derived lowpass filter

Let f_1 be the highest frequency of maximum transmission (in hertz) for a series m-derived lowpass LC filter as shown in Fig. 4.7. Let f_2 be the lowest frequency of maximum attenuation (in hertz); let R be the load resistance (in ohms). Then the *filter constant* is denoted m and is given by:

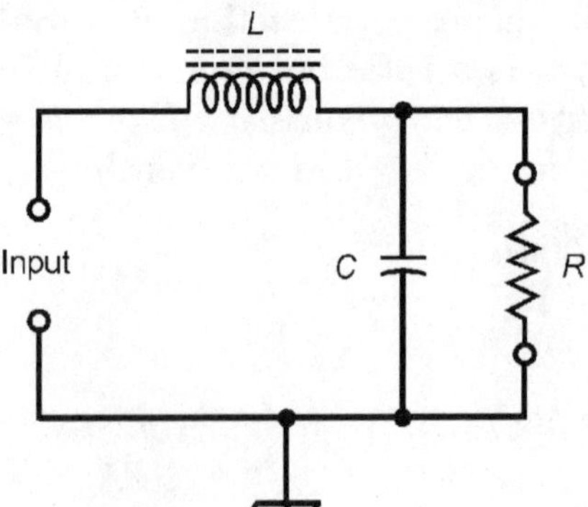

Figure 4.6 Constant-k lowpass filter.

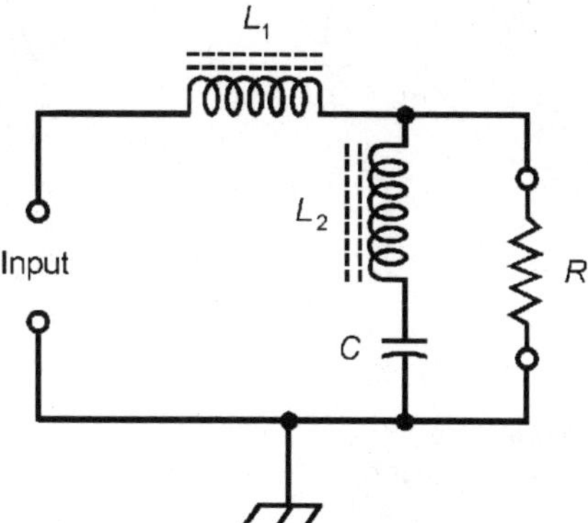

Figure 4.7 Series m-derived lowpass filter.

$$m = (1 - f_1^2/f_2^2)^{1/2}$$

The optimum inductance L_1 (in henrys) for the filter shown in Fig. 4.7 is given by:

$$L_1 = mR/(\pi f_1)$$

The optimum inductance L_2 (in henrys) for the filter shown in Fig. 4.7 is given by:

$$L_2 = R\,(1 - m^2)/(4\pi m f_1)$$

The optimum capacitance C (in farads) for the filter shown in Fig. 4.7 is given by:

$$C = (1 - m^2)/(\pi R f_1)$$

Shunt *m*-derived lowpass filter

Let f_1 be the highest frequency of maximum transmission (in hertz) for a shunt m-derived lowpass LC filter as shown in Fig. 4.8. Let f_2 be the lowest frequency of maximum attenuation (in hertz); let R be the load resistance (in ohms). Then the filter constant m is given by:

$$m = (1 - f_1^2/f_2^2)^{1/2}$$

The optimum inductance L (in henrys) for the filter shown in Fig. 4.8 is given by:

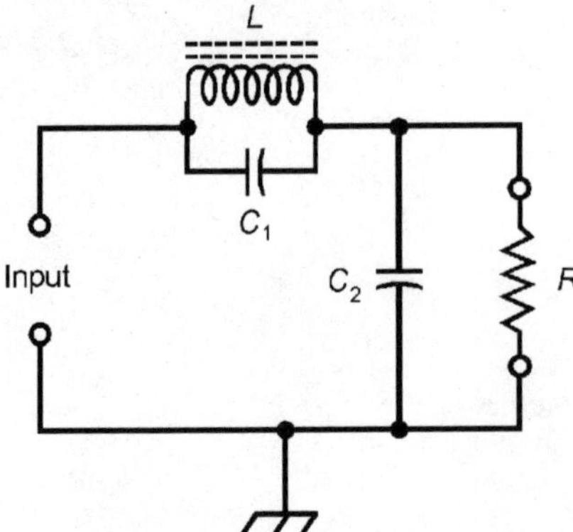

Figure 4.8 Shunt m-derived lowpass filter.

$$L = mR/(\pi f_1)$$

The optimum capacitance C_1 (in farads) for the filter shown in Fig. 4.8 is given by:

$$C_1 = (1 - m^2)/(4\pi R m f_1)$$

The optimum capacitance C_2 (in farads) for the filter shown in Fig. 4.8 is given by:

$$C_2 = m/(\pi R f_2)$$

Highpass filter

A *highpass filter* offers little or no attenuation of signals at frequencies greater than the *cutoff*, and significant attenuation of signals at frequencies less than the cutoff.

Constant-*k* highpass filter

Let f be the cutoff frequency (in hertz) for a constant-k highpass LC filter as shown in Fig. 4.9. Let R be the load resistance (in ohms). Then the optimum inductance L (in henrys) is given by:

$$L = R/(4\pi f)$$

The optimum capacitance C (in farads) for the filter shown in Fig. 4.9 is given by:

$$C = 1/(4\pi f R)$$

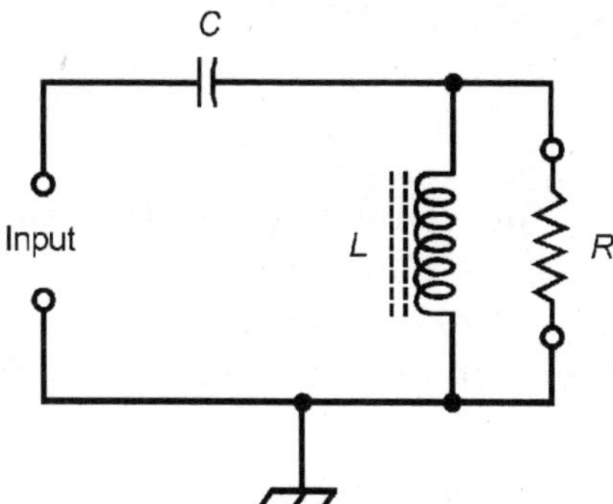

Figure 4.9 Constant-k highpass filter.

Series m-derived highpass filter

Let f_1 be the highest frequency of maximum attenuation (in hertz) for a series m-derived highpass LC filter as shown in Fig. 4.10. Let f_2 be the lowest frequency of maximum transmission (in hertz); let R be the load resistance (in ohms). Then the filter constant m is given by:

$$m = (1 - f_1^2/f_2^2)^{1/2}$$

The optimum inductance L (in henrys) for the filter shown in Fig. 4.10 is given by:

$$L = R/(4\pi m f_2)$$

The optimum capacitance C_1 (in farads) for the filter shown in Fig. 4.10 is given by:

$$C_1 = 1/(4\pi m f_2 R)$$

The optimum capacitance C_2 (in farads) for the filter shown in Fig. 4.10 is given by:

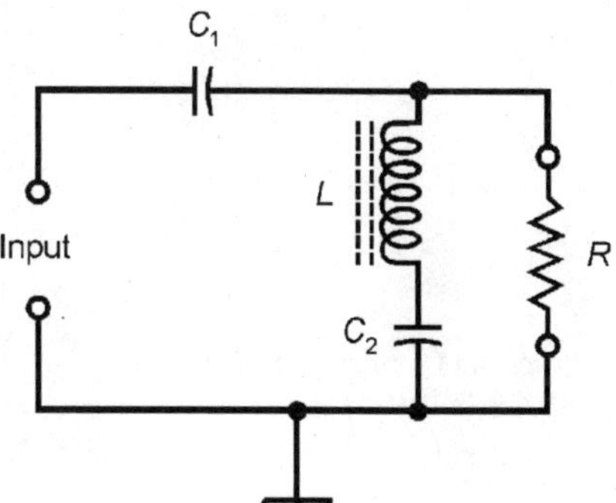

Figure 4.10 Series m-derived highpass filter.

$$C_2 = m/(\pi f_2 R(1 - m^2))$$

Shunt *m*-derived highpass filter

Let f_1 be the highest frequency of maximum attenuation (in hertz) for a shunt m-derived highpass LC filter as shown in Fig. 4.11. Let f_2 be the lowest frequency of maximum transmission (in hertz); let R be the load resistance (in ohms). Then the filter constant m is given by:

$$m = (1 - f_1^2/f_2^2)^{1/2}$$

The optimum inductance L_1 (in henrys) for the filter shown in Fig. 4.11 is given by:

$$L_1 = mR/(\pi f_2(1 - m^2))$$

The optimum inductance L_2 (in henrys) for the filter shown in Fig. 4.11 is given by:

$$L_2 = R/(4\pi m f_2)$$

The optimum capacitance C (in farads) for the filter shown in Fig. 4.11 is given by:

$$C = 1/(4\pi m f_2 R)$$

Bandpass filter

A *bandpass filter* offers little or no attenuation of signals whose frequencies are between a *lower cutoff* and an *upper cutoff*, and

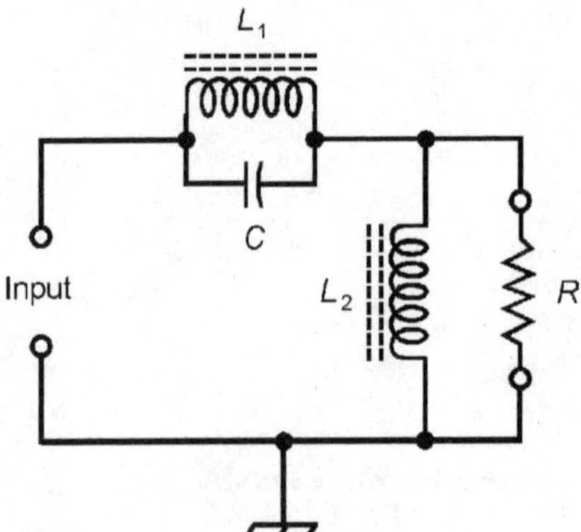

Figure 4.11 Shunt m-derived highpass filter.

significant attenuation of signals whose frequencies are outside this range.

Constant-*k* bandpass filter

Let f_1 be the lower cutoff frequency (in hertz) for a constant-k bandpass LC filter as shown in Fig. 4.12. Let f_2 be the upper cutoff frequency (in hertz). Let R be the load resistance (in ohms). Then the optimum inductances (in henrys) are given by:

$$L_1 = R/(\pi(f_2 - f_1))$$

$$L_2 = (f_2 - f_1)R/(4\pi f_1 f_2)$$

The optimum capacitances (in farads) for the filter shown in Fig. 4.12 are given by:

$$C_1 = (f_2 - f_1)/(4\pi f_1 f_2 R)$$

$$C_2 = 1/(\pi R(f_2 - f_1))$$

Series *m*-derived bandpass filter

Let frequencies f_1, f_2, f_3, and f_4 be expressed in hertz and defined as shown in Fig. 4.13. Let R be the load resistance (in ohms). Define the quantities x, m (the filter constant), y, and z as follows:

$$x = ((1 - f_2^2/f_3^2)(1 - f_3^2/f_4^2))^{1/2}$$

$$m = x/(1 - f_2 f_3/f_4^2)$$

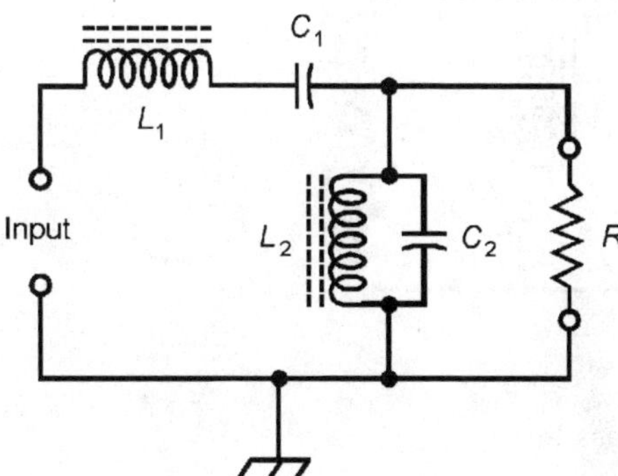

Figure 4.12 Constant-*k* bandpass filter.

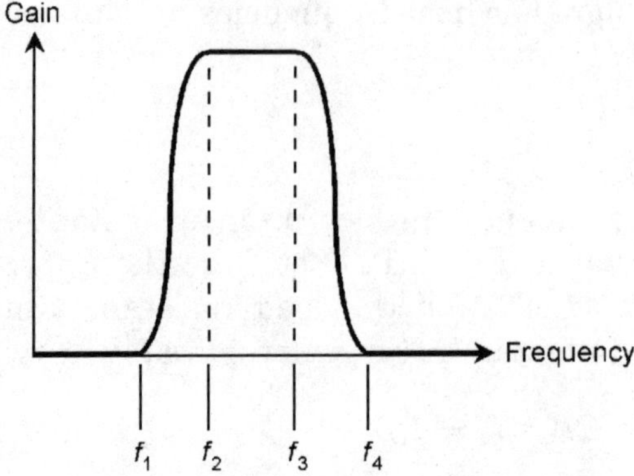

Figure 4.13 Bandpass filter response curve.

$$y = (1 - m^2)(1 - f_1^2/f_4^2)f_2f_3/(4xf_1^2)$$

$$z = (1 - m^2)(1 - f_1^2/f_4^2)/(4x)$$

The optimum inductances (in henrys) for the filter shown in Fig. 4.14 are given by:

$$L_1 = mR/(\pi(f_3 - f_2))$$

$$L_2 = zR/(\pi(f_3 - f_2))$$

$$L_3 = yR/(\pi(f_3 - f_2))$$

The optimum capacitances (in farads) for the filter shown in Fig. 4.14 are given by:

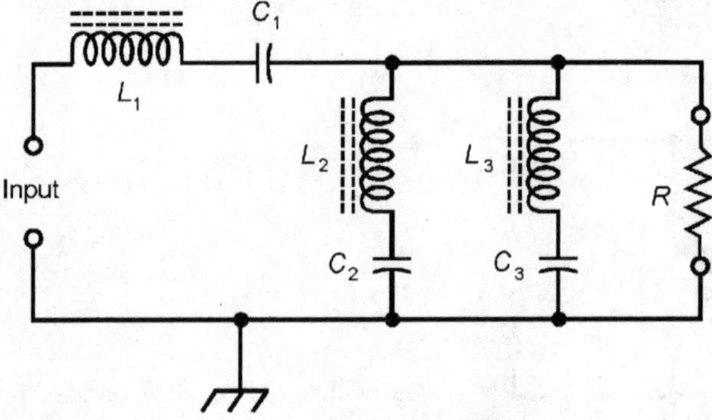

Figure 4.14 Series *m*-derived bandpass filter.

$$C_1 = (f_3 - f_2)/(4\pi mRf_2f_3)$$

$$C_2 = (f_3 - f_2)/(4\pi yRf_2f_3)$$

$$C_3 = (f_3 - f_2)/(4\pi zRf_2f_3)$$

Shunt *m*-derived bandpass filter

Let frequencies f_1, f_2, f_3, and f_4 be expressed in hertz and defined as shown in Fig. 4.13. Let R be the load resistance (in ohms). Define the quantities x, m (the filter constant), y, and z as in the preceding section for the series m-derived bandpass filter. Then the optimum inductances (in henrys) for the filter shown in Fig. 4.15 are given by:

$$L_1 = (f_3 - f_2)R/(4\pi zf_2f_3)$$

$$L_2 = (f_3 - f_2)R/(4\pi yf_2f_3)$$

$$L_3 = (f_3 - f_2)R/(4\pi mf_2f_3)$$

The optimum capacitances (in farads) for the filter shown in Fig. 4.15 are given by:

$$C_1 = z/(\pi R(f_3 - f_2))$$

$$C_2 = y/(\pi R(f_3 - f_2))$$

$$C_3 = m/(\pi R(f_3 - f_2))$$

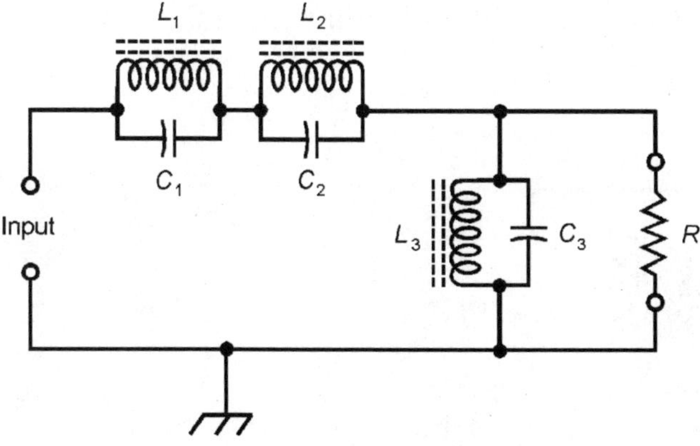

Figure 4.15 Shunt m-derived bandpass filter.

Bandstop filter

A *bandstop filter*, also called a *band-rejection filter*, offers significant attenuation of signals whose frequencies are between a *lower cutoff* and an *upper cutoff,* and little or no attenuation of signals whose frequencies are outside this range.

Constant-*k* bandstop filter

Let f_1 be the lower cutoff frequency (in hertz) for a constant-*k* bandstop *LC* filter as shown in Fig. 4.16. Let f_2 be the upper cutoff frequency (in hertz). Let *R* be the load resistance (in ohms). Then the optimum inductances (in henrys) are given by:

$$L_1 = R(f_2 - f_1)/(\pi f_1 f_2)$$

$$L_2 = R/(4\pi(f_2 - f_1))$$

The optimum capacitances (in farads) for the filter shown in Fig. 4.16 are given by:

$$C_1 = 1/((4\pi R(f_2 - f_1))$$

$$C_2 = (f_2 - f_1)/(\pi R f_1 f_2)$$

Series *m*-derived bandstop filter

Let frequencies f_1, f_2, f_3, and f_4 be expressed in hertz and defined as shown in Fig. 4.17. Let *R* be the load resistance (in ohms). Define the quantities *m* (the filter constant), *x*, and *y* as follows:

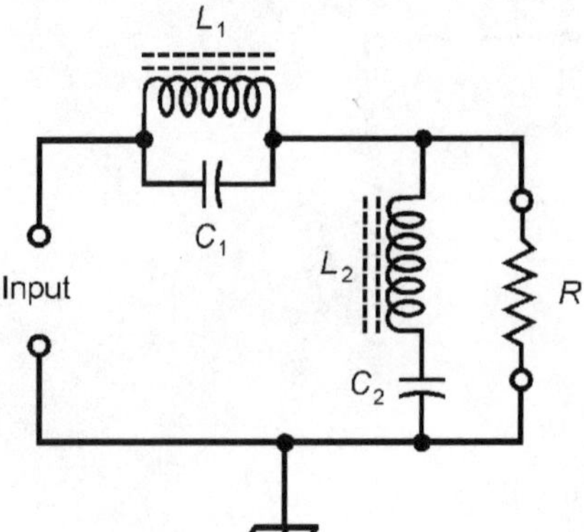

Figure 4.16 Constant-*k* bandstop filter.

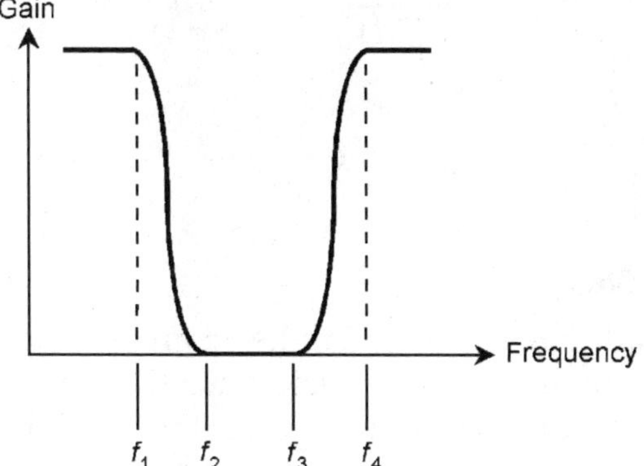

Figure 4.17 Bandstop filter response curve.

$$m = ((1 - f_1^2/f_3^2)(1 - f_3^2/f_4^2)/(1 - f_1/f_4))^{1/2}$$

$$x = (1/m)(1 + f_1 f_4/f_3^2)$$

$$y = (1/m)(1 + f_3^2/(f_1 f_4))$$

The optimum inductances (in henrys) for the filter shown in Fig. 4.18 are given by:

$$L_1 = mR(f_4 - f_1)/(\pi f_1 f_4)$$

$$L_2 = R/(4\pi m(f_4 - f_1))$$

$$L_3 = yR/(4\pi(f_4 - f_1))$$

The optimum capacitances (in farads) for the filter shown in Fig. 4.18 are given by:

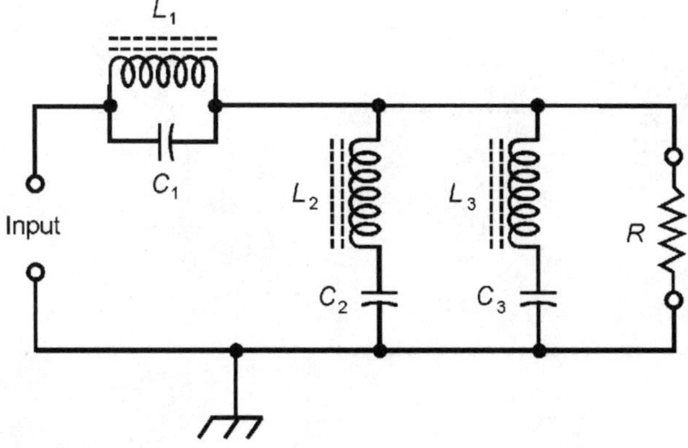

Figure 4.18 Series m-derived bandstop filter.

$$C_1 = 1/(4\pi m R(f_4 - f_1))$$

$$C_2 = (f_4 - f_1)/(\pi y R f_1 f_4)$$

$$C_3 = (f_4 - f_1)/(\pi x R f_1 f_4)$$

Shunt *m*-derived bandstop filter

Let frequencies f_1, f_2, f_3, and f_4 be expressed in hertz and defined as shown in Fig. 4.17. Let R be the load resistance (in ohms). Define the quantities m (the filter constant), x, and y as in the preceding section for the series m-derived bandstop filter. The optimum inductances (in henrys) for the filter shown in Fig. 4.19 are given by:

$$L_1 = (f_4 - f_1)R/(\pi y f_1 f_4)$$

$$L_2 = (f_4 - f_1)R/(\pi x f_1 f_4)$$

$$L_3 = R/(4\pi m(f_4 - f_1))$$

The optimum capacitances (in farads) for the filter shown in Fig. 4.19 are given by:

$$C_1 = x/(4\pi R(f_4 - f_1))$$

$$C_2 = y/(4\pi R(f_4 - f_1))$$

$$C_3 = m(f_4 - f_1)/(\pi R f_1 f_4)$$

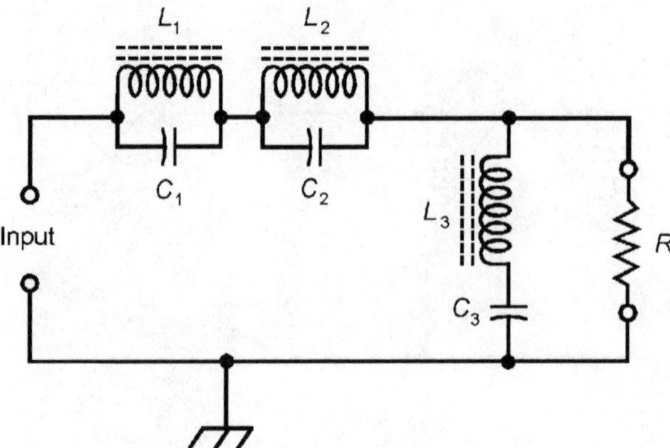

Figure 4.19 Shunt m-derived bandstop filter.

Thermal noise power

Let k represent Boltzmann's constant (approximately 1.3807×10^{-23} joules per degree Kelvin); let T represent the absolute temperature (in degrees Kelvin); let B represent the bandwidth (in hertz). Then the *thermal noise power* (in watts), P_{nt}, is given by:

$$P_{nt} = kTB$$

Thermal noise voltage

Let R represent the resistance of a noise source (in ohms); let P_{nt} represent the thermal noise power (in watts). Then the *thermal noise voltage* (in volts), V_{nt}, is given by:

$$V_{nt} = (P_{nt}R)^{1/2}$$

Signal-to-noise ratio

Let P_n represent the noise power (in watts) at the output of a circuit; let P_s represent the signal power (in watts) at the output of the same circuit. Then the *signal-to-noise ratio* (in decibels) is denoted $S{:}N$ and is given by:

$$S{:}N = 10 \log_{10} (P_s/P_n)$$

The value of $S{:}N$ can also be calculated in terms of voltages or currents. Let V_n represent the noise voltage (in volts) at the output of a circuit; let I_n represent the noise current (in amperes) at that point; let V_s represent the signal voltage (in volts) at that point; let I_s represent the signal current (in amperes) at that point. Then the signal-to-noise ratio $S{:}N$ (in decibels), assuming constant impedance, is given by either of these formulas:

$$S{:}N = 20 \log_{10} (V_s/V_n)$$

$$S{:}N = 20 \log_{10} (I_s/I_n)$$

Signal-plus-noise-to-noise ratio

Let P_n represent the noise power (in watts) at the output of a circuit; let P_s represent the signal power (in watts) at the output of the same circuit. Then the *signal-plus-noise-to-noise ratio* (in decibels) is denoted $(S{+}N){:}N$ and is given by:

$$(S{+}N){:}N = 10 \log_{10} ((P_s + P_n)/P_n)$$

The value of $(S{+}N){:}N$ can also be calculated in terms of voltages or currents. Let V_n represent the noise voltage (in volts) at the output of a circuit; let I_n represent the noise current (in amperes) at that point; let V_s represent the signal voltage (in volts) at that point; let I_s represent the signal current (in amperes) at that point. Then the signal-plus-noise-to-noise ratio $(S{+}N){:}N$ (in decibels), assuming constant impedance, is given by either of these formulas:

$$(S{+}N){:}N = 20 \log_{10} ((V_s + V_n)/V_n)$$

$$(S{+}N){:}N = 20 \log_{10} ((I_s + I_n)/I_n)$$

Noise figure

Let P_i represent the noise power (in watts) at the output of an ideal circuit; let P_a represent the noise power (in watts) at the output of an actual circuit. Then the *noise figure* (in decibels) of the actual circuit is denoted N and is given by:

$$N = 10 \log_{10} (P_a/P_i)$$

The noise figure can also be calculated in terms of $S{:}N$ ratios. Let $S{:}N_i$ be the $S{:}N$ ratio (in decibels) at the output of an ideal circuit; let $S{:}N_a$ be the $S{:}N$ ratio (in decibels) at the output of an actual circuit receiving the same signal. Then the noise figure N (in decibels) of the actual circuit is given by:

$$N = S{:}N_i - S{:}N_a$$

Semiconductor Diodes

A *diode* exhibits a nonlinear relationship between voltage and current. This relationship differs in the forward direction compared with the reverse direction. It also differs in the dynamic (changing) sense compared with the static (unchanging) sense. The following several formulas are relevant to semiconductor diodes.

Forward current

Let I_{rs} represent the reverse saturation current (in amperes) for a particular diode. Let q represent the charge on an electron (approximately 1.602×10^{-19} coulomb); let V_f represent the forward voltage (in volts); let k represent Boltzmann's constant (approximately 1.3807×10^{-23} joules per degree Kelvin); let T represent the absolute temperature (in degrees Kelvin); let e represent the exponential constant (approximately 2.718). Consider x to be defined as follows:

$$x = qV_f/(kT)$$

Then the *forward current*, I_f (in amperes), is given by:

$$I_f = I_{rs}\,(e^x - 1)$$

Static resistance

Let V_{DC} represent the DC voltage drop (in volts) across a diode; let I_{DC} represent the direct current (in amperes) through the diode. Then the *static resistance*, R_s (in ohms), of the diode is given by:

$$R_s = V_{DC}/I_{DC}$$

Dynamic resistance

Let V represent the instantaneous voltage drop (in volts) across a diode; let I represent the instantaneous current (in amperes) through the diode. Then the *dynamic resistance*, R_d (in ohms), of the diode is given by:

$$R_d = dV\,/\,dI$$

That is, R_d is the derivative of the voltage with respect to the current, or the slope of the characteristic curve V versus I (Fig. 4.20) at a specified point.

Rectification efficiency

Let V_{DC} represent the DC output voltage of a diode rectifier (in volts); let V_{pk} represent the peak AC input voltage (in volts). Then the *rectification efficiency*, η (as a ratio), is given by:

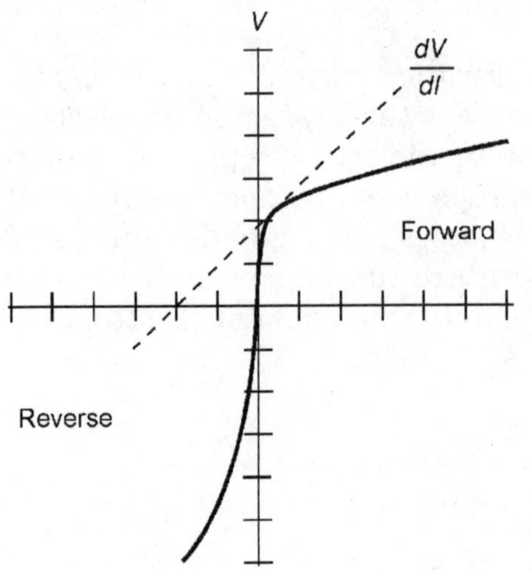

Figure 4.20 Characteristic curve for a semiconductor diode.

$$\eta = V_{DC}/V_{pk}$$

The rectification efficiency, $\eta_{\%}$ (in percent), is given by:

$$\eta_{\%} = 100V_{DC}/V_{pk}$$

Bipolar Transistors

The following several formulas are relevant to *bipolar transistors,* both the NPN type and the PNP type.

Static forward current transfer ratio

Assume the collector voltage, V_c, in a common-emitter circuit (Fig. 4.21) is constant. Let I_c represent the collector current (in amperes); let I_b represent the base current (in amperes). Then the *static forward current transfer ratio,* H_{FE}, is given by:

$$H_{FE} = I_c/I_b$$

Dynamic base resistance

Assume the collector voltage, V_c, is constant. Let V_b represent the base voltage (in volts); let I_b represent the base current (in amperes). Then the *dynamic base resistance,* R_b (in ohms), is:

$$R_b = dV_b/dI_b$$

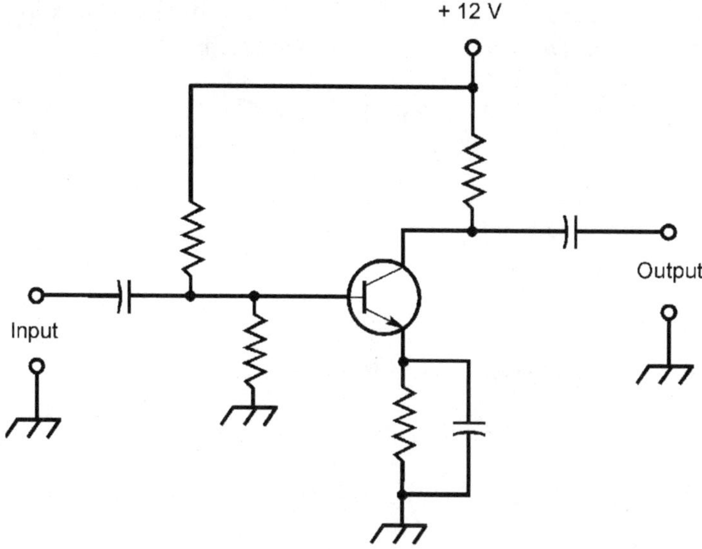

Figure 4.21 Common-emitter bipolar transistor circuit.

Dynamic emitter resistance

Assume the collector voltage, V_c, applied to a transistor is constant. Let V_e represent the emitter voltage (in volts); let I_e represent the emitter current (in amperes). Then the *dynamic emitter resistance*, R_e (in ohms), is given by:

$$R_e = dV_e/dI_e$$

Dynamic collector resistance

Assume the emitter current, I_e, through a transistor is constant. Let V_c represent the collector voltage (in volts); let I_c represent the collector current (in amperes). Then the *dynamic collector resistance*, R_c (in ohms), is given by:

$$R_c = dV_c/dI_c$$

Dynamic emitter feedback conductance

Assume the emitter voltage, V_e, applied to a transistor is constant. Let I_e represent the emitter current (in amperes); let V_c

represent the collector voltage (in volts). Then the *dynamic emitter feedback conductance,* G_{ec} (in siemens), is given by:

$$G_{ec} = dI_e/dV_c$$

Alpha

Assume the collector voltage, V_c, in a common-base circuit (Fig. 4.22) is constant. Let I_c represent the collector current (in amperes); let I_e represent the emitter current (in amperes). Then the *dynamic current amplification in common-base arrangement,* or *alpha* (symbolized α), is given by:

$$\alpha = dI_c/dI_e$$

This quantity is always greater than 0 but less than 1.

Beta

Assume the collector voltage, V_c, applied to a transistor in a common-emitter circuit (Fig. 4.21) is constant. Let I_c represent the collector current (in amperes); let I_b represent the base current (in amperes). Then the *dynamic current amplification in common-emitter arrangement,* or *beta* (symbolized β), is:

$$\beta = dI_c/dI_b$$

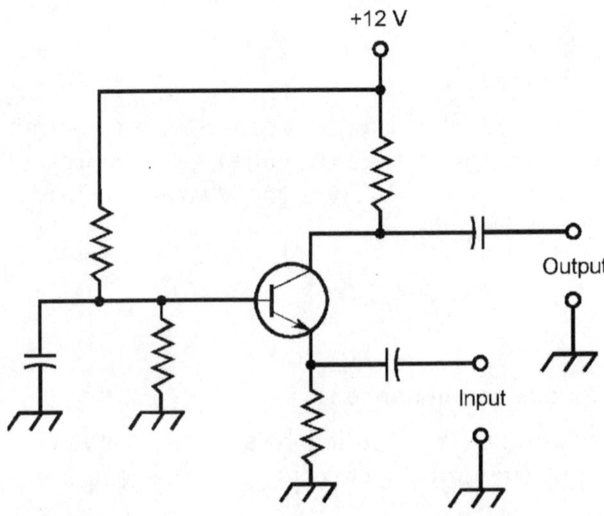

Figure 4.22 Common-base bipolar transistor circuit.

Alpha as a function of beta

Suppose the beta (β) of a bipolar transistor is known. The alpha (α) of that transistor, assuming the collector voltage V_c remains constant, is given by:

$$\alpha = \beta/(1 + \beta)$$

Beta as a function of alpha

Suppose the alpha (α) of a bipolar transistor is known. The beta (β) of that transistor, assuming the collector voltage V_c remains constant, is given by:

$$\beta = \alpha/(1 - \alpha)$$

Dynamic stability factor

Let I_c represent the collector current through a bipolar transistor (in amperes); let I_ω represent the collector leakage current (in amperes). Then the *dynamic stability factor, S*, of the transistor is given by:

$$S = dI_c/dI_\omega$$

Resistance parameters (common base)

Let α represent the dynamic current amplification of a bipolar transistor in the common-base arrangement. Let the following symbols represent resistances (in ohms):

R_b = dynamic base resistance
R_c = dynamic collector resistance
R_e = dynamic emitter resistance
R_{in} = input resistance
R_{rt} = reverse transfer resistance
R_{ft} = forward transfer resistance
R_{out} = output resistance

Then the following equations hold:

$$R_{\text{in}} = R_e + R_b$$

$$R_{\text{rt}} = R_b$$

$$R_{\text{ft}} = R_b + \alpha R_c$$

$$R_{\text{out}} = R_c + R_b$$

Resistance parameters (common emitter)

Let α represent the dynamic current amplification of a bipolar transistor in the common-base arrangement. Let the following symbols represent resistances (in ohms) in a common-emitter circuit:

R_b = dynamic base resistance
R_c = dynamic collector resistance
R_e = dynamic emitter resistance
R_{in} = input resistance
R_{rt} = reverse transfer resistance
R_{ft} = forward transfer resistance
R_{out} = output resistance

Then the following equations hold:

$$R_{\text{in}} = R_e + R_b$$

$$R_{\text{rt}} = R_e$$

$$R_{\text{ft}} = R_e - \alpha R_c$$

$$R_{\text{out}} = R_c + R_e - \alpha \mathrm{R}_c$$

Resistance parameters (common collector)

Let α represent the dynamic current amplification of a bipolar transistor in the common-base arrangement. Let the following symbols represent resistances (in ohms) in a common-collector circuit (Fig. 4.23):

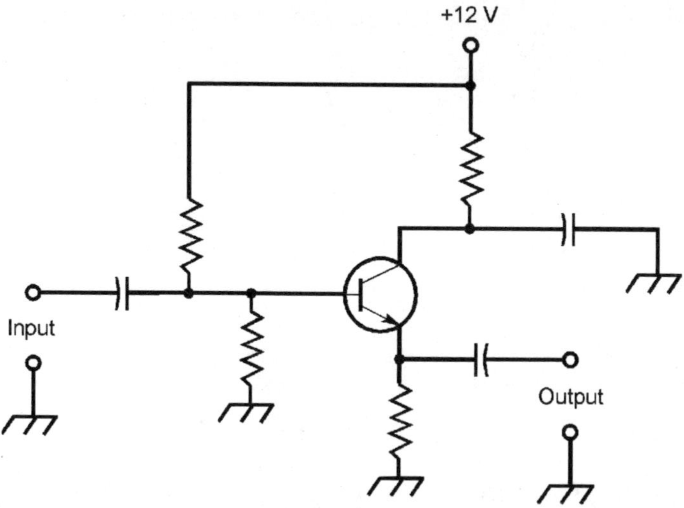

Figure 4.23 Common-collector bipolar transistor circuit, also known as emitter follower.

R_b = dynamic base resistance

R_c = dynamic collector resistance

R_e = dynamic emitter resistance

R_{in} = input resistance

R_{rt} = reverse transfer resistance

R_{ft} = forward transfer resistance

R_{out} = output resistance

Then the following equations hold:

$$R_{in} = R_b + R_c$$

$$R_{rt} = R_c - \alpha R_c$$

$$R_{ft} = R_e (1 - \alpha)$$

$$R_{out} = R_e + R_c - \alpha R_c$$

Hybrid parameters (common emitter)

In a common-emitter bipolar-transistor circuit, let the following symbols represent the indicated parameters. Currents are in amperes, resistances are in ohms, conductances are in siemens, and voltages are in volts.

I_b = base current
I_c = collector current
V_{eb} = emitter-base voltage
V_{ce} = collector-emitter voltage
R_{in} = input resistance for constant V_{ce}
G_{out} = output conductance for constant I_b
h_f = forward transfer characteristic for constant V_{ce}
h_r = reverse transfer characteristic for constant I_b

Then the following equations hold:

$$R_{in} = dV_{eb}/dI_b$$

$$G_{out} = dI_c/dV_{ce}$$

$$h_f = dI_c/dI_b$$

$$h_r = dV_{eb}/dV_{ce}$$

Hybrid parameters (common base)

In a common-base bipolar-transistor circuit, let the following symbols represent the indicated parameters. Currents are in amperes, resistances are in ohms, conductances are in siemens, and voltages are in volts.

I_e = emitter current
I_c = collector current
V_{cb} = collector-base voltage
V_{eb} = emitter-base voltage
R_{in} = input resistance for constant V_{cb}
G_{out} = output conductance for constant I_e
h_f = forward transfer characteristic for constant V_{cb}
h_r = reverse transfer characteristic for constant I_e

Then the following equations hold:

$$R_{in} = dV_{eb}/dI_e$$

$$G_{out} = dI_c/dV_{cb}$$

$$h_f = dI_c/dI_e$$

$$h_r = dV_{eb}/dV_{cb}$$

Hybrid parameters (common collector)

In a common-collector bipolar-transistor circuit, let the following symbols represent the indicated parameters. Currents are in amperes, resistances are in ohms, conductances are in siemens, and voltages are in volts.

I_b = base current
I_e = emitter current
V_{ec} = emitter-collector voltage
V_{bc} = base-collector voltage
R_{in} = input resistance for constant V_{ec}
G_{out} = output conductance for constant I_b
h_f = forward transfer characteristic for constant V_{ec}
h_r = reverse transfer characteristic for constant I_b

Then the following equations hold:

$$R_{in} = dV_{ec}/dI_e$$

$$G_{out} = dI_e/dV_{ec}$$

$$h_f = dV_{bc}/dI_b$$

$$h_r = dV_{bc}/dV_{ec}$$

Field-effect Transistors

The following several formulas are relevant to *field-effect transistors* (FETs), both the N-channel type and the P-channel type.

Forward transconductance (common source)

Let I_d represent drain current (in amperes) in a common-source FET circuit (Fig. 4.24 or 4.25). Let V_g represent gate voltage (in volts). *Forward transconductance,* G_{fs} (in siemens), is:

$$G_{fs} = dI_d/dV_g$$

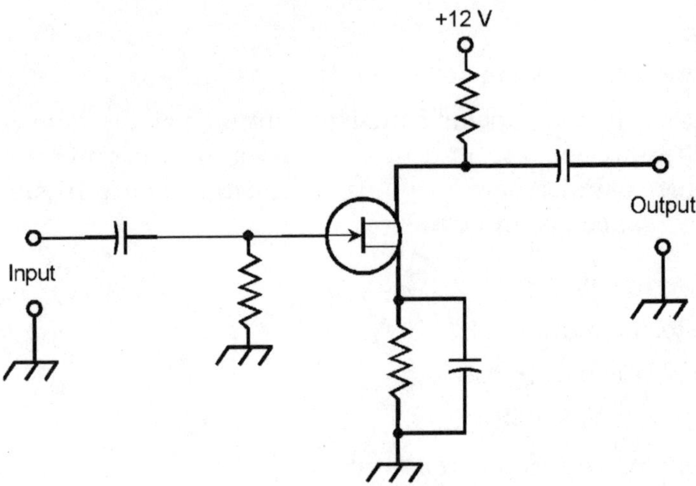

Figure 4.24 Common-source FET circuit with bypassed source resistor.

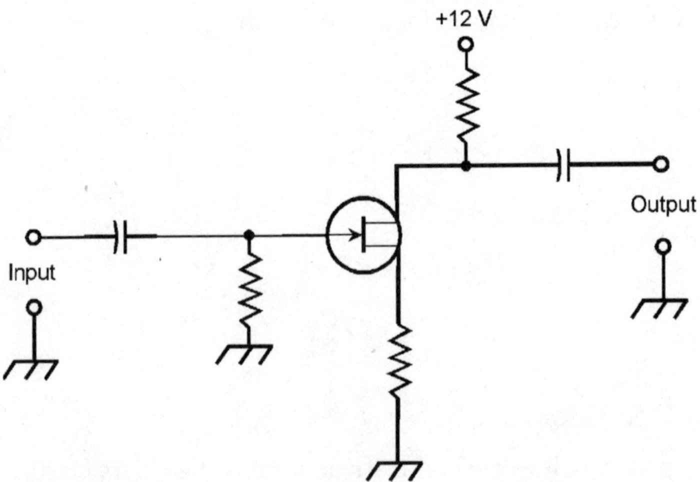

Figure 4.25 Common-source FET circuit with unbypassed source resistor.

Voltage amplification (common source)

Let G_{fs} represent the forward transconductance (in siemens) for a common-source circuit with an unbypassed source resistor (Fig. 4.25). Let R_d represent the resistance (in ohms) of an external drain resistor. Let R_s represent the resistance (in ohms) of an external source resistor. Then the *voltage amplification,* A_V (as a ratio), is given by:

$$A_V = G_{fs}R_d/(1 + G_{fs}\,R_s)$$

If the source resistor is bypassed (Fig. 4.24), then:

$$A_V = G_{fs}\,R_d$$

Voltage amplification (source follower)

Let G_{fs} represent the forward transconductance (in siemens) for a source-follower FET circuit (Fig. 4.26). Let R_d represent the resistance (in ohms) of an external drain resistor. Let R_s represent the resistance (in ohms) of an external source resistor. Then the *voltage amplification,* A_V (as a ratio), is given by:

$$A_V = G_{fs}R_s/(1 + G_{fs}\,R_s)$$

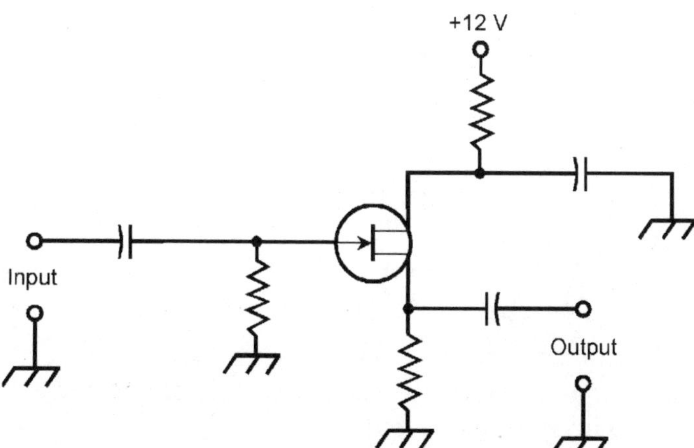

Figure 4.26 Common-drain FET circuit, also known as source follower.

Output impedance (source follower)

Let G_{fs} represent the forward transconductance (in siemens) for a source-follower FET circuit (Fig. 4.26). Let R_s represent the resistance (in ohms) of an external source resistor. Then the *output impedance,* Z_{out} (in ohms), is given by:

$$Z_{out} = R_s/(1 + G_{fs} R_s)$$

Electron Tubes

This section contains formulas relevant to *electron tubes* (often called simply *tubes,* or *valves* in England).

Diode tube perveance

Let A_p represent the surface area of the anode (plate) of a diode tube (in square centimeters); let s_{cp} represent the separation between the cathode and the plate (in centimeters). Then the *diode perveance,* G_d, is given by:

$$G_d = 2.3 \times 10^{-6} (A_p/s_{cp})$$

Triode tube perveance

Let A_p represent the surface area of the anode (plate) of a diode tube (in square centimeters); let s_{cg} represent the separation between the cathode and the grid (in centimeters). Then the *triode perveance,* G_t, is given by:

$$G_t = 2.3 \times 10^{-6} (A_p/s_{cg})$$

3/2-power law for diode tube

Let V_p represent the plate voltage (in volts) in a diode tube; let G_d represent the diode perveance. Then the *plate current,* I_p (in amperes) is approximately given by:

$$I_p = V_p^{3/2} G_d$$

3/2-power law for triode tube

Let μ represent the amplification factor of a triode tube; let V_g represent the grid voltage (in volts); let V_p represent the plate voltage (in volts); let G_t represent the triode perveance. Then the plate current, I_p (in amperes) is approximately given by:

$$I_p = (\mu V_g + V_p)^{3/2} G_t$$

Plate current vs perveance in triode

Let V_g represent the grid voltage (in volts) in a triode; let V_p represent the plate voltage (in volts); let G_t represent the triode perveance; let μ represent the amplification factor. Then the plate current, I_p (in amperes) is approximately given by:

$$I_p = ((\mu V_g + V_p)/(\mu + 1))^{3/2} G_t$$

DC internal plate resistance

Let V_p represent the DC plate-cathode voltage (in volts) in a vacuum tube; let I_p represent the DC flowing in the plate circuit (in amperes). Then the *DC internal plate resistance,* R_p (in ohms), is given by:

$$R_p = V_p/I_p$$

Dynamic internal plate resistance

Let V_p represent the instantaneous plate-cathode voltage (in volts); let I_p represent the instantaneous current flowing in the plate circuit (in amperes). Assume the control-grid voltage, V_g, is constant. Then the *dynamic internal plate resistance,* R_{pd} (in ohms), is given by:

$$R_{pd} = dV_p/dI_p$$

DC internal screen resistance

Let V_s represent the DC screen-grid voltage (in volts) in a tetrode or pentode tube; let I_s represent the DC flowing in the screen circuit (in amperes). Then the *DC screen resistance,* R_s (in ohms), is given by:

$$R_s = V_s/I_s$$

Dynamic internal screen resistance

Let V_s represent the instantaneous screen-grid voltage (in volts) in a tetrode or pentode tube; let I_s represent the instantaneous current flowing in the screen circuit (in amperes). Then the *dynamic internal screen resistance,* R_{sd} (in ohms), is given by:

$$R_{sd} = dV_s/dI_s$$

Transconductance of tube

Let V_g represent the instantaneous control-grid voltage (in volts); let I_p represent the instantaneous current flowing in the plate circuit. Assume the DC plate voltage, V_p, is constant. Then the *transconductance,* g_m (in siemens), is given by:

$$g_m = dI_p/dV_g$$

Plate amplification factor

Let V_p represent the instantaneous plate voltage (in volts); let V_g represent the instantaneous control-grid voltage (in volts); let g_m represent the transconductance (in siemens); let R_{pd} represent the dynamic internal plate resistance (in ohms). Assume that the plate current, I_p, is constant. Then the *plate amplification factor,* μ_p (as a ratio), is given by either of the following formulas:

$$\mu_p = dV_p/dV_g$$

$$\mu_p = R_{pd}g_m$$

Screen amplification factor

Let V_s represent the instantaneous screen-grid voltage (in volts) in a tetrode or pentode tube; let V_g represent the instantaneous control-grid voltage (in volts). Assume that the screen-grid current, I_s, is constant. Then the *screen amplification factor,* μ_s (as a ratio), is given by:

$$\mu_s = dV_s/dV_g$$

Output resistance in cathode follower

Let g_m represent the transconductance of a tube (in siemens) connected in a cathode-follower arrangement as shown in Fig. 4.27; let R_k represent the value of the external cathode resistor (in ohms). Then the *output resistance,* R_{out} (in ohms) is given by:

$$R_{out} = R_k/(1 + g_m R_k)$$

Input capacitance of tube

Let C_{gk} represent the capacitance between the control grid and the cathode (in picofarads); let C_{gp} represent the capacitance between the control grid and the plate (in picofarads); let μ represent the amplification factor (as a ratio). Then the *input capacitance,* C_{in} (in picofarads), is given by:

$$C_{in} = C_{gk} + C_{gp} (\mu + 1)$$

Required DC supply voltage for tube

Refer to Fig. 4.28. Let V_k represent the required cathode voltage (in volts); let R_k represent the value of the external cathode resistor (in ohms); let R_L represent the value of the external

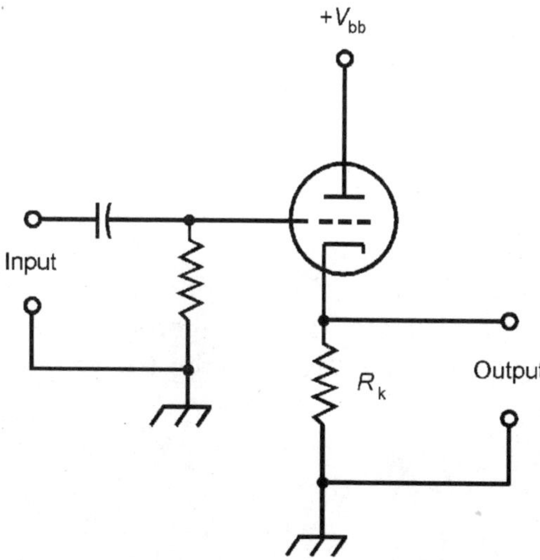

Figure 4.27 Common-plate electron-tube circuit, also known as cathode follower.

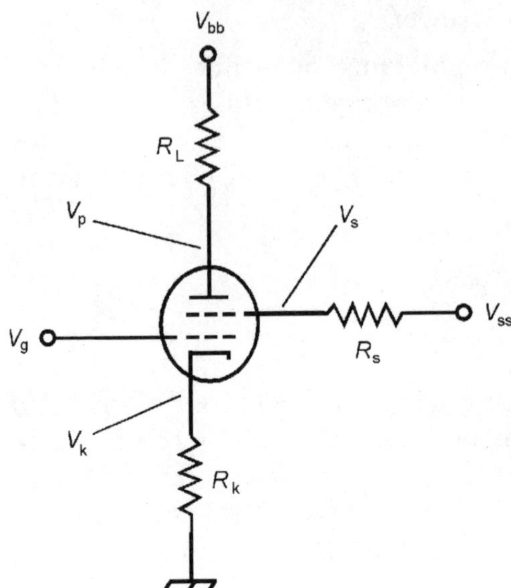

Figure 4.28 General electron-tube circuit.

plate resistor (in ohms); let V_L represent the voltage (in volts) across the external plate resistor; let I_k represent the cathode current (in amperes); let I_p represent the plate current (in amperes); let V_p represent the required plate-cathode voltage (in volts). Then the *required DC supply voltage,* V_{bb} (in volts), is given by either of the following two formulas:

$$V_{bb} = V_p + V_k + V_L$$

$$V_{bb} = V_p + I_k R_k + I_p R_L$$

DC plate-cathode voltage

Refer to Fig. 4.28. Let V_{bb} represent the supply voltage (in volts); let V_k represent the voltage (in volts) on the cathode relative to ground; let I_p represent the plate current (in amperes); let R_L represent the value of the external plate resistor (in ohms). Then the *DC plate-cathode voltage,* V_p, is given by:

$$V_p = V_{bb} - (I_p R_L + V_k)$$

DC screen voltage

Refer to Fig. 4.28; let V_{ss} represent the DC screen-grid supply voltage (in volts); let I_s represent the screen current (in amperes); let R_s represent the value of the external screen-circuit resistor (in ohms). Then the *DC screen voltage,* V_s (in volts), is given by:

$$V_s = V_{ss} - I_s R_s$$

Screen current

Refer to Fig. 4.28; let V_g represent the DC voltage on the control grid (in volts); let V_s represent the DC screen voltage (in volts); let G represent the perveance of the electron tube; let μ_s represent the screen amplification factor. Then the *screen current,* I_s (in amperes), is given by:

$$I_s = G\,(V_g + V_s/\mu_s)$$

Required external plate resistance

Refer to Fig. 4.28. Let V_{bb} represent the DC supply voltage (in volts); let V_k represent the required cathode voltage (in volts) relative to ground; let V_p represent the required plate-cathode voltage (in volts); let I_p represent the required plate current (in amperes). Then the *required external plate resistance,* R_L (in ohms), is given by:

$$R_L = (V_{bb} - (V_p + V_k))/I_p$$

Required external cathode resistance

Refer to Fig. 4.28. Let V_{bb} represent the supply voltage (in volts); let V_g represent the grid voltage (in volts); let I_k represent the cathode current (in amperes). Then the *required external cathode resistance,* R_k (in ohms), is given by:

$$R_k = (V_{bb} - V_g)/I_k$$

Required external screen resistance

Refer to Fig. 4.28. Let V_s represent the required screen voltage (in volts); let V_{ss} represent the screen supply voltage (in volts); let I_s represent the required screen current (in amperes). Then the *required external screen resistance,* R_s (in ohms), is given by:

$$R_s = (V_{ss} - V_s)/I_s$$

Voltage amplification and gain for tube

Let g_m represent the transconductance of an electron tube; let μ represent the amplification factor; let R_p represent the internal plate resistance (in ohms); let R_L represent the value of the external plate resistor (in ohms). Then the *voltage amplification,* A_V (as a ratio), is given by either of these formulas:

$$A_V = g_m R_p R_L/(R_p + R_L)$$

$$A_V = \mu R_L/(R_p + R_L)$$

The *voltage gain,* G_V (in decibels), is given by either of the following formulas, assuming constant impedance:

$$G_V = 20 \log_{10} (g_m R_p R_L/(R_p + R_L))$$

$$G_V = 20 \log_{10} (\mu R_L/(R_p + R_L))$$

Power amplification and gain for tube

Refer to Fig. 4.28. Let P_{in} represent the input signal power (in watts) applied to the control grid of an electron tube; let P_{out} represent the output signal power in the plate circuit. Then the *power amplification,* A_P (as a ratio), is given by:

$$A_P = P_{out}/P_{in}$$

The *power gain,* G_P (in decibels), is given by:

$$G_P = 10 \log_{10} (P_{out}/P_{in})$$

Filament power demand

Let V_f represent the effective filament voltage in an electron tube (in volts rms); let I_f represent the effective filament current (in amperes rms); let R_f represent the filament resistance (in ohms). Then the *filament power demand,* P_f (in watts), is given by any of the following formulas:

$$P_f = V_f I_f$$

$$P_f = I_f^2 R_f$$

$$P_f = V_f^2/R_f$$

DC screen power

Refer to Fig. 4.28. Let V_s represent the DC screen voltage (in volts); let I_s represent the direct current in the screen circuit (in amperes). Then the *DC screen power,* P_s (in watts), is given by:

$$P_s = V_s I_s$$

DC plate input power

Refer to Fig. 4.28. Let V_p represent the DC plate voltage (in volts); let I_p represent the direct current in the plate circuit (in amperes). Then the *DC plate input power,* $P_{p\text{-in}}$ (in watts), is:

$$P_{p\text{-in}} = V_p I_p$$

Signal output power from tube

Refer to Fig. 4.28. Let V_{max} represent the maximum instantaneous plate voltage (in volts); let V_{min} represent the minimum instantaneous plate voltage (in volts); let I_{max} represent the maximum instantaneous plate current (in amperes); let I_{min} represent the minimum instantaneous plate current (in amperes). Then the *signal output power,* $P_{s\text{-out}}$ (in watts), is given by:

$$P_{s\text{-out}} = 0.125\,(V_{max} I_{max} - V_{max} I_{min} - V_{min} I_{max} + V_{min} I_{min})$$

Plate power dissipation

Refer to Fig. 4.28. Let $P_{p\text{-in}}$ represent DC plate power input (in watts); let $P_{s\text{-out}}$ represent signal output power (in watts). Then the *plate power dissipation,* $P_{p\text{-dis}}$ (in watts), is given by:

$$P_{p\text{-dis}} = P_{p\text{-in}} - P_{s\text{-out}}$$

Plate efficiency

Refer to Fig. 4.28. Let $P_{p\text{-in}}$ represent DC plate input power (in watts); let $P_{s\text{-out}}$ represent signal output power (in watts). Then the *plate efficiency,* η_p (as a ratio), is given by:

$$\eta_p = P_{s\text{-out}}/P_{p\text{-in}}$$

As a percentage, the plate efficiency $\eta_{p\%}$ is given by:

$$\eta_{p\%} = 100\,P_{s\text{-out}}/P_{p\text{-in}}$$

Input power sensitivity for tube

Refer to Fig. 4.28. Let $V_{g\text{-in}}$ represent the grid input signal voltage (in volts rms); let $P_{s\text{-out}}$ represent the signal power output (in watts). Then the *input power sensitivity,* S_P (in watts per volt), is given by:

$$S_P = P_{s\text{-out}}/V_{g\text{-in}}$$

Electromagnetic Fields

An *electromagnetic (EM) field* is generated whenever charged particles are accelerated. In most practical situations, this acceleration is alternating and periodic.

Frequency vs wavelength

Let the frequency (in hertz) of an EM wave be represented by f; let the wavelength (in meters) be represented by λ; let the speed of propagation (in meters per second) be represented by c. Then the following formula holds:

$$c = f\lambda$$

In free space, c is approximately 2.99792×10^8 meters per second. For most practical applications this is rounded off to 3.00×10^8 meters per second.

Free-space wavelength

The *free-space wavelength* of an RF field depends on the frequency. In general, the higher the frequency, the shorter the free-space wavelength. Let:

λ_{ft} = free-space wavelength (in feet)
λ_{in} = free-space wavelength (in inches)
λ_m = free-space wavelength (in meters)
λ_{cm} = free-space wavelength (in centimeters)
f_{MHz} = frequency (in megahertz)
f_{GHz} = frequency (in gigahertz)

Then the following equations hold:

$$\lambda_{ft} = 984/f_{MHz}$$

$$\lambda_{ft} = 0.984/f_{GHz}$$

$$\lambda_{in} = 11.8/f_{GHz}$$

$$\lambda_m = 300/f_{MHz}$$

$$\lambda_m = 0.300/f_{GHz}$$

$$\lambda_{cm} = 30.0/f_{GHz}$$

Angular frequency

Let f be the frequency of an EM field (in hertz). Then the *angular frequency,* ω (in radians per second), is given by:

$$\omega = 2\pi f \approx 6.28f$$

The angular frequency in degrees per second is given by:

$$\omega = 360f$$

Period

Let f be the frequency of an EM field (in hertz). The the period, T (in seconds), is given by:

$$T = 1/f$$

For an angular frequency ω in radians per second:

$$T = 2\pi/\omega$$

$$T \approx 6.28/\omega$$

For an angular frequency ω in degrees per second:

$$T = 360/\omega$$

RF Transmission Lines

The following formulas apply to RF transmission lines, also called feed lines, used wireless transmitting and receiving antenna systems.

Characteristic impedance of coaxial cable

Let d_1 represent the outside diameter of the center conductor in an air-dielectric coaxial transmission line; let d_2 represent the inside diameter of the shield or braid (in the same units as d_1). Then the *characteristic impedance,* Z_0, of the line (in ohms), is given by:

$$Z_0 = 138 \log_{10} (d_2/d_1)$$

Characteristic impedance of two-wire line

Let d represent the outside diameter of either wire in a two-wire, air-dielectric transmission line; assume both wires have the same diameter. Let s represent the spacing between the centers of the two conductors in the line; assume s is the same at all points along the line, and is specified in the same units as d. Then the characteristic impedance, Z_0, of the line (in ohms), is given by:

$$Z_0 = 276 \log_{10} (2s/d)$$

Velocity factor

Let c_0 represent the speed at which an EM disturbance propagates along a transmission line (in meters per second). Then the *velocity factor,* v, of the line (as a ratio), is given by:

$$v = c_0/(3.00 \times 10^8)$$

The velocity factor as a percentage is denoted $v_{\%}$ and is given by:

$$v_{\%} = c_0/(3.00 \times 10^6)$$

Table 4.1 lists approximate velocity factors for common types of RF transmission line. The value of v is always positive, but it can never be greater than 1 (100 percent). In wire lines, v depends primarily on the nature of the dielectric separating the conductors.

TABLE 4.1 Velocity factors for RF transmission lines. These figures are approximate.

General Description	Velocity Factor
Coaxial cable, solid polyethylene dielectric	0.66
Coaxial hard line, solid polyethylene dielectric	0.66
Coaxial cable, foamed polyethylene dielectric	0.75–0.85
Coaxial hard line, foamed polyethylene dielectric	0.75–0.85
Coaxial hard line, solid polyethylene disk spacers	0.85–0.90
TV "twin-lead" ribbon, 75-ohm	0.70–0.80
TV "twin-lead" ribbon, 300-ohm	0.80–0.90
Parallel-wire "window" ribbon	0.85–0.90
Parallel-wire "ladder line" with plastic spacers	0.90–0.95
Open-wire line without spacers	0.95
Single-wire line	0.95

Electrical wavelength

In a medium other than free space, the wavelength depends on the frequency and also on the velocity factor (v) of the medium in which the field propagates. Let:

λ_{ft} = electrical wavelength (in feet)

λ_{in} = electrical wavelength (in inches)

λ_{m} = electrical wavelength (in meters)

λ_{cm} = electrical wavelength (in centimeters)

f_{MHz} = frequency (in megahertz)

f_{GHz} = frequency (in gigahertz)

Then the following equations hold:

$$\lambda_{ft} = 984v/f_{MHz}$$

$$\lambda_{ft} = 0.984v/f_{GHz}$$

$$\lambda_{in} = 11.8v/f_{GHz}$$

$$\lambda_{m} = 300v/f_{MHz}$$

$$\lambda_{m} = 0.300v/f_{GHz}$$

$$\lambda_{cm} = 30.0v/f_{GHz}$$

Let $v_{\%}$ represent the velocity factor as a percentage between 0 and 100. Then:

$$\lambda_{ft} = 9.84 v_{\%} / f_{MHz}$$

$$\lambda_{ft} = (9.84 \times 10^{-3}) v_{\%} / f_{GHz}$$

$$\lambda_{in} = 0.118 v_{\%} / f_{GHz}$$

$$\lambda_{m} = 3.00 v_{\%} / f_{MHz}$$

$$\lambda_{m} = (3.00 \times 10^{-3}) v_{\%} / f_{GHz}$$

$$\lambda_{cm} = 0.300 v_{\%} / f_{GHz}$$

Length of 1/4-wave matching section

The length of a quarter-wave section of transmission line, commonly used for impedance matching, depends on the frequency, and also on the velocity factor of the line. Let:

s_{ft} = section length (in feet)
s_{in} = section length (in inches)
s_{m} = section length (in meters)
s_{cm} = section length (in centimeters)
f_{MHz} = frequency (in megahertz)
f_{GHz} = frequency (in gigahertz)
v = velocity factor (as a ratio between 0 and 1)

Then the following equations hold:

$$s_{cm} = 7.50 v / f_{GHz}$$

$$s_{ft} = 246 v / f_{MHz}$$

$$s_{ft} = 0.246 v / f_{GHz}$$

$$s_{in} = 2.95 v / f_{GHz}$$

$$s_{m} = 75.0 v / f_{MHz}$$

$$s_{m} = 0.0750 v / f_{GHz}$$

Let $v_{\%}$ represent the velocity factor as a percentage between 0 and 100. Then:

$$s_{\text{cm}} = 0.0750 v_{\%} / f_{\text{GHz}}$$

$$s_{\text{ft}} = 2.46 v_{\%} / f_{\text{MHz}}$$

$$s_{\text{ft}} = (2.46 \times 10^{-3}) v_{\%} / f_{\text{GHz}}$$

$$s_{\text{in}} = 0.0295 v_{\%} / f_{\text{GHz}}$$

$$s_{\text{m}} = 0.750 v_{\%} / f_{\text{MHz}}$$

$$s_{\text{m}} = (7.50 \times 10^{-4}) v_{\%} / f_{\text{GHz}}$$

Characteristic impedance of 1/4-wave matching section

In an optimally designed antenna feed system, the signal input and output impedances must both be purely resistive, and the *characteristic impedance* of a quarter-wave matching section must be equal to the geometric mean of the input and output impedances. Let:

Z_0 = characteristic impedance of matching section (in ohms)
Z_{in} = input impedance (in ohms)
Z_{out} = output impedance (in ohms)

The following formulas apply:

$$Z_0 = (Z_{\text{in}} Z_{\text{out}})^{1/2}$$

$$Z_{\text{out}} = Z_0^2 / Z_{\text{in}}$$

$$Z_{\text{in}} = Z_0^2 / Z_{\text{out}}$$

If the above mentioned criteria are not met, then *standing waves* will exist along the transmission line. This is not always a serious practical concern, but in some cases it can result in degradation of system efficiency and/or physical damage to the transmission line.

Standing wave ratio (SWR)

Suppose an RF transmission line is terminated in a load whose impedance (in ohms) is a pure resistance, R_{load}. Let Z_0 represent the characteristic impedance of the line (in ohms). If $R_{load} > Z_0$, the *standing-wave ratio,* abbreviated SWR, is given by:

$$\text{SWR} = R_{load}/Z_0$$

If $R_{load} < Z_0$, then:

$$\text{SWR} = Z_0/R_{load}$$

If $R_{load} = Z_0$, then:

$$\text{SWR} = R_{load}/Z_0 = Z_0/R_{load} = 1{:}1 = 1$$

When a transmission line is terminated with a load whose impedance is not a pure resistance, the SWR is determined according to the maximum and minimum voltage or current in the line.

Voltage standing wave ratio (VSWR)

Let V_{max} represent the maximum RF voltage (in volts) between the conductors of a transmission line; let V_{min} represent the minimum RF voltage (in volts) between the conductors of the line. Points at which V_{max} and V_{min} occur are separated by ¼ electrical wavelength. The *voltage standing-wave ratio,* abbreviated VSWR, is given by:

$$\text{VSWR} = V_{max}/V_{min}$$

Current standing wave ratio (ISWR)

Let I_{max} represent the maximum RF current (in amperes) in a transmission line; let I_{min} represent the minimum RF current in the line (in amperes). Points at which I_{max} and I_{min} occur are separated by ¼ electrical wavelength. Current maxima normally exist at the same points on a transmission line as voltage minima; current minima normally exist at the same points as voltage maxima. The *current standing-wave ratio,* abbreviated ISWR, is given by:

$$\text{ISWR} = I_{max}/I_{min}$$

Relation among SWR, VSWR, and ISWR

In theory, assuming zero loss in a transmission line, the following equation holds:

$$SWR = VSWR = ISWR$$

In practice, when a line has significant loss, these quantities differ slightly depending on the points where current and voltage are measured. In transmitting antenna systems, the ratios are lower toward the equipment (transmitter) end of the line, and higher toward the antenna (load) end.

Reflection coefficient vs SWR

Let s represent the SWR, VSWR, or ISWR measured at the antenna (load) end of a transmitting RF transmission line. Then the *reflection coefficient, k*, is given by:

$$k = (s - 1)/s$$

Reflection coefficient vs load resistance

Suppose an RF transmission line is terminated in a load whose impedance (in ohms) is a pure resistance, R_{load}. Let Z_0 represent the characteristic impedance of the line (in ohms). Then the reflection coefficient, k, is given by:

$$k = (R_{load} - Z_0)/(R_{load} + Z_0)$$

Loss in matched lines

Table 4.2 gives the approximate loss (in decibels per 100 feet and per 100 meters) for various types of transmission line under

TABLE 4.2A Approximate loss in decibels per 100 feet for various transmission lines under conditions of 1:1 SWR.

Line type	1 MHz	10 MHz	100 MHz
600-ohm ladder line	0.05	0.1	0.5
300-ohm TV ribbon	0.1	0.5	1.5
RG-8/U coaxial cable	0.15	0.6	2.0
RG-59/U coaxial cable	0.3	1.0	4.0
RG-58/U coaxial cable	0.3	1.4	5.0

TABLE 4.2B Approximate loss in decibels per 100 meters for various transmission lines under conditions of 1:1 SWR.

Line Type	1 MHz	10 MHz	100 MHz
600-ohm ladder line	0.16	0.33	1.6
300-ohm TV ribbon	0.33	3.3	4.9
RG-8/U coaxial cable	0.49	2.0	6.5
RG-59/U coaxial cable	1.0	3.3	13
RG-58/U coaxial cable	0.3	1.4	5.0

conditions of 1:1 SWR (a perfect impedance match between the line and the load). Dielectrics are assumed to be solid polyethylene, except for ladder line in which the dielectric is dry air with plastic spacers.

SWR loss

Figure 4.29 shows the approximate loss (in decibels) that occurs in addition to the matched-line loss in a transmission line when the SWR is not 1:1. This additional loss is called *SWR loss* and

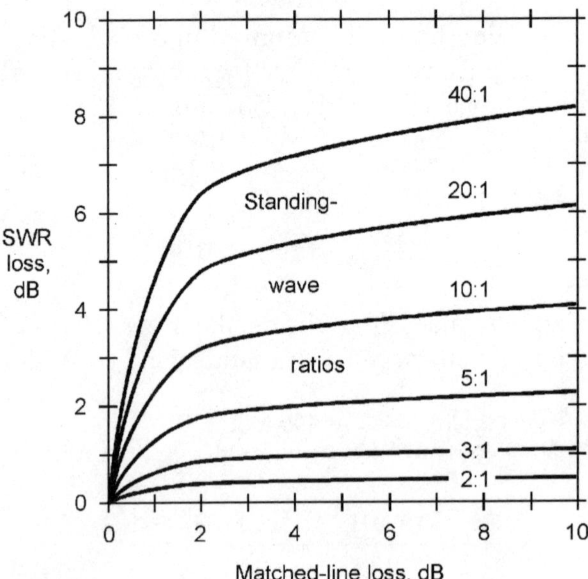

Figure 4.29 Approximate SWR loss, as a function of both the matched-line loss and the SWR at the load end of the line.

is insignificant in practical terms unless the SWR is more than 2:1. In severely mismatched, long lines at high frequencies, the SWR loss can be considerable.

Antennas

The physical size of a resonant antenna depends on the electrical wavelength, which in turn depends on the frequency.

Radiation resistance

Let P_{rad} represent the power radiated from a resonant antenna (in watts); let I_{rad} represent the RF current (in amperes) that would flow in a non-reactive resistor inserted at the feed point, if the use of that resistor in place of the antenna would produce the same feed-line current distribution as does the antenna. Then the *radiation resistance,* R_{rad} (in ohms), of the resonant antenna is given by:

$$R_{rad} = P_{rad}/I_{rad}{}^2$$

Let P_{rad} represent the power radiated from a resonant antenna (in watts); let V_{rad} represent the RF voltage (in volts) that would appear across a non-reactive resistor inserted at the feed point, if the use of that resistor in place of the antenna would produce the same feed-line voltage distribution as does the antenna. Then the *radiation resistance,* R_{rad} (in ohms), of the resonant antenna is given by:

$$R_{rad} = V_{rad}{}^2/P_{rad}$$

Antenna efficiency

Let R_{rad} represent the radiation resistance of an antenna (in ohms); let R_{loss} represent the *loss resistance* in the antenna and associated components such as loading coils, traps, ground system, etc. Then the *antenna efficiency,* η (as a ratio), is given by:

$$\eta = R_{rad}/(R_{rad} + R_{loss})$$

The efficiency as a percentage, $\eta_{\%,}$ is given by:

$$\eta_{\%} = 100\ R_{rad}/(R_{rad} + R_{loss})$$

Length of 1/2-wave dipole antenna

For a half-wave dipole antenna fed at the center, placed at least 1/4 wavelength above effective ground and constructed of common wire, let:

s_{ft} = end-to-end length (in feet)
s_{in} = end-to-end length (in inches)
s_{m} = end-to-end length (in meters)
s_{cm} = end-to-end length (in centimeters)
f_{MHz} = frequency (in megahertz)
f_{GHz} = frequency (in gigahertz)

Then the following formulas apply:

$$s_{ft} \approx 468/f_{MHz}$$

$$s_{ft} \approx 0.468/f_{GHz}$$

$$s_{in} \approx 5.62/f_{GHz}$$

$$s_{m} \approx 143/f_{MHz}$$

$$s_{m} \approx 0.143/f_{GHz}$$

$$s_{cm} \approx 14.3/f_{GHz}$$

For antennas constructed of metal tubing, the above values should be multiplied by approximately 0.95 (95 percent). However, the exact optimum antenna length in any given case must be determined by experimentation, because it depends on the ratio of tubing diameter to wavelength, and also on the surrounding environment.

Height of 1/4-wave vertical antenna

For a quarter-wave vertical antenna constructed of common wire and placed over perfectly conducting ground, let:

h_{ft} = radiating-element height (in feet)
h_{in} = radiating-element height (in inches)

h_m = radiating-element height (in meters)

h_{cm} = radiating-element height (in centimeters)

f_{MHz} = frequency (in megahertz)

f_{GHz} = frequency (in gigahertz)

Then the following formulas apply:

$$h_{ft} \approx 234/f_{MHz}$$

$$h_{ft} \approx 0.234/f_{GHz}$$

$$h_{in} \approx 2.81/f_{GHz}$$

$$h_m \approx 71.5/f_{MHz}$$

$$h_m \approx 0.0715/f_{GHz}$$

$$h_{cm} \approx 7.15/f_{GHz}$$

For antennas constructed of metal tubing, the above values should be multiplied by approximately 0.95 (95 percent). However, the exact optimum antenna height in any given case must be determined by experimentation, because it depends on the ratio of tubing diameter to wavelength, and also on the surrounding environment.

Length of harmonic antenna

For a resonant harmonic antenna fed at integral multiples of ¼ wavelength from either end, placed at least ¼ wavelength above effective ground and constructed of common wire, let:

s_{ft} = end-to-end length (in feet)

s_{in} = end-to-end length (in inches)

s_m = end-to-end length (in meters)

s_{cm} = end-to-end length (in centimeters)

f_{MHz} = frequency (in megahertz)

f_{GHz} = frequency (in gigahertz)

n = harmonic at which antenna is operated (a positive integer)

Then the following formulas apply:

$$s_{ft} \approx 492\ (n - 0.05)/f_{MHz}$$

$$s_{ft} \approx 0.492\ (n - 0.05)/f_{GHz}$$

$$s_{in} \approx 5.90\ (n - 0.05)/f_{GHz}$$

$$s_{m} \approx 150\ (n - 0.05)/f_{MHz}$$

$$s_{m} \approx 0.150\ (n - 0.05)/f_{GHz}$$

$$s_{cm} \approx 15.0\ (n - 0.05)/f_{GHz}$$

Length of resonant unterminated long wire

For a resonant unterminated long wire antenna fed at either end, placed at least 1/4 wavelength above effective ground and constructed of common wire, let:

s_{ft} = end-to-end length (in feet)
s_{in} = end-to-end length (in inches)
s_{m} = end-to-end length (in meters)
s_{cm} = end-to-end length (in centimeters)
f_{MHz} = frequency (in megahertz)
f_{GHz} = frequency (in gigahertz)
n = length of wire in wavelengths

Then the following formulas apply:

$$s_{ft} \approx 984\ (n - 0.025)/f_{MHz}$$

$$s_{ft} \approx 0.984\ (n - 0.025)/f_{GHz}$$

$$s_{in} \approx 11.8\ (n - 0.025)/f_{GHz}$$

$$s_{m} \approx 300\ (n - 0.025)/f_{MHz}$$

$$s_{m} \approx 0.300\ (n - 0.025)/f_{GHz}$$

$$s_{cm} \approx 30.0\ (n - 0.025)/f_{GHz}$$

Bridge Circuits

A *bridge circuit* is used to determine unknown resistances, reactances, impedances, and/or frequencies. Variable components are adjusted until a condition of balance (zero output) occurs, at which time the unknown values can be calculated.

Anderson bridge

Let L_x and R_x represent an unknown inductance (in henrys) and an unknown resistance (in ohms) in series. Assume they are inserted in the *Anderson bridge* configuration of Fig. 4.30, and the variable components are adjusted for balance. Let C_s represent a precision standard capacitance (in farads). The following formulas apply:

$$L_x = C_s\,(R_3(1 + R_2/R_4) + R_2)$$

$$R_x = R_1R_2/R_4$$

Hay bridge

Let L_x and R_x represent an unknown inductance (in henrys) and an unknown resistance (in ohms) in series. Assume they are

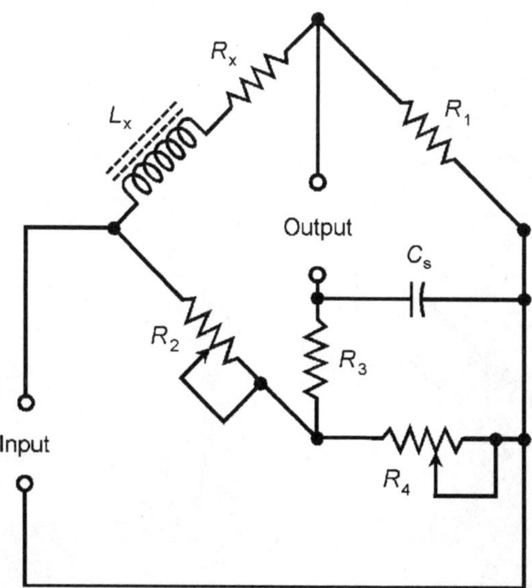

Figure 4.30 Anderson bridge for determining the value of an inductance (L_x) and resistance (R_x) in series.

inserted in the *Hay bridge* configuration of Fig. 4.31, and the variable components are adjusted for balance. Let C_s represent a precision standard capacitance (in farads). Let f represent the frequency (in hertz). The following formulas apply:

$$L_x = C_s R_1 R_2$$

$$R_x = (4\pi^2 f^2 C_s^{\,2} R_1 R_2 R_3)/(1 + 4\pi^2 f^2 C_s^{\,2} R_3^{\,2})$$

Maxwell bridge

Let L_x and R_x represent an unknown inductance (in henrys) and an unknown resistance (in ohms) in series. Assume they are inserted in the *Maxwell bridge* configuration of Fig. 4.32, and the variable components are adjusted for balance. Let C_s represent a precision standard capacitance (in farads). The following formulas apply:

$$L_x = C_s R_1 R_2$$

$$R_x = R_1 R_2 / R_3$$

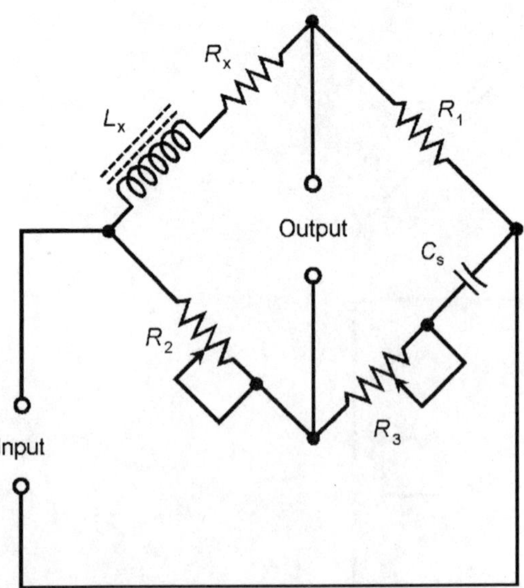

Figure 4.31 Hay bridge for determining the value of an inductance (L_x) and resistance (R_x) in series.

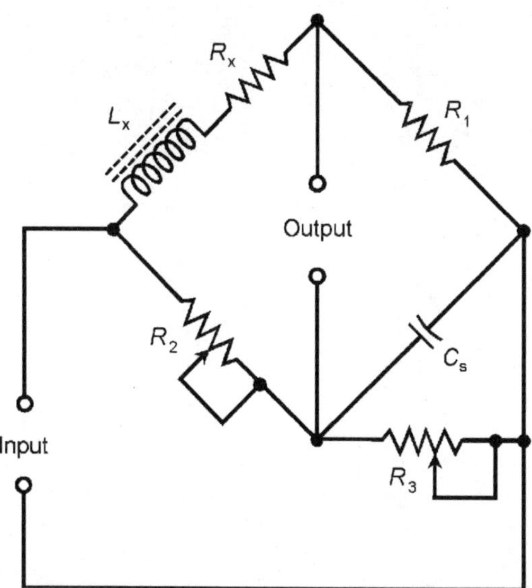

Figure 4.32 Maxwell bridge for determining the value of an inductance (L_x) and resistance (R_x) in series.

Owen bridge

Let L_x and R_x represent an unknown inductance (in henrys) and an unknown resistance (in ohms) in series. Assume they are inserted in the *Owen bridge* configuration of Fig. 4.33, and the variable components are adjusted for balance. The following formulas apply:

$$L_x = C_2 R_1 R_2$$

$$R_x = R_1 C_2 / C_1$$

Schering bridge

Let C_x and R_x represent an unknown capacitance (in farads) and an unknown resistance (in ohms) in series. Assume they are inserted in the *Schering bridge* configuration of Fig. 4.34, and the variable components are adjusted for balance. Let C_s represent a precision standard capacitance (in farads). The following formulas apply:

$$C_x = C_s R_2 / R_1$$

$$R_x = R_1 C_1 / C_s$$

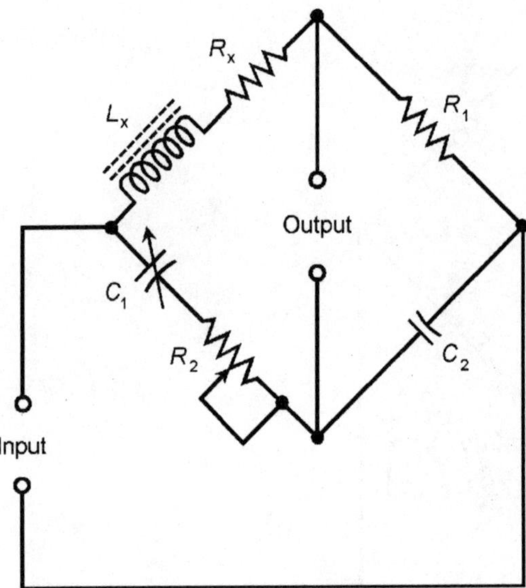

Figure 4.33 Owen bridge for determining the value of an inductance (L_x) and resistance (R_x) in series.

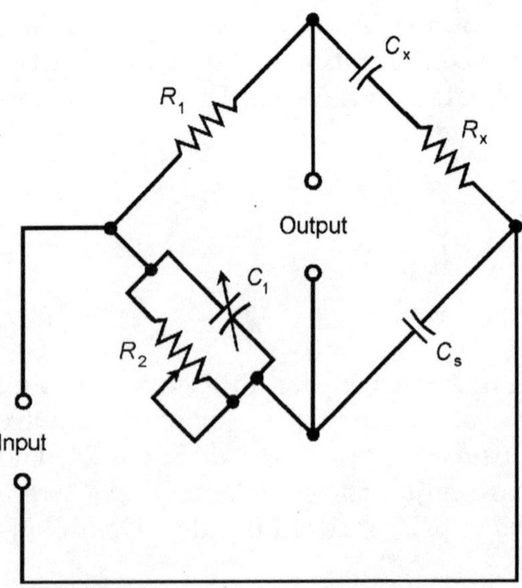

Figure 4.34 Schering bridge for determining the value of a capacitance (C_x) and resistance (R_x) in series.

Wheatstone bridge

Let R_x represent an unknown resistance (in ohms). Assume it is inserted in the *Wheatstone bridge* configuration of Fig. 4.35, and potentiometer R_2 is adjusted for balance. The following formula applies:

$$R_x = R_1 R_2 / R_3$$

Wien bridge

Let the resistances (in ohms) and capacitances (in farads) of the *Wien bridge* circuit of Fig. 4.36 be related as follows:

$$R_2 = 2R_1$$

$$C_1 = C_2$$

$$R_3 = R_4$$

Then the input frequency, f (in hertz), that results in zero output (balance) is given by:

$$f = 1/(2\pi R_3 C_1)$$

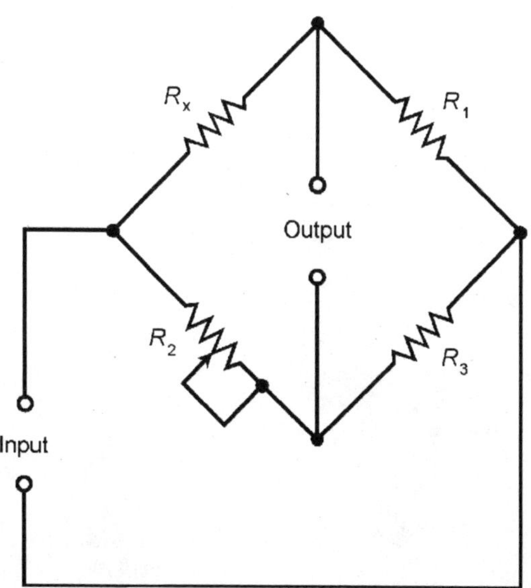

Figure 4.35 Wheatstone bridge for determining the value of a resistance (R_x).

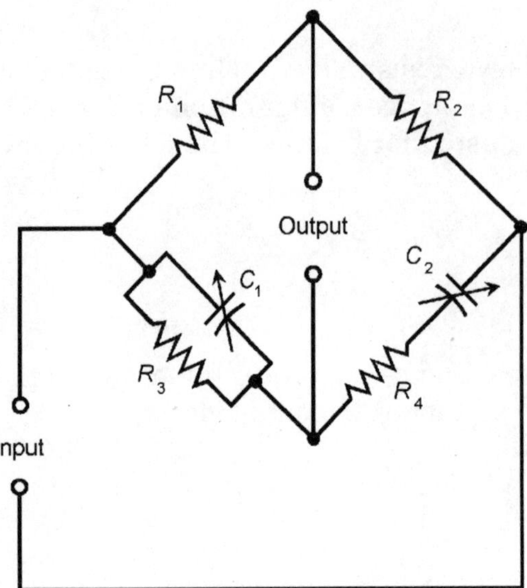

Figure 4.36 Wien bridge for measurement of frequency.

Null Networks

A *null network* produces zero output at a specific frequency that is determined by the values of the inductances, capacitances, and resistances in the circuit.

LC bridged T

Suppose an inductance L (in henrys), capacitances C_1 and C_2 (in farads), and a resistance R (in ohms) are connected in the configuration shown in Fig. 4.37. Further suppose that $C_1 = C_2$,

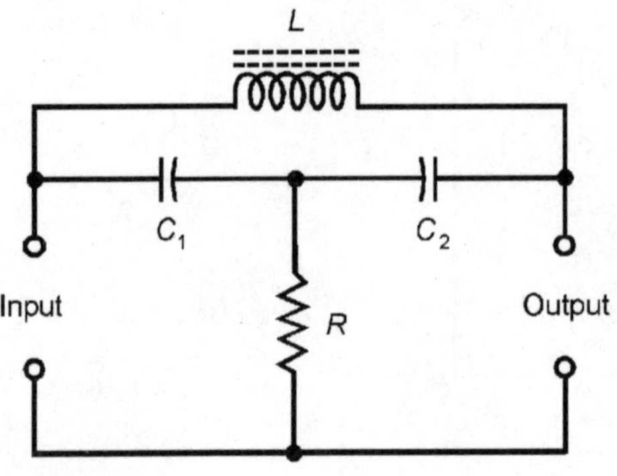

Figure 4.37 Inductance-capacitance (LC) bridged T null network.

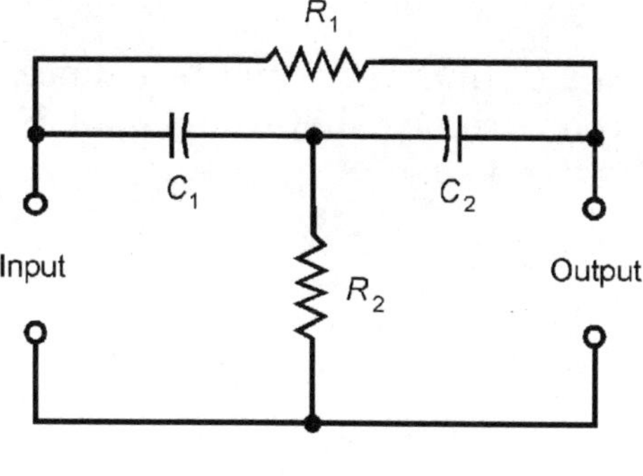

Figure 4.38 Resistance-capacitance (RC) bridged T null network.

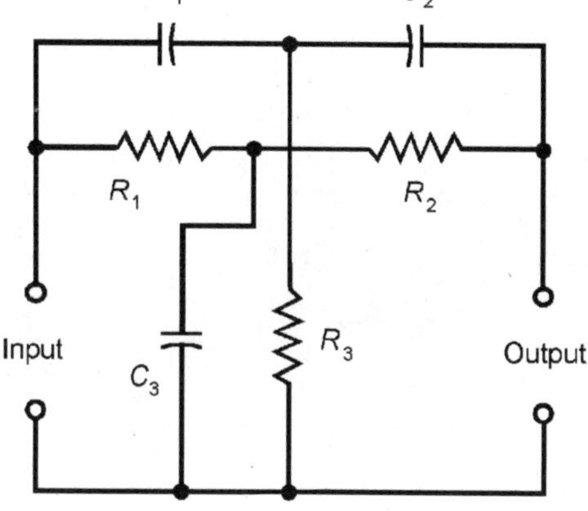

Figure 4.39 Resistance-capacitance (RC) parallel T null network.

so the network is symmetrical. Then the null frequency, f (in hertz), is given by either of the following:

$$f = 1/(\pi(2LC_1)^{1/2})$$

$$f = 1/(\pi(2LC_2)^{1/2})$$

RC bridged T

Suppose capacitances C_1 and C_2 (in farads) and resistances R_1 and R_2 (in ohms) are connected as shown in Fig. 4.38. Further suppose $C_1 = C_2$, so the network is symmetrical. Then the null frequency, f (in hertz), is given by either of the following:

$$f = 1/(2\pi C_1(R_1R_2)^{1/2})$$

$$f = 1/(2\pi C_2(R_1R_2)^{1/2})$$

RC parallel T

Suppose capacitances C_1, C_2, and C_3 (in farads) and resistances R_1, R_2, and R_3 (in ohms) are connected as shown in Fig. 4.39. Further suppose that the following hold:

$$C_3 = 2C_1 = 2C_2$$

$$R_1 = R_2 = 2R_3$$

Then the null frequency, f (in hertz), is given by:

$$f = 1/(2\pi R_1 C_1)$$

Chapter

5

Physical and Chemical Data

This chapter contains information and formulas relevant to classical physics, relativistic physics, general chemistry, and related disciplines.

Units

The *Standard International (SI) System of Units,* formerly called the *Meter/Kilogram/Second* (*MKS*) *System,* consists of seven fundamental quantities in nature. These are the first seven units defined below. Most other units are derived from these.

Displacement

One *meter* (1 m) is equivalent to 1.65076373×10^6 wavelengths in a vacuum of the radiation corresponding to the transition between the two levels of the krypton-86 atom. It was originally defined as 10^{-7} of the distance from the North Geographic Pole to the equator as measured over the surface of the earth at the Greenwich Meridian. Displacement is represented in equations by the italicized lowercase letters d or s.

Mass

One *kilogram* (1 kg) is the mass of 1000 cubic centimeters (1.000×10^3 cm^3), or one liter, of pure liquid water at the temperature of its greatest density (approximately 281 degrees Kelvin). Mass is represented by the italicized lowercase letter m.

Time

One *second* (1 s) is 1/86,400 = 1.1574×10^{-5} part of a solar day. It is also defined as the time required for a beam of visible light to propagate over a distance of 2.99792×10^{8} m in a vacuum. Time is represented in equations by the italicized lowercase letter t.

Temperature

One *degree Kelvin* (1°K) is 3.66086×10^{-3} part of the difference between absolute zero (the absence of all heat) and the freezing point of pure water at standard atmospheric temperature and pressure. Temperature is represented in equations by the italicized uppercase letter T, and occasionally by the italicized lowercase letter t.

Electric current

One *ampere* (1 A) represents the movement of 6.24×10^{18} charge carriers (usually electrons) past a specific fixed point in an electrical conductor over a time span of 1 s. Current is represented in equations by the uppercase letter I, and occasionally by the italicized lowercase letter i.

Luminous intensity

One *candela* (1 cd) represents the radiation from a surface area of 1.667×10^{-6} m^2 of a blackbody at the solidification temperature of pure platinum. Luminous intensity is represented in equations by the italicized uppercase letters B, F, I, or L.

Material quantity

One *mole* (1 mol) is the number of atoms in precisely 0.012 kg of carbon-12. This is approximately 6.022169×10^{23}, also known as *Avogadro's number*. Material quantity is represented in equations by the italicized uppercase letter N.

Area

The standard unit of area is the *square meter* (sq m) or *meter squared* (m^2). Area is represented in equations by the italicized uppercase letter A.

Volume

The standard unit of volume is the *cubic meter* (cu m) or *meter cubed* (m^3). Volume is represented in equations by the italicized uppercase letter V.

Plane angular measure

The standard unit of angular measure is the *radian* (rad). It is the angle subtended by an arc on a circle, whose length, as measured on the circle, is equal to the radius of the circle. Angles are represented in equations by italicized lowercase Greek letters, usually ϕ (phi) or θ (theta).

Solid angular measure

The standard unit of solid angular measure is the *steradian* (sr). A solid angle of 1 sr is represented by a cone with its apex at the center of a sphere, intersecting the surface of the sphere in a circle such that, within the circle, the enclosed area on the sphere is equal to the square of the radius of the sphere. Solid angles are represented in equations by italicized lowercase Greek letters, usually ϕ (phi) or θ (theta).

Velocity

The standard unit of linear speed is the *meter per second* (m/s). The unit of velocity requires two specifications: speed and direction. Direction is indicated in radians clockwise from geographic north on the earth's surface, and counterclockwise from the positive x axis in the coordinate xy-plane. In three dimensions, direction can be specified in rectangular, spherical, or cylindrical coordinates. Speed and velocity are represented in equations by the italicized lowercase letter v, and sometimes by the italicized lowercase letters u or w.

Angular velocity

The standard unit of angular velocity is the *radian per second* (rad/s). Angular velocity is represented in equations by the italicized lowercase Greek letter ω (omega).

Acceleration

The standard unit of acceleration is the *meter per second per second* (m/s/s) or *meter per second squared* (m/s^2). Linear acceleration is represented in equations by the italicized lowercase letter a.

Angular acceleration

The standard unit of angular acceleration is the *radian per second per second* (rad/s/s) or *radian per second squared* (rad/s^2). Angular acceleration is represented in equations by the italicized lowercase Greek letter α (alpha).

Force

The standard unit of force is the *newton* (N). It is the impetus required to cause the linear acceleration of a 1-kg mass at a rate of 1 m/s^2. Force is represented in equations by the italicized uppercase letter F.

Unit electric charge

The *unit electric charge* is the charge contained in a single electron. This charge is also contained in a hole (electron absence within an atom), a proton, a positron, and an anti-proton. Charge quantity in terms of unit electric charges is represented in equations by the italicized lowercase letter e.

Electric charge quantity

The standard unit of charge quantity is the *coulomb* (C), which represents the total charge contained in 6.24×10^{18} electrons. Charge quantity is represented in equations by the italicized uppercase letter Q or the italicized lowercase letter q.

Energy

The standard SI unit of energy is the *joule* (J). Mathematically, it is expressed in terms of unit mass multiplied by unit distance squared per unit time squared:

$$1 \text{ J} = 1 \text{ kg} \cdot \text{m}^2/\text{s}^2$$

Energy is represented in equations by the italicized uppercase letter E. Occasionally it is represented by the italicized uppercase H, T, or V.

Electromotive force

The standard unit of electromotive force (EMF), also called electric potential or potential difference, is the *volt* (V). It is equivalent to 1 J/C. Electromotive force is represented in equations by the italicized uppercase letters E or V.

Resistance

The standard unit of resistance is the *ohm* (Ω). It is the resistance that results in 1 A of electric current with an applied EMF of 1 V. Resistance is represented in equations by the italicized uppercase letter R, and occasionally by the italicized lowercase letter r.

Resistivity

The standard unit of resistivity is the *ohm-meter* (Ω · m). If a length of material measuring 1 m carries 1 A of current when a potential difference of 1 V is applied, then it has a resistivity of 1 Ω · m. Resistivity is represented in equations by the italicized lowercase Greek letter ρ (rho).

Conductance

The standard unit of conductance is the *siemens* (S), formerly called the *mho*. Mathematically, conductance is the reciprocal of resistance. Conductance is represented in equations by the italicized uppercase letter G. If R is the resistance of a component in ohms, and G is the conductance of the component in siemens, then

$$G = 1/R$$

$$R = 1/G$$

Conductivity

The standard unit of conductivity is the *siemens per meter* (S/m). If a length of material measuring 1 m carries 1 A of current when a potential difference of 1 V is applied, then it has a conductivity of 1 S/m. Conductivity is represented in equations by the italicized lowercase Greek letter σ (sigma).

Power

The standard unit of power is the *watt* (W), equivalent to 1 J/s. Power is represented in equations by the italicized uppercase letters P or W. In electrical and electronic circuits containing no reactance, if P is the power in watts, V is the voltage in volts, I is the current in amperes, and R is the resistance in ohms, then the following holds:

$$P = VI = I^2R = V^2/R$$

Period

The standard unit of alternating-current (AC) cycle period is the *second* (s). This is a large unit in practice; typical signals have periods on the order of thousandths, millionths, billionths, or trillionths of a second. Period is represented in equations by the italicized uppercase letter T.

Frequency

The standard unit of frequency is the *hertz* (Hz), formerly called the *cycle per second* (cps). This is a small unit in practice; typical signals have frequencies on the order of thousands, millions, billions, or trillions of hertz. Frequency, which is the mathematical reciprocal of period, is represented in equations by the italicized lowercase letter *f*. If *T* is the period of a wave disturbance in seconds, then the frequency *f* in hertz is given by:

$$f = 1/T$$

Capacitance

The standard unit of capacitance is the *farad* (F), which is equal to 1 C/V. This is a large unit in practice. In electronic circuits, most values of capacitance are on the order of millionths, billionths, or trillionths of a farad. Capacitance is represented in equations by the italicized uppercase letter *C*.

Inductance

The standard unit of inductance is the *henry* (H), which is equal to 1 V · s/A. This is a large unit in practice. In electronic circuits, most values of inductance are on the order of thousandths or millionths of a henry. Inductance is represented in equations by the italicized uppercase letter *L*.

Reactance

The standard unit of reactance is the *ohm* (Ω). Reactance is represented in equations by the italicized uppercase letter *X*. It can be positive (inductive) and symbolized by X_L, or negative (capacitive) and symbolized by X_C. Reactance is dependent on frequency. Let *f* represent frequency in hertz, *L* represent inductance in henrys, and *C* represent capacitance in farads. Then the following formulas hold:

$$X_L = 2\pi f L$$

$$X_C = -1/(2\pi f C)$$

Complex impedance

In the determination of complex impedance, there are two components: resistance (*R*) and reactance (*X*). The reactive component is multiplied by the unit imaginary number known as the *j operator*. Mathematically, *j* is equal to the positive square root of −1, so the following formulas hold:

$$j = (-1)^{1/2}$$

$$j^2 = j \times j = -1$$

$$j^3 = j^2 \times j = -1 \times j = -j$$

$$j^4 = j^3 \times j = -j \times j = 1$$

For powers of j beyond 4, the cycle repeats, so in general, for integers $n > 4$:

$$j^n = j^{(n-4)}$$

Let Z represent complex impedance, R represent resistance, and X represent reactance (either inductive or capacitive). Then:

$$Z = R + jX$$

Absolute-value impedance

Complex impedance can be represented as a vector in a rectangular coordinate plane, where resistance is plotted on the abscissa (horizontal axis) and reactance is plotted on the ordinate (vertical axis). The length of this vector is called the *absolute-value impedance,* symbolized by the italicized uppercase letter Z and expressed in ohms. This impedance is usually discussed only when $jX = 0$, that is, when the impedance is a pure resistance ($Z = R$). In the broad sense, if Z is the absolute-value impedance, then

$$Z = (R^2 + X^2)^{1/2}$$

There are, in theory, infinitely many combinations of R and X that can result in a given absolute-value impedance Z.

Electric field strength

The standard unit of electric field strength is the *volt per meter* (V/m). An electric field of 1 V/m is represented by a potential difference of 1 V existing between two points displaced by 1 m. Electric field strength is represented in equations by the italicized uppercase letter E.

Electromagnetic field strength

The standard unit of electromagnetic (EM) field strength is the *watt per square meter* (W/m^2). An EM field of 1 W/m^2 is represented by 1 W of power impinging perpendicularly on a flat surface whose area is 1 m^2.

Electric susceptibility

The standard unit of electric susceptibility is the *coulomb per volt-meter,* abbreviated $C/(V \cdot m)$. This quantity is represented in equations by the italicized lowercase Greek letter η (eta).

Permittivity

The standard unit of permittivity is the *farad per meter* (F/m). Permittivity is represented in equations by the italicized lowercase Greek letter ϵ (epsilon).

Charge-carrier mobility

The standard unit of charge-carrier mobility, also called carrier mobility or simply mobility, is the *meter squared per volt-second,* abbreviated $m^2/(V \cdot s)$. Mobility is represented in equations by the italicized lowercase Greek letter μ (mu).

Magnetic flux

The standard unit of magnetic flux is the *weber* (Wb). This is a large unit in practice, equivalent to 1 $A \cdot H$, represented by a constant, direct current of 1 A flowing through a coil having an inductance of 1 H. Magnetic flux is represented in equations by the italicized uppercase Greek letter Φ (phi).

Magnetic flux density

The standard unit of magnetic flux density is the *tesla* (T), equivalent to 1 Wb/m^2. Sometimes, magnetic flux density is spoken of in terms of the number of lines of flux per unit area; this is an imprecise terminology.

Magnetic field intensity

The standard unit of magnetic field intensity is the *oersted* (Oe), equivalent to 79.6 A/m. Magnetic field intensity is represented in equations by the italicized uppercase letter H.

Magnetic pole strength

The standard unit of magnetic pole strength is the *ampere-meter* (A · m). Pole strength is represented in equations by the italicized lowercase letter p or uppercase P.

Magnetomotive force

The standard unit of magnetomotive force is the *ampere-turn*, produced by a constant, direct current of 1 A flowing in a single-turn, air-core coil. Magnetomotive force is independent of the radius of the coil.

Classical Mechanics

The following paragraphs summarize basic Newtonian equations relating to the motions of masses in Euclidean space.

Displacement, velocity, and time

Let v_{av} represent the average velocity of an object (in meters per second); let t represent the elapsed time (in seconds). Then the displacement s (in meters) is given by:

$$s = v_{av}t$$

If the displacement s (in meters) and the time t (in seconds) are known, then the average velocity v_{av} (in meters per second) is:

$$v_{\mathrm{av}} = s/t$$

If the displacement s (in meters) and the average velocity v_{av} (in meters per second) are known, then the elapsed time t (in seconds) is given by:

$$t = s/v_{\mathrm{av}}$$

Instantaneous velocity

Let v_{ins} represent the instantaneous velocity of an object (in meters per second). If s_{ins} represents the instantaneous displacement (in meters) and t represents the time (in seconds), then the instantaneous velocity is given by the derivative of the instantaneous displacement with respect to time:

$$v_{ins} = ds_{ins}/dt$$

Average acceleration

Let v_1 represent the velocity of an object (in meters per second) at the beginning of a specified period of time of duration t (in seconds); let v_2 represent the velocity of the object (in meters per second) at the end of that period. Then the average acceleration, a_{av} (in meters per second squared) during the period is given by:

$$a_{av} = (v_2 - v_1)/t$$

Displacement, average acceleration, and time

Let a_{av} represent the average acceleration of an object (in meters per second squared); let t represent the elapsed time (in seconds). Then the displacement s (in meters) is given by:

$$s = (a_{av}t^2)/2$$

If the displacement s (in meters) and the time t (in seconds) are known, then the average acceleration a_{av} (in meters per second squared) is given by:

$$a_{av} = 2s/t^2$$

If the displacement s (in meters) and the average acceleration a_{av} (in meters per second squared) are known, then the elapsed time t (in seconds) is given by:

$$t = (2s/a_{av})^{1/2}$$

Instantaneous acceleration

Let a_{ins} represent the instantaneous acceleration of an object (in meters per second squared). If v_{ins} represents the instantaneous velocity (in meters per second) and t represents the time (in seconds), then the instantaneous acceleration is given by the derivative of the instantaneous velocity with respect to time:

$$a_{ins} = dv_{ins}/dt$$

Newton's First Law

This law consists of two parts:

- A mass at rest will remain at rest unless acted on by an outside force
- A mass in motion will continue to move with constant velocity unless acted on by an outside force.

Newton's Second Law

Consider an object of mass m (in kilograms). Suppose this mass is acted upon by a constant outside force F (in newtons). The object will experience a constant acceleration a (in meters per second squared) equal to the ratio of the force to the mass:

$$a = F/m$$

The foregoing can be more generally stated by considering the force as a vector **F** (in newtons in a specific direction) and the acceleration as a vector **a** (representing a change in the velocity vector **v**):

$$\mathbf{a} = \mathbf{F}/m$$

The following relation derives from the above:

$$\mathbf{F} = m\mathbf{a}$$

In addition, the following relation holds when $\mathbf{a} \neq 0$, although this is primarily of theoretical interest:

$$m = \mathbf{F}/\mathbf{a}$$

Newton's Third Law

This law in its most familiar form states that for every action, there is an equal and opposite reaction. If **F** represents a force vector acting on a given body, then the resultant reaction force vector **G** is related to **F** as follows:

$$\mathbf{G} = -\mathbf{F}$$

Law of Universal Gravitation

Let m_1 and m_2 represent the masses (in kilograms) of two objects M_1 and M_2. Let d be the distance (in meters) separating the mass centers. Then a gravitational force vector **F** (whose magnitude is expressed in newtons) acts on M_1 in the direction of M_2, and an equal but opposite gravitational force vector $-\mathbf{F}$ acts on M_2 in the direction of M_1, such that:

$$|\mathbf{F}| = Gm_1m_2/d^2$$

where G is a number known as the *gravitational constant*, and

$$G \approx 6.673 \times 10^{-11}$$

This constant is expressed in newton-meters squared per kilogram squared.

Coefficient of static friction

Suppose two objects, M_1 and M_2, are in physical contact along a common flat surface S. Suppose there is friction caused by the contact. Let F_n represent the normal (perpendicular) force at S (in newtons) with which M_1 pushes against M_2. Let F_m represent the maximum force (in newtons) that can be applied to M_1 relative to M_2 parallel to S, such that the two objects remain stationary with respect to each other. The *coefficient of static friction*, denoted μ_s, is defined as follows:

$$\mu_s = F_m/F_n$$

Coefficient of kinetic friction

Suppose two objects, M_1 and M_2, are in physical contact along a common flat surface S, and that the two objects are in relative motion at a constant linear velocity. Suppose there is friction caused by the contact. Let F_n represent the normal (perpendicular) force at S (in newtons) with which M_1 pushes against M_2. Let F_f represent the frictional force (in newtons) between M_1 and M_2 parallel to S. The *coefficient of kinetic friction*, denoted μ_k, is defined as follows:

$$\mu_k = F_f / F_n$$

Torque

Let r represent the radial distance (in meters) from a pivot point P to some point Q at which a force vector $\mathbf{F}$ (whose magnitude is expressed in newtons) is applied, as shown in Fig. 5.1. Let θ represent the angle (in radians) between $\mathbf{F}$ and line segment PQ. The *torque*, also called the *moment* (denoted τ and expressed in newton-meters) is defined as follows:

$$\tau = r|\mathbf{F}| \sin \theta$$

Work

Let $\mathbf{F}$ be a force vector (whose magnitude is expressed in newtons) that acts on an object M, causing M to be displaced by a

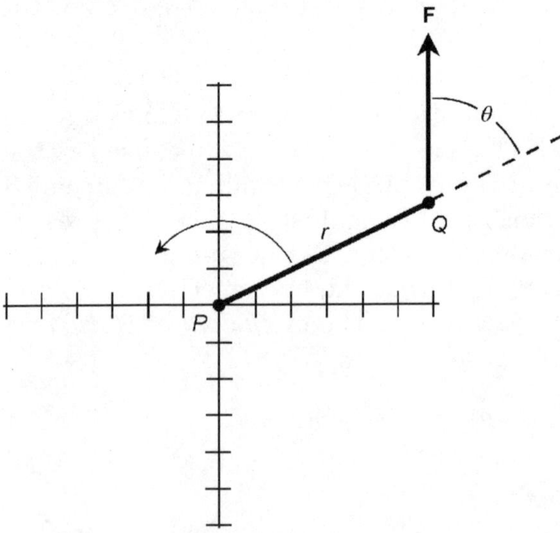

Figure 5.1 Determination of torque, also known as moment, about a rotational axis.

vector **s** (whose magnitude is expressed in meters). Let θ be the angle between vectors **F** and **s**, as shown in Fig. 5.2. Then the *work* done by **F** acting on M is denoted W, is expressed in newton-meters or joules, and is given by:

$$W = |\mathbf{F}||\mathbf{s}| \cos \theta$$

Potential energy

Let m be the mass (in kilograms) of an object M. Let h be the altitude (in meters) of M above a reference level, through which M will freely fall if released. Let g be the gravitational acceleration (in meters per second squared) in which M is placed. Then the *potential energy* of M is denoted E_p, is expressed in newton-meters or joules, and is given by:

$$E_p = gmh$$

In the earth's gravitational field at the surface, $g \approx 9.8067$.

Kinetic energy

Let m be the mass (in kilograms) of an object M moving in a straight line at constant speed. Let v be the constant linear velocity (in meters per second) at which M moves. Then the kinetic energy of M is denoted E_k, is expressed in newton-meters or joules, and is given by:

$$E_k = mv^2/2$$

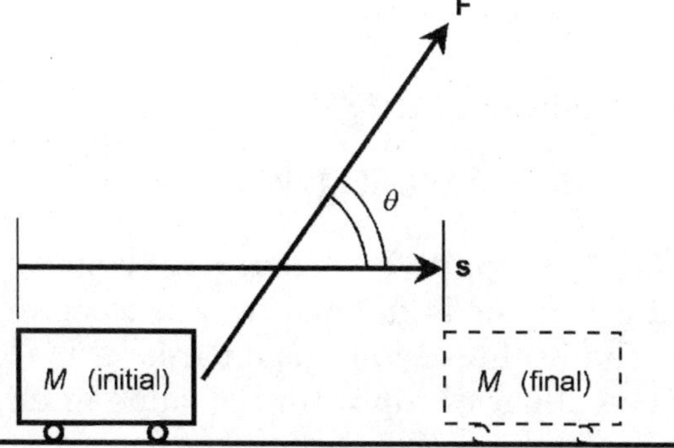

Figure 5.2 Determination of work in straight-line motion.

Principle of work

Let W_{in} be the total work input to a machine; let W_{out} be the work output from that machine in useful or intended form; let W_f be the work required to overcome friction or loss in the machine. Further, let all these quantities be expressed in the same units (usually joules). Then the following equation holds:

$$W_{out} = W_{in} - W_f$$

Displacement ratio

Let s_{in} be the displacement over which an input force moves an object M. Let s be the actual displacement of M, in the same units as s_{in}. The *displacement ratio*, R_s, is given by:

$$R_s = s_{in}/s$$

Force ratio

Let F_{load} be the force exerted on an object M by a machine. Let F be the force actually used to operate the machine, in the same units as F_{load}. The *force ratio*, R_F, is given by:

$$R_F = F_{load}/F$$

Efficiency of machine

Let W_{out} represent the useful work output from a machine; let W_{in} represent the useful work input (in the same units as W_{out}). The *efficiency* of the machine, *Eff*, is:

$$Eff = W_{out}/W_{in}$$

As a percentage, the efficiency $Eff_{\%}$ is:

$$Eff_{\%} = 100\ W_{out}/W_{in}$$

The above formulas also apply for useful power output versus power input, because power is defined as work per unit time. Let P_{out} represent the useful power output from a machine; let P_{in} represent the useful power input (in the same units as P_{out}). The *efficiency* of the machine, *Eff*, is:

$$Eff = P_{out}/P_{in}$$

As a percentage, the efficiency $Eff_{\%}$ is:

$$Eff_{\%} = 100\ P_{out}/P_{in}$$

Linear impulse

Let **F** be a force (whose magnitude is expressed in newtons) applied to an object M for a length of time t (in seconds). The *linear impulse* (whose magnitude is expressed in kilogram-meters per second, or in newton-seconds) is denoted **I** and is given by:

$$\mathbf{I} = \mathbf{F}t$$

Linear momentum

Let m be the mass (in kilograms) of an object M. Let **v** be the velocity (whose magnitude is expressed in meters per second). Then the *linear momentum* (whose magnitude is expressed in kilogram-meters per second) is denoted **p** and is given by:

$$\mathbf{p} = m\mathbf{v}$$

Conservation of linear momentum

Let $S = \{M_1, M_2, M_3, \ldots M_n\}$ be a system of objects; let their masses (in kilograms) be denoted $m_1, m_2, m_3, \ldots m_n$. Let the velocity of each object (the magnitudes of which are expressed in meters per second) be denoted $\mathbf{v}_1, \mathbf{v}_2, \mathbf{v}_3, \ldots \mathbf{v}_n$ respectively. Let the net external force acting on S be denoted **F** (whose magnitude is expressed in newtons). Let the momentum of each object (the magnitudes of which are expressed in kilogram-meters per second) be denoted $\mathbf{p}_1, \mathbf{p}_2, \mathbf{p}_3, \ldots \mathbf{p}_n$ respectively. Let the vector sum of the momenta of the objects in S be given by:

$$\mathbf{p} = \mathbf{p}_1 + \mathbf{p}_2 + \mathbf{p}_3 + \ldots \mathbf{p}_n$$

$$= m_1\mathbf{v}_1 + m_2\mathbf{v}_2 + m_3\mathbf{v}_3 + \ldots + m_n\mathbf{v}_n$$

Suppose the magnitude and direction of **F** both remain constant. Then the magnitude and direction of **p** will both remain constant. This holds true even if the individual momenta, $\mathbf{p}_i$, the individual masses, $\mathbf{m}_i$, and/or the individual velocities, $\mathbf{v}_i$, do not all remain constant.

Collisions and momentum

Let M_1 and M_2 be two separate objects having masses (in identical units) m_1 and m_2, respectively. Let $\mathbf{v}_1$ and $\mathbf{v}_2$ be their respective velocities (whose magnitudes are expressed in identical units) before the objects collide. Let $\mathbf{w}_1$ and $\mathbf{w}_2$ be their respective velocities after they collide (whose magnitudes are expressed in the same units as $\mathbf{v}_1$ and $\mathbf{v}_2$). Then the following holds true:

$$m_1\mathbf{v}_1 + m_2\mathbf{v}_2 = m_1\mathbf{w}_1 + m_2\mathbf{w}_2$$

The *total system momentum* before the collision is equal to the total system momentum after the collision.

Elastic collisions and kinetic energy

Let M_1 and M_2 be two separate objects having masses (in identical units) m_1 and m_2, respectively. Let v_1 and v_2 be their respective linear speeds before the objects undergo an *elastic collision*. Let w_1 and w_2 be their respective linear speeds after they undergo the elastic collision (expressed in the same units as v_1 and v_2). Then the following holds true:

$$m_1 v_1^2/2 + m_2 v_2^2/2 = m_1 w_1^2/2 + m_2 w_2^2/2$$

The total system kinetic energy before the elastic collision is equal to the total system energy after the elastic collision.

Average angular speed

Let M be an object that is rotating about a fixed axis L. Let θ_1 be the angular displacement of M (in radians, relative to some reference axis K) at an instant of time t_1, and let θ_2 be the angular displacement of M (in radians relative to K) at some later instant of time t_2, as shown in Fig. 5.3. The *average angular speed* (in radians per second) is denoted ω_{av}, and is given by:

$$\omega_{av} = (\theta_2 - \theta_1)/(t_2 - t_1)$$

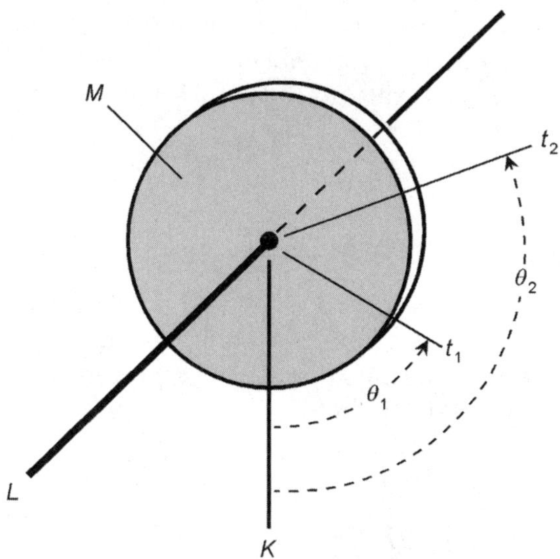

Figure 5.3 Determination of average angular speed.

Angular frequency

Let M be an object that is rotating about a fixed axis L at a constant rate. Let f be the number of complete rotations that occur per second. Then the *angular frequency* (in radians per second) is denoted ω, and is given by:

$$\omega = 2\pi f$$

Angular acceleration

Let M be an object that is rotating about a fixed axis L at a rate that changes uniformly over a time interval starting at time t_1 and ending at time t_2 ($t_2 - t_1$ seconds later). Let the angular speed (in radians per second) at time t_1 be ω_1, and let the angular speed at time t_2 be ω_2. Then the *angular acceleration* (in radians per second squared) during the time interval $[t_1,t_2]$ is denoted α, and is given by:

$$\alpha = (\omega_2 - \omega_1)/(t_2 - t_1)$$

Moment of inertia

Let an object M be comprised of smaller mass components m_1, m_2, m_3, . . . , m_n (in kilograms). Suppose these components are at distances $r_1, r_2, r_3, \ldots, r_n$ respectively (in meters) from an axis L about which M rotates. The *moment of inertia* (denoted I and expressed in kilogram-meters squared) of M is given by:

$$I = m_1 r_1^2 + m_2 r_2^2 + m_3 r_3^2 + \ldots + m_n r_n^2$$

Torque vs angular acceleration

Suppose a torque τ (in newton-meters) acts on an object M that has a moment of inertia I (in kilogram-meters squared). Let α be the angular acceleration (in radians per second squared) that results. Then the following relations hold:

$$\tau = I\alpha$$

$$\alpha = \tau / I$$

Angular momentum

Suppose an object M of mass m (in kilograms) rotates about an axis L at an angular speed ω (in radians per second). The *angular momentum* (in kilogram-meters squared per second) is a vector directed along L with a magnitude equal to $I\omega$, the product of the angular speed and the moment of inertia.

Angular impulse

Suppose a constant torque τ (in newton-meters) acts on an object M for a period of time $t = t_2 - t_1$ (expressed in seconds). Let I be the moment of inertia of M. Let ω_1 be the angular speed (in radians per second) of M at time t_1; let ω_2 be the angular speed of M at some later time t_2. Then the *angular impulse* is a vector $\mathbf{I}_\theta$ (whose magnitude is expressed in kilogram-second-meters squared):

$$|\mathbf{I}_\theta| = \tau (t_2 - t_1) = I\omega_2 - I\omega_1$$

Conservation of angular momentum

If the net torque on an object M is zero, the angular momentum vector of M will remain constant in magnitude and constant in direction.

Kinetic energy of rotation

Let M be an object rotating about an axis L with an angular speed ω (in radians per second). Then the *rotational kinetic energy* (denoted E_{kr} and expressed in joules) of M is given by:

$$E_{kr} = I\omega^2/2$$

Work on rotating object

Let M be an object rotating about an axis L as a result of constant torque τ (in newton-meters). Suppose M rotates through an angle θ (in radians). Then the *rotational work* (denoted W_r and expressed in joules) performed on M is given by:

$$W_r = \tau\theta$$

Power on rotating object

Let M be an object rotating about an axis L as a result of constant torque τ (in newton-meters). Suppose M rotates at an angular speed ω (in radians per second). Then the *rotational power* (denoted P_r and expressed in watts) performed on M is given by:

$$P_r = \tau\omega$$

Centripetal acceleration

Let M be an object revolving around a central point Q at a constant rate. Let v be the tangential speed of M (in meters per second); let ω be the angular speed (in radians per second); let r be the radius of revolution (in meters), as shown in Fig. 5.4. Then the magnitude of the *centripetal acceleration*, denoted a_c (in meters per second squared), is given by:

$$a_c = v^2/r = \omega^2 r$$

The centripetal-acceleration vector, $\mathbf{a}_c$, is directed from M toward Q.

Centripetal force

Let M be an object revolving around a central point Q at a constant rate. Let v be the tangential speed of M (in meters per

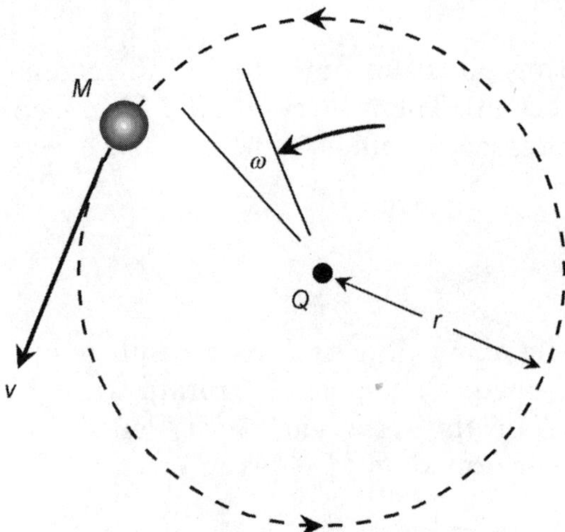

Figure 5.4 Determination of centripetal acceleration.

second); let ω be the angular speed (in radians per second); let r be the radius of revolution (in meters), and let m be the mass of M (in kilograms), as shown in Fig. 5.4. Then the magnitude of the *centripetal force*, denoted F_c (in kilogram-meters per second squared), is given by:

$$F_c = mv^2/r = m\omega^2 r$$

The centripetal-force vector, $\mathbf{F}_c$, is directed from M toward Q.

Hooke's Law

Let M be an object in an elastic system. Suppose M is moved from its equilibrium position so the elastic medium is stretched over a displacement s (in meters), defined such that:

$$s > 0 \text{ represents expansion}$$

$$s < 0 \text{ represents compression}$$

Let the stiffness of the elastic medium be expressed as a constant k (in newtons per meter). Then the magnitude of the *external force* (denoted F_x and expressed in newtons) is given by:

$$F_x = ks$$

The magnitude of the *restoring force* (denoted F_r and expressed in newtons) is given by:

$$F_r = -ks$$

The direction of the external-force vector, $\mathbf{F}_x$, is the same as the direction of the displacement of M; the direction of the restoring-force vector, $\mathbf{F}_r$, is opposite to the direction of the displacement of M.

Elastic potential energy

Let M be an object in an elastic system. Suppose M is moved from its equilibrium position so the elastic medium is stretched over a displacement s (in meters), defined such that:

$$s > 0 \text{ represents expansion}$$

$$s < 0 \text{ represents compression}$$

Let the stiffness of the elastic medium be expressed as a constant k (in newtons per meter). Then the *elastic potential energy* (denoted E_{px} and expressed in joules) stored in the stretched or compressed medium is given by:

$$E_{px} = ks^2/2$$

Suppose M is released and allowed to oscillate as a result of the elasticity of the medium. This oscillation is known as *simple harmonic motion* (SHM). Let s_{inst} be the instantaneous displacement of M from the midpoint (the point of zero expansion or compression) at a given instant in time. Then:

$$E_{px} = ks_{inst}{}^2/2$$

At the midpoint of the oscillation, when s_{inst} is zero and the medium is not stretched,

$$E_{px} = 0$$

Elastic kinetic energy

Let M be an object in an elastic system. Suppose M is moved from its equilibrium position so the elastic medium is stretched over a displacement s (in meters), defined such that:

$$s > 0 \text{ represents expansion}$$

$$s < 0 \text{ represents compression}$$

Let the stiffness of the elastic medium be expressed as a constant k (in newtons per meter). Suppose M is released and allowed to oscillate in SHM as a result of the elasticity of the medium. Let v_{inst} be the instantaneous speed of M at some point in the oscillation (in meters per second). Let v be the maximum speed attained by M during the oscillation. The *elastic kinetic energy* (denoted E_{kx} and expressed in joules) manifested by M at any given instant of time is given by:

$$E_{kx} = kv_{inst}{}^2/2$$

At the midpoint of the oscillation, when $v_{inst} = v$ and the medium is not stretched,

$$E_{kx} = kv^2/2 = ks^2/2$$

At either endpoint of the oscillation, when v_{inst} is zero and the medium is fully expanded or compressed:

$$E_{kx} = 0$$

Elastic kinetic vs potential energy in SHM

Refer to the preceding two paragraphs. At all times in an oscillating system that exhibits SHM such as that described, the combined total of the elastic kinetic energy manifested by M and the elastic potential energy stored in M is constant:

$$E_{px} + E_{kx}$$

$$= ks_{inst}{}^2/2 + kv_{inst}{}^2/2$$

Period of oscillation in SHM

Let M be an object of mass m (in kilograms) in an elastic system. Suppose M is moved from its equilibrium position as described in the preceding paragraphs, causing M to oscillate in SHM. Let the stiffness of the elastic medium be expressed as a constant k (in newtons per meter). Then the *period of oscillation* (denoted T and expressed in seconds) is given by:

$$T = 2\pi(m/k)^{1/2}$$

Instantaneous speed in SHM

Let M be an object in an elastic system. Suppose M is moved from its equilibrium position so the elastic medium is stretched over a displacement s (in meters), defined such that:

$s > 0$ represents expansion

$s < 0$ represents compression

Let the stiffness of the elastic medium be expressed as a constant k (in newtons per meter). Suppose M is released and allowed to oscillate in SHM as a result of the elasticity of the medium. Let s_{inst} be the instantaneous displacement of M from the midpoint (the point of zero expansion or compression) at a given instant in time. Let v_{inst} be the *instantaneous speed* of M at some point in the oscillation (in meters per second). Then:

$$v_{inst} = (ks^2 - ks_{inst}{}^2)/m)^{1/2}$$

Instantaneous acceleration in SHM

Let M be an object in an elastic system. Suppose M is moved from its equilibrium position so the elastic medium is stretched over displacement s (in meters) such that:

$s > 0$ represents expansion

$s < 0$ represents compression

Let the stiffness of the elastic medium be expressed as a constant k (in newtons per meter). Suppose M is released and allowed to oscillate in SHM as a result of the elasticity of the medium. Let s_{inst} be the instantaneous displacement of M from the midpoint (the point of zero expansion or compression) at a given instant in time. Let a_{inst} be the *instantaneous acceleration* of M (in meters per second squared). Then

$$a_{inst} = -ks_{inst}/m$$
$$= -4\pi^2 s_{inst}/T^2$$

Period of oscillation for pendulum

Let M be a mass that is hung from a rigid wire and allowed to swing over a small arc as a simple pendulum. Let s be the length of the wire (in meters); let g represent the gravitational acceleration (approximately 9.8067 meters per second squared at the earth's surface). Then the approximate *period of oscillation* for M (denoted T and expressed in seconds) is given by:

$$T = 2\pi(s/g)^{1/2}$$

Mass density

Let M be an object whose mass is m (in kilograms). Let V be the volume of M (in cubic meters). Then the *mass density* of M (denoted ρ and expressed in kilograms per meter cubed) is given by:

$$\rho = m/V$$

Specific gravity for solid

Let M be an object whose mass is m (in kilograms). Let V be the volume of M (in liters). Then the *specific gravity* of M (denoted sp gr) is given by:

$$\text{sp gr} = m/V$$

Specific gravity for liquid

Let L be a sample of liquid whose mass is m (in kilograms). Let V be the volume of L (in liters). Then the *specific gravity* of L (denoted sp gr) is given by:

$$\text{sp gr} = m/V$$

Specific gravity for gas

Let G be a sample of gas whose mass is m (in grams). Let V be the volume of G (in liters). Then the *specific gravity* of G (denoted sp gr) is given by:

$$\text{sp gr} = 0.7735m/V$$

Pressure

Let M be an object whose surface area is A (in square meters). Let F be a compressive force (in newtons) directed perpendicular to the surface of M at all points. Then the *pressure* (denoted P and expressed in pascals) on M is given by:

$$P = F/A$$

Stress

Let M be a solid object. Let A be the area (in square meters) of a cross section S of M. Let F be a force (in newtons) applied over S. Then the *stress* (denoted σ and expressed in pascals) on M is given by:

$$\sigma = F/A$$

Strain

Let X be a specified linear dimensional measure (in meters) of an object M with no force applied. Let X_F be the measure of the same linear dimension (in meters) with a force F (in newtons) applied. Then the *strain* (denoted ε) on M is given by:

$$\varepsilon = |X_F - X|/X$$

Modulus of elasticity

Let σ be the stress (in pascals) on an object M; let ε be the strain on M. Then the *modulus of elasticity* (expressed in pascals and denoted Y) is given by:

$$Y = \sigma/\varepsilon$$

This quantity is also known as *Young's modulus*. In general, low values of Y represent flexible or highly elastic substances; high values of Y represent comparatively inflexible or rigid substances.

Fluidics and Thermodynamics

This section deals with the behavior of fluids, gases, heat, thermal expansion and contraction, and entropy.

Hydrostatic pressure

Let C be a column of fluid whose mass density (in kilograms per meter cubed, or grams per liter) is ρ, and whose height (in meters) is h. Let g represent the gravitational acceleration (approximately 9.8067 meters per second squared at the earth's surface). Then the *hydrostatic pressure* (denoted P and expressed in pascals) at the bottom of C is given by:

$$P = \rho \mathrm{hg}$$

Archimedes' Principle

Let M be an object of density ρ_M completely immersed in a fluid having density ρ (in kilograms per meter cubed). Let V be the volume of M (in cubic meters). Let g be the gravitational acceleration (g $\approx$ 9.8067 meters per second squared at the earth's surface). The *buoyant force* vector, $\mathbf{F}_B$, is directed upward with magnitude (in newtons) given by:

$$|\mathbf{F}_B| = \rho \mathrm{Vg} - \rho_M Vg$$

When an object is partially immersed in fluid, the buoyant force vector is directed upward, and has a magnitude equal to the weight of the fluid displaced.

Pascal's Principle

When the pressure changes on a parcel of confined fluid or gas, then the pressure on every other parcel of the same confined fluid or gas changes to the same extent.

Fluid flow rate

Suppose a fluid C passes along a pipe. Let s be the average speed of C relative to the pipe (in meters per second); let A be the cross-sectional area of the pipe (in meters squared). The *fluid flow rate* (denoted J and expressed in meters cubed per second) is:

$$J = As$$

Continuity principle

Suppose a non-compressible fluid C passes along a pipe. Let s_1 be the speed of C relative to the pipe (in meters per second) at some point P_1 along the length of the pipe; let A_1 be the cross-sectional area of the pipe (in meters squared) at P_1. Let s_2 be the speed of C relative to the pipe (in meters per second) at some point P_2 along the length of the pipe; let A_2 be the cross-sectional area of the pipe (in meters squared) at P_2. Then the following equation holds:

$$A_1 s_1 = A_2 s_2 = J$$

That is, the flow rate (denoted J and expressed in meters cubed per second) of a non-compressible fluid moving through a pipe is the same at all points along the pipe.

The so-called *continuity principle* can also be stated as follows:

$$A_1/A_2 = s_2/s_1$$

That is, the speed of a noncompressible fluid flowing through a pipe at a given point is inversely proportional to the cross-sectional area of the pipe at that point.

Shear stress

Let M be an object having the shape of a rectangular prism. Suppose that a force F (in newtons) acts in one direction on one face of M whose area (in square meters) is A. Suppose the same force F acts in the opposite direction on the opposite face of M, whose area is also A. Suppose the application of these forces does not change the volume of M relative to its volume with no forces applied. The *shear stress* (denoted σ_s and expressed in pascals) is given by:

$$\sigma_s = F/A$$

Shear strain

Let M, F, and A be defined as in the preceding paragraph. Let s be the shearing displacement (in meters); let d be the distance (in meters) between the opposite faces to which the opposing forces F are applied. The *shear strain* (denoted ϵ_s) is given by:

$$\varepsilon_s = s/d$$

Shear rate

Let M, F, and A be defined as in the preceding paragraphs. Let s be the shearing displacement (in meters); let d be the distance (in meters) between the opposite faces to which the opposing forces F are applied. The *shear rate* is defined as the rate of change of shear strain. Let t represent time (in seconds) and the shear rate (per second) be denoted R_s. Then:

$$R_s = d\varepsilon_s/dt$$

Shear modulus

Let M, F, A, s, and d be defined as in the preceding two paragraphs. The *shear modulus* (denoted S_s and expressed in pascals) is the ratio of the shear stress to the shear strain, and is given by:

$$S_s = \sigma_s/\varepsilon_s = (Fd)/(As)$$

Fluid viscosity

Let σ_s be the shear stress of a fluid C (in pascals); let R_s be the shear rate of C (per second). Then the *fluid viscosity* (denoted η and expressed in pascal-seconds) of C is given by:

$$\eta = \sigma_s/R_s$$

Reynolds number for fluids

Let η be the viscosity (in pascal-seconds) of a fluid C flowing through a pipe. Let ρ be the density of C (in kilograms per meter cubed, or grams per liter). Let s be the speed (in meters per second) at which C flows through the pipe whose radius is r (in meters). Then the *Reynolds number*, denoted N_R, of C is defined as:

$$N_R = 2\rho sr/\eta$$

Poiseuille's Law for fluids

Let η be the viscosity (in pascal-seconds) of a fluid C flowing through a pipe. Let P_{in} be the pressure (in pascals) on C at the input end of the pipe; let P_{out} be the pressure (in pascals) on C at the output end of the pipe. Let m be the length of the pipe (in meters); let r be the radius of the pipe (in meters). Then the fluid flow rate (denoted J and expressed in meters cubed per second) is given by:

$$J = \pi r^4(P_{in} - P_{out})/(8\eta m)$$

Bernoulli's Law for fluids

Suppose a non-compressible, non-viscous fluid C flows at a steady rate J (in meters cubed per second) along a path or through a pipe. Consider two points X_1 and X_2, at heights h_1 and h_2 (in meters) respectively above a specified reference level. Let s_1 and s_2 be the fluid speeds (in meters per second) at X_1 and X_2 respectively. Let ρ be the fluid density (in kilograms per meter cubed, or grams per liter) at X_1 and X_2; this parameter does not change because C is incompressible. Let P_1 and P_2 be the fluid pressures (in pascals) at X_1 and X_2 respectively. Let g be the gravitational acceleration. Then the following equation holds:

$$P_1 + s_1^2\rho/2 + h_1 g\rho = P_2 + s_2^2\rho/2 + h_2 g\rho$$

In the earth's gravitational field at the surface, $g \approx 9.8067$.

Torricelli's Law for fluids

Suppose a volume of a non-compressible, non-viscous fluid C is held by a stationary container open at the top. Suppose there is a hole in the tank at a distance a (in meters) below the surface of the liquid. Let s be the speed (in meters per second) at which C flows out of the hole as a result of the effects of gravitational acceleration g (in meters per second squared). Then the following equation holds:

$$s = (2ga)^{1/2}$$

In the earth's gravitational field at the surface, $g \approx 9.8067$.

Linear thermal expansion

Let T_1 be the initial temperature (in degrees Kelvin) of a solid object M; let T_2 be the final temperature (in degrees Kelvin). Let s_1 be the initial measure of a specific linear dimension of M (in meters); let s_2 be the final measure of the same linear dimension (in meters). Then the following approximation holds:

$$s_2 - s_1 \approx \alpha s_1 (T_2 - T_1)$$

where α is a dimensionless constant known as the *coefficient of linear expansion* for the substance comprising M.

Area thermal expansion

Let T_1 be the initial temperature (in degrees Kelvin) of a solid object M; let T_2 be the final temperature (in degrees Kelvin). Let A_1 be the initial area of a specific cross section of M (in meters squared); let A_2 be the final area of the same cross section (in meters squared). Then the following approximation holds:

$$A_2 - A_1 \approx \beta A_1 (T_2 - T_1)$$

where β is a dimensionless constant known as the *coefficient of area expansion* for the substance comprising M.

Volume thermal expansion

Let T_1 be the initial temperature (in degrees Kelvin) of a solid object M; let T_2 be the final temperature (in degrees Kelvin). Let V_1 be the initial volume of M (in meters cubed); let V_2 be the final volume of M (in meters cubed). Then the following approximation holds:

$$V_2 - V_1 \approx \gamma V_1 (T_2 - T_1)$$

where γ is a dimensionless constant known as the *coefficient of volume expansion* for the substance comprising M.

Isotropic substance

Let M be a solid object that expands to the same extent in all linear dimensions when the temperature increases. Then the solid substance comprising M is called *isotropic*. Let α be the coefficient of linear expansion, let β be the coefficient of area expansion, and let γ be the coefficient of volume expansion for an isotropic solid. Then the following approximations hold:

$$\beta \approx 2\alpha$$

$$\gamma \approx 3\alpha$$

STP for a gas

For a substance in the gaseous state, *standard temperature and pressure* (STP) are defined as follows, in degrees Kelvin and pascals respectively:

$$T_0 = 273.15$$

$$P_0 = 1.013 \times 10^5$$

This corresponds to one atmosphere at the freezing point of pure water.

Ideal Gas Law

Suppose a quantity N (in moles) of a gas C is confined to a container having volume V (in meters cubed). Let P be the pressure of the gas (in pascals); let T be the absolute temperature (in degrees Kelvin). Suppose the following relation holds under conditions of varying pressure, temperature, and volume:

$$P = R_0 NT/V$$

where $R_0 \approx 8.3145$. If this property holds for C, then C is considered an *ideal gas*. The value $R_0 \approx 8.3145$ is known as the *ideal gas constant*, and is expressed in joules per mole degree Kelvin.

This law is sometimes expressed in another way. Suppose a confined volume of a gas C undergoes compression or expansion. Let P_1 represent the initial pressure of C (in pascals); let T_1 represent the initial temperature (in degrees Kelvin); let V_1 represent the initial volume (in meters cubed). Let P_2 represent the final pressure (in pascals); let T_2 represent the final temperature (in degrees Kelvin); let V_2 represent the final volume (in meters cubed). Then C is an ideal gas if and only if the following equation holds:

$$P_1V_1/T_1 = P_2V_2/T_2$$

Dalton's Law

Let C be a mixture of ideal, non-reactive gases $C_1, C_2, C_3, \ldots, C_n$, occupying a container of volume V (in meters cubed). Let P be the pressure (in pascals) exerted by C; let $P_1, P_2, P_3, \ldots, P_n$ be the *partial pressures* (in pascals) exerted by $C_1, C_2, C_3, \ldots,$ and C_n respectively. Then the following equation holds:

$$P = P_1 + P_2 + P_3 + \ldots + P_n$$

Mass of atom

Let M be the *atomic mass* of a chemical element. Let m be the mass (in kilograms) of an individual atom of that element. Then the following equation holds:

$$m = M/N$$

where N is a constant known as *Avogadro's number*, expressed in particles per mole:

$$N \approx 6.022169 \times 10^{23}$$

Mass of molecule

Let M be the *molecular mass* of a chemical compound. Let m be the mass (in kilograms) of an individual atom of that compound. Then the following equation holds:

$$m = M/N$$

Average translational kinetic energy of gas particle

Let T be the absolute temperature (in degrees Kelvin) of a gas. The *average translational kinetic energy* (denoted E_{kt-av} and expressed in joules) of an atom or molecule in that gas is given by:

$$E_{kt-av} = 3kT/2$$

where k is *Boltzmann's constant*, expressed in joules per degree Kelvin:

$$k \approx 1.3807 \times 10^{-23}$$

Root-mean-square (rms) speed of gas particle

Let m be the mass (in kilograms) of an atom or molecule in a gas. Let T be the absolute temperature (in degrees Kelvin). Let k be Boltzmann's constant, as defined in the preceding paragraph. The *root-mean-square speed* (denoted v_{rms} and expressed in meters per second) of the particle is given by:

$$v_{rms} = (3kT/m)^{1/2}$$

Length of mean free path (mfp) for gas particle

Suppose an ideal gas is confined to a chamber of volume V (in meters cubed). Let N be the number of gas particles in the enclosure. Suppose each atom or molecule undergoes numerous, repeated collisions with other atoms or molecules of the same ideal gas. Further suppose that all the particles are perfect spheres with identical radius r (in meters). The length of the *mean free path* for each particle (denoted L_{mfp} and expressed in meters) is given by:

$$L_{mfp} = V/((32\pi^2)^{1/2}r^2N)$$

Specific heat capacity

Suppose a sample X of a substance has mass m (in kilograms). Let T_1 be the temperature (in degrees Kelvin) of X before the application of an amount q (in joules) of heat energy. Let T_2 be the temperature (in degrees Kelvin) of X following the application of the heat energy. The *specific heat capacity* (denoted c and expressed in joules per kilogram degree Kelvin) of X is given by:

$$c = q/(mT_2 - mT_1)$$

Heat gain

Suppose a sample X of a substance has mass m (in kilograms) and a specific heat capacity c (in joules per kilogram degree Kelvin). Let T_1 be the initial temperature (in degrees Kelvin) of X; let T_2 be the final temperature (in degrees Kelvin) of X. Suppose $T_2 > T_1$ (that is, X heats up), and the state of X (solid, liquid, or gas) remains the same. Then the *heat gain* (denoted q_+ and expressed in joules) of X is given by:

$$q_+ = mcT_2 - mcT_1$$

Heat loss

Suppose a sample X of a substance has mass m (in kilograms) and a specific heat capacity c (in joules per kilogram degree Kelvin). Let T_1 be the initial temperature (in degrees Kelvin) of X; let T_2 be the final temperature (in degrees Kelvin) of X. Suppose $T_2 < T_1$ (that is, X cools down), and the state of X (solid, liquid, or gas) remains the same. Then the *heat loss* (denoted q_- and expressed in joules) of X is given by:

$$q_- = mcT_1 - mcT_2$$

Emissivity

Let M be an object whose surface area (in meters squared) is A. Let M_b be a blackbody of identical proportions and surface area to M. Let T be the absolute temperature (in degrees Kelvin) of both M and M_b. Suppose M_b emits energy E_b (in joules) while M emits energy E (also in joules). Then the *emissivity* (denoted ε, a dimensionless quantity) of M is given by:

$$\varepsilon = E/E_b$$

Emissivity can also be expressed in terms of power. Suppose

M_b emits power P_b (in watts) while M emits power P (also in watts). Then:

$$\varepsilon = P/P_b$$

Emissivity is sometimes expressed as a percentage $\varepsilon_{\%}$ rather than as a ratio. For energy and power, respectively, the following equations apply:

$$\varepsilon_{\%} = 100E/E_b$$

$$\varepsilon_{\%} = 100P/P_b$$

Power radiated by surface

Let M be an object with surface area A (in meters squared). Let T be the temperature of M (in degrees Kelvin); let ϵ be the emissivity (expressed as a ratio) of M. Let P be the power (in watts) radiated by M. Then the following equation holds:

$$P = T^4 \sigma \varepsilon A$$

where σ represents the *Stefan-Boltzmann constant*, expressed in watts per meter squared degree Kelvin to the fourth power:

$$\sigma \approx 5.6697 \times 10^{-8}$$

First Law of Thermodynamics

Let X be a physical system such as a machine. Suppose X absorbs or accepts a quantity of heat input energy E (in joules). Let E_w represent the work energy (in joules) expended by X in the desired form; let E_{int} represent the increased internal energy (in joules) of X resulting from the input energy E. Then the following equation holds:

$$E = E_w + E_{int}$$

Stated another way, the useful work energy expended by a physical system is always equal to the difference between the input energy and the increased internal energy:

$$E_w = E - E_{int}$$

Entropy gain

Let X be a system whose temperature (in degrees Kelvin) is T. Suppose an amount q_+ of heat energy (in joules) is introduced into X. Then the *entropy gain* (denoted S_+ and expressed in joules per degree Kelvin) is given by:

$$S_+ = q_+/T$$

Entropy loss

Let X be a system whose temperature (in degrees Kelvin) is T. Suppose an amount q_- of heat energy (in joules) leaves X. Then the *entropy loss* (denoted S_- and expressed in joules per degree Kelvin) is given by:

$$S_- = q_-/T$$

Second Law of Thermodynamics

This law can be stated in three parts.

- Let M_1 and M_2 be objects having absolute temperatures T_1 and T_2. If $T_1 > T_2$, then heat energy can spontaneously flow from T_1 to T_2, but not from T_2 to T_1. If $T_1 < T_2$, then heat energy can spontaneously flow from T_2 to T_1, but not from T_1 to T_2.
- The efficiency of a continuously cycling heat engine is always less than 100 percent. Let E represent the heat energy input to the engine; let E_w represent the useful work energy output. Then $E_w < E$.
- When a system changes spontaneously, its entropy cannot decrease.

A corollary to this law is the fact that if there are several different ways in which a system can exist, the most probable state is the state in which the entropy is greatest, and the least probable state is the state in which the entropy is smallest.

Waves and Optics

This section deals with the physics of wave motion, interference, refraction, reflection, and diffraction.

Wavelength/frequency/speed

Let λ denote the wavelength (in meters) of a wave disturbance. Let f denote the frequency (in hertz); let c be the speed (in meters per second) at which the disturbance propagates. Then the following relations hold:

$$c = f\lambda$$

$$f = c/\lambda$$

$$\lambda = c/f$$

Engineers generally use the italicized lowercase English letter f to represent frequency; physicists more often use the italicized lowercase Greek ν (nu).

Wavelength/period/speed

Let λ denote the wavelength (in meters) of a wave disturbance. Let T denote the period (in seconds); let c be the speed (in meters per second) at which the disturbance propagates. Then the following relations hold:

$$\lambda = cT$$

$$c = \lambda/T$$

$$T = \lambda/c$$

Resonance

Suppose a wire, waveguide, antenna element, or cavity M has length L (in meters). Suppose a disturbance propagates along or through M at a speed c (in meters per second). Let f be the frequency (in hertz) of the disturbance. Let T be the period (in seconds). Let λ be the wavelength of the disturbance (in meters).

Let n be any positive integer. Then M will exhibit resonance if and only if the following relations hold:

$$\lambda = 2L/n$$

$$f = nc/(2L)$$

$$T = 2L/(nc)$$

Speed of acoustic waves in ideal gas

Let C be an ideal gas whose atomic or molecular mass is M. Suppose the absolute temperature (in degrees Kelvin) is T. Let c_p represent the specific heat capacity (in joules per kilogram degree Kelvin) of C at constant pressure; let c_v represent the specific heat capacity of C at constant volume. Let R_0 represent the ideal gas constant (approximately 8.3145 joules per mole degree Kelvin). Then the *speed of acoustic waves* in C (denoted v and expressed in meters per second) is given by:

$$v = (c_p R_0 T/(c_v M))^{1/2}$$

Speed of acoustic waves versus temperature

Let C be an ideal gas. Let T_1 be the initial temperature of C (in degrees Kelvin); let T_2 be the temperature (in degrees Kelvin) after heating or cooling. Let v_1 be the speed (in meters per second) of acoustic waves in C at temperature T_1; let v_2 be the speed (in meters per second) of acoustic waves in C at temperature T_2. The following relation holds:

$$v_2^2/v_1^2 = T_2/T_1$$

This can also be expressed as a statement that the speed of acoustic-wave propagation in an ideal gas is proportional to the square root of the absolute temperature:

$$v_2/v_1 = (T_2/T_1)^{1/2}$$

Acoustic wave intensity

Suppose an acoustic wave disturbance propagates through an ideal gas C at speed v (in meters per second). Let ρ be the density of C (in grams per liter or kilograms per meter cubed); let d be the maximum displacement (in meters) of the atoms or molecules in the gas caused by the acoustic disturbance. Let f be the frequency (in hertz) of the wave disturbance. Then the *acoustic wave intensity* (denoted I and expressed in watts per meter squared) is given by:

$$I = 2\rho v(\pi f d)^2$$

Threshold of audibility

The *threshold of audibility* is the minimum acoustic wave intensity that can be detected by the average human listener when the listener is anticipating the sound. By convention, this is denoted I_0 and is defined as 1.000×10^{-12} watt per meter squared, or 1.000 picowatt per meter squared.

Loudness

Let I be the intensity of an acoustic disturbance (in picowatts per meter squared). The *loudness*, also called the *volume*, of the sound (denoted β and expressed in decibels) is given by:

$$\beta = 10 \log_{10} I$$

When I is expressed in microwatts per meter squared:

$$\beta = 60 + 10 \log_{10} I$$

When I is expressed in milliwatts per meter squared:

$$\beta = 90 + 10 \log_{10} I$$

When I is expressed in watts per meter squared:

$$\beta = 120 + 10 \log_{10} I$$

Doppler effect for acoustic waves

Let v be the speed (in meters per second) of acoustic waves in a medium C. Let L be a line at rest in C. Suppose that a sound source, S, and an observer, O, both move along L in such a way that they approach one another, and both move in opposite directions relative to C. Let s_s be the speed of the acoustic wave source (in meters per second) relative to C; let s_o be the speed of the observer (in meters per second) relative to C. Let f_s be the frequency (in hertz) of the sound emitted by S. Then the frequency f_o (in hertz) of the acoustic disturbance perceived by O is given by:

$$f_o = (f_s v + f_s s_o)/(v - s_s)$$

If S and O both move along L in such a way that they recede from each other, and both move in opposite directions relative to C, then:

$$f_o = (f_s v - f_s s_o)/(v + s_s)$$

Reflection from flat surface

Refer to Fig. 5.5. Let M be a flat reflective surface. Let N be a line normal to M that intersects M at a point P. Suppose a ray of light R_1 strikes M at P, subtending an angle θ_1 (in radians)

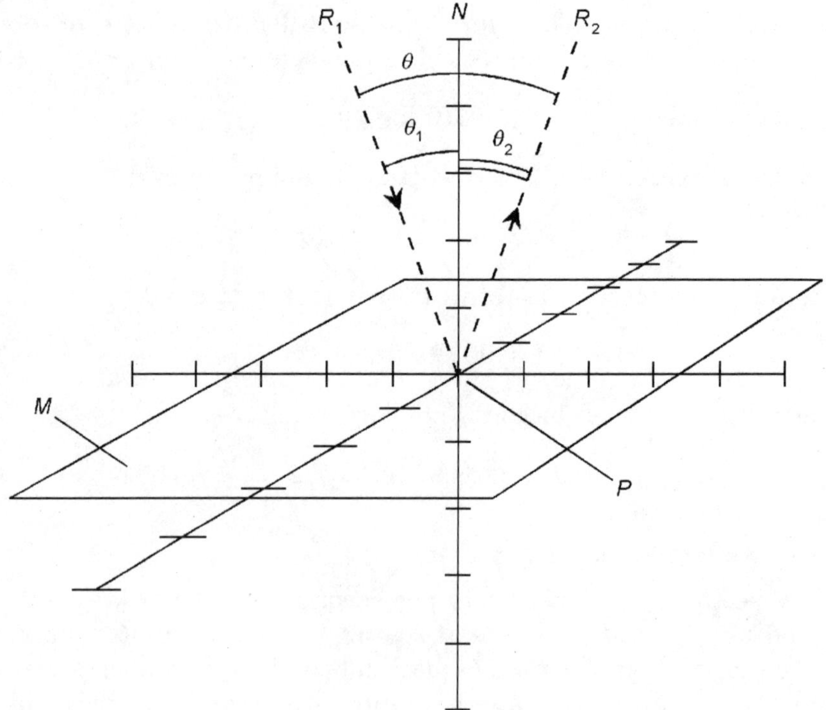

Figure 5.5 Reflection from a flat surface.

relative to N. Let θ_2 be the angle (in radians) that the reflected ray R_2, emanating from P, subtends relative to N. Let θ be the angle (in radians) between R_1 and R_2. Then the measure of the angle of incidence equals the measure of the angle of reflection, and the sum of the measures of the two angles is equal to twice the measure of either. The following equations hold:

$$\theta_1 = \theta_2$$

$$\theta_1 + \theta_2 = \theta$$

Reflection from concave spherical surface

Refer to Fig. 5.6. Let M be a concave spherical reflective surface whose radius of curvature is r (in meters). Let N be a line pass-

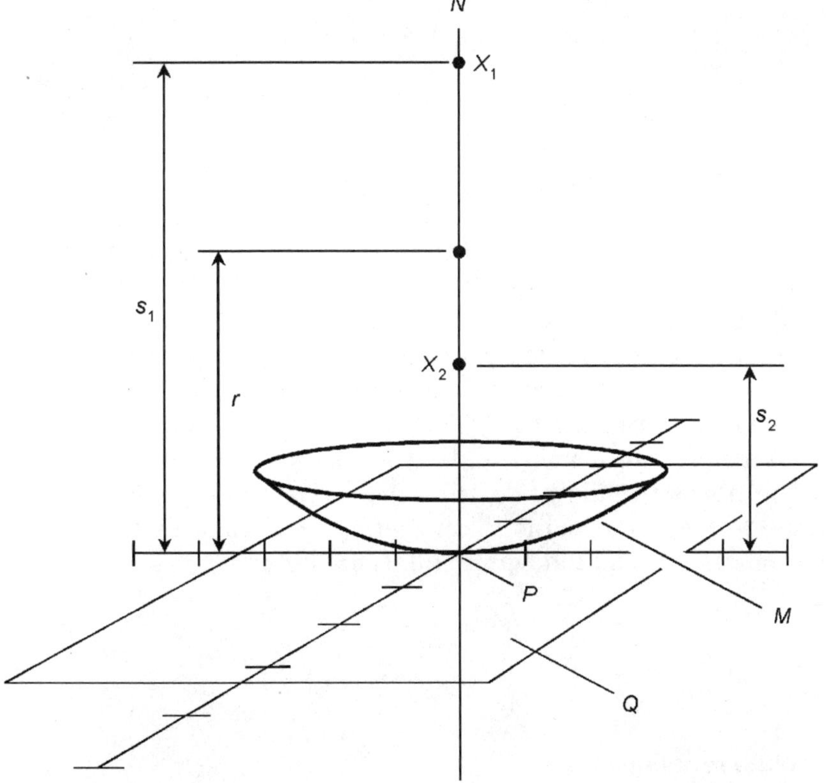

Figure 5.6 Reflection from concave spherical surface.

ing through some point P on M, such that N is normal to the plane Q tangent to M at P. Let X_1 be a point source of light located on line N, at a distance s_1 (in meters) from point P. The light reflected from M will converge toward a point X_2 that is also located on line N. Let s_2 be the distance (in meters) of X_2 from point P. Then the following equations hold:

$$s_2 = 2s_1r/(2s_1 - r)$$

$$s_1 = 2s_2r/(2s_2 - r)$$

Note that if $s_1 = r/2$, then point X_2 vanishes because the reflected rays are parallel. In addition, if point X_1 is located arbitrarily far away so that its arriving rays are parallel, then $s_2 = r/2$. The distance $r/2$ is defined as the *focal length* of the concave spherical reflector M.

The above equations can be rearranged to obtain:

$$1/s_1 + 1/s_2 = 2/r$$

$$r = 2s_1s_2/(s_1 + s_2)$$

Reflection from convex spherical surface

Refer to Fig. 5.7. Let M be a convex spherical reflective surface whose radius of curvature is r (in meters), and is defined as a negative radius. Let N be a line passing through some point P on M, such that N is normal to the plane Q tangent to M at P. Let X_1 be a point source of light located on line N, at a distance s_1 (in meters) from point P. The light reflected from M will emanate away from a point X_2 that is also located on line N, but on the opposite side of M from X_1. Let s_2 be the distance (in meters) of X_2 from point P, and let s_2 be defined as a negative distance. Then the following equations hold:

$$1/s_1 + 1/s_2 = 2/r$$

$$r = 2s_1s_2/(s_1 + s_2)$$

Absolute refractive index

Let v be the speed (in meters per second) of an electromagnetic wave disturbance D in some medium M. The *absolute refractive*

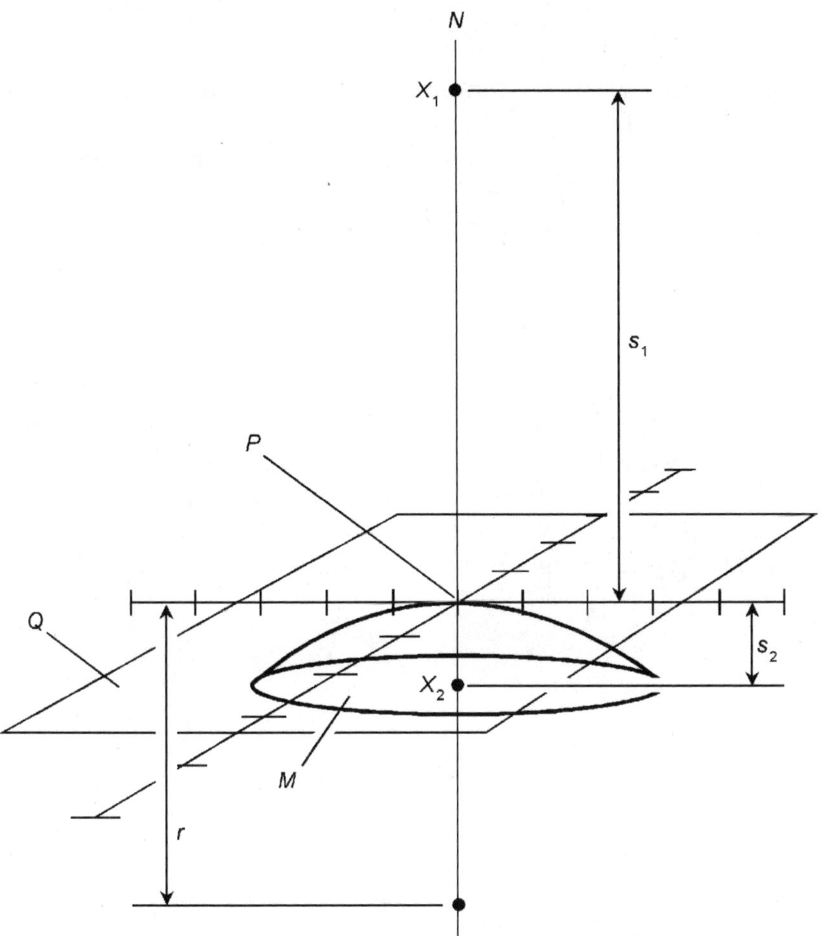

Figure 5.7 Reflection from convex spherical surface.

index of M (a dimensionless quantity, usually symbolized n) for D is approximately equal to:

$$n \approx 2.99792 \times 10^8/v$$

Relative refractive index

Let v_1 be the speed of an electromagnetic wave disturbance D in some medium M_1. Let v_2 be the speed (in the same units as v_1) of D in some other medium M_2. The *relative refractive index* (a dimensionless quantity, symbolized $n_{2:1}$) of M_2 with respect to M_1 for D is given by:

$$n_{2:1} = v_1/v_2$$

The relative refractive index can also be stated in terms of the absolute refractive indices of two media. Let n_1 be the absolute refractive index of medium M_1 for an electromagnetic wave disturbance D; let n_2 be the absolute refractive index of medium M_2 for D. Then the following equation holds:

$$n_{2:1} = n_2/n_1$$

Snell's Law for $n_2 > n_1$

Refer to Fig. 5.8. Let B be a flat boundary between two media M_1 and M_2 whose absolute indices of refraction are n_1 and n_2, respectively. Suppose that $n_2 > n_1$, that is, the ray passes from a medium having a relatively lower refractive index to a medium having a relatively higher refractive index. Let N be a line

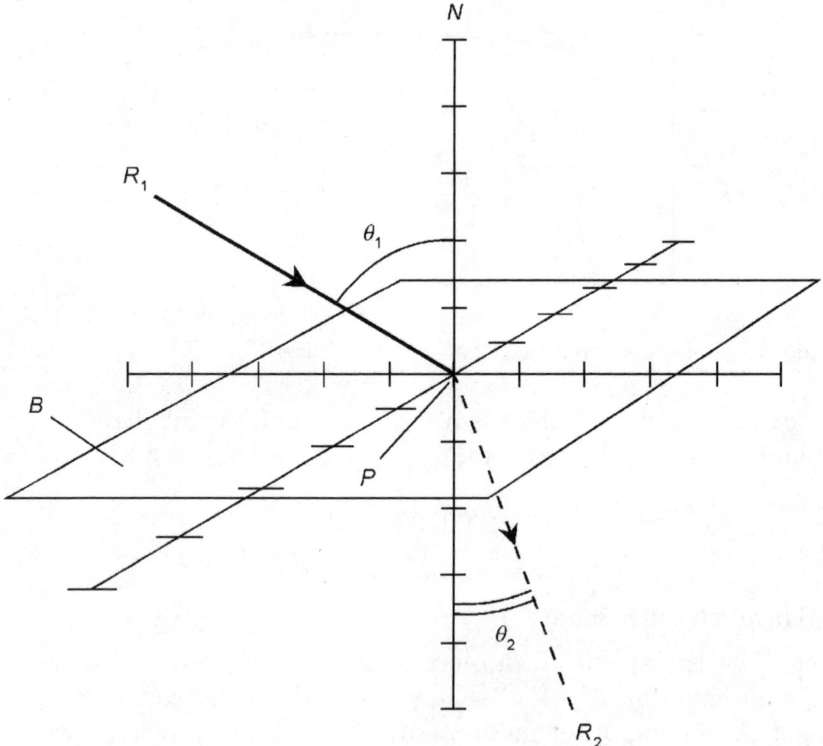

Figure 5.8 A ray passing from a medium with a relatively lower refractive index to a medium with a relatively higher refractive index.

passing through some point P on B, such that N is normal to B at P. Let R_1 be a ray of some electromagnetic disturbance D, traveling through M_1, that strikes B at P. Let θ_1 be the angle (in radians) that R_1 subtends relative to N at P. Let R_2 be the ray of D that emerges from P into M_2. Let θ_2 be the angle (in radians) that R_2 subtends relative to N at P. Then line N, ray R_1, and ray R_2 lie in a common plane, and $\theta_2 \leq \theta_1$. (The two angles are equal if and only if both measure zero; that is, if and only if R_1 coincides with N.) The following equation, known as *Snell's Law*, holds for D traveling through M_1 and M_2:

$$\sin \theta_2 / \sin \theta_1 = n_1 / n_2$$

Snell's Law for $n_2 < n_1$

Refer to Fig. 5.9. Let B be a flat boundary between two media M_1 and M_2 whose absolute indices of refraction are n_1 and n_2, respectively. Suppose that $n_2 < n_1$, that is, the ray passes from

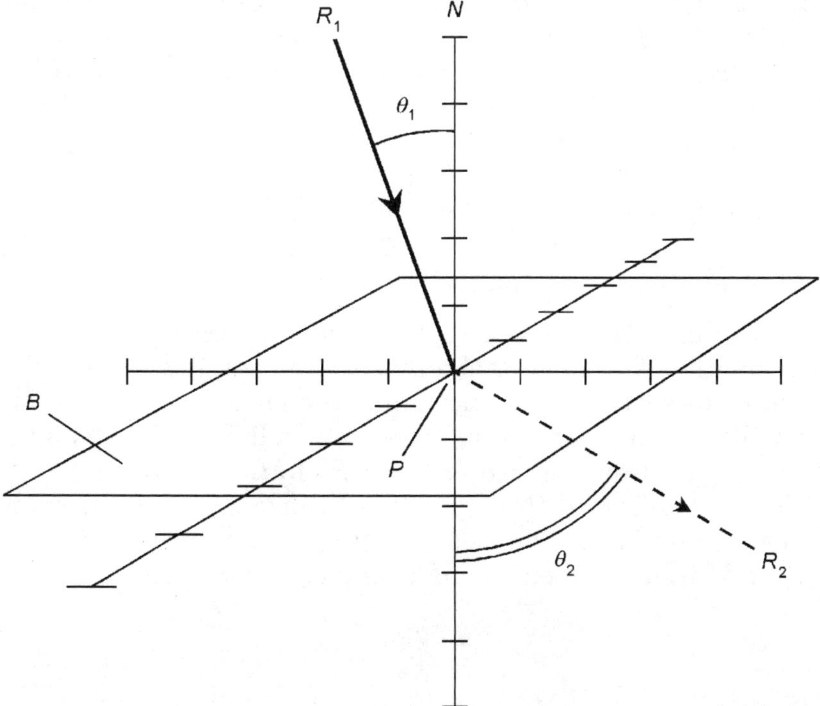

Figure 5.9 A ray passing from a medium with a relatively higher refractive index to a medium with a relatively lower refractive index.

a medium having a relatively higher refractive index to a medium having a relatively lower refractive index. Let N be a line passing through some point P on B, such that N is normal to B at P. Let R_1 be a ray of some electromagnetic disturbance D, traveling through M_1, that strikes B at P. Let θ_1 be the angle (in radians) that R_1 subtends relative to N at P. Let R_2 be the ray of D that emerges from P into M_2. Let θ_2 be the angle (in radians) that R_2 subtends relative to N at P. Then line N, ray R_1, and ray R_2 lie in a common plane, and $\theta_2 \geq \theta_1$. (The two angles are equal if and only if both measure zero; that is, if and only if R_1 coincides with N.) Snell's law holds in this case, just as in the situation described previously:

$$\sin \theta_2/\sin \theta_1 = n_1/n_2$$

Critical angle

Consider the situation depicted by Fig. 5.9. As θ_1 increases, $\theta_2 \rightarrow \pi/2$. In the case where $n_2 < n_1$, $\theta_2 = \pi/2$ for some $\theta_1 < \pi/2$. The *critical angle* for R_1, denoted θ_c, is the smallest θ_1 for which $\theta_2 = \pi/2$. When $\theta_c \leq \theta_1 < \pi/2$, B behaves as a reflective surface, and R_1 obeys the rule for reflection from a flat surface. The value of θ_c (in radians) is equal to the arcsine of the ratio of the indices of refraction:

$$\theta_c = \sin^{-1} (n_2/n_1)$$

Refraction through convex lens

Refer to Fig. 5.10. Let M be a symmetrical convex lens whose focal length is f (in meters). Let N be a line passing through the center of M, such that N is normal to the plane Q defined by M. Let X_1 be a point source of light located on line N, at a distance s_1 (in meters) from the center point P of M. Assume that $s_1 > f$. The light refracted through M will converge toward a point X_2 that is also located on line N. Let s_2 be the distance (in meters) of X_2 from P. Then the following equations hold:

$$s_2 = s_1 f/(s_1 - f)$$

$$s_1 = s_2 f/(s_2 - f)$$

Note that if $s_1 = f$, then point X_2 vanishes because the re-

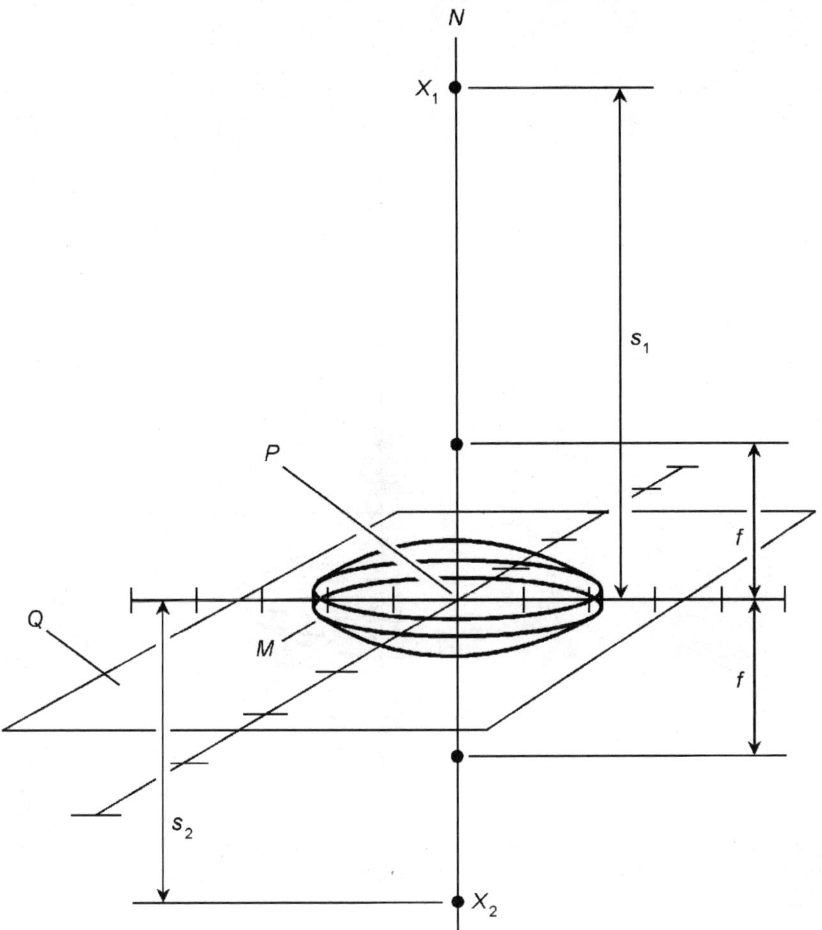

Figure 5.10 Refraction through a symmetrical convex lens when the distance from the source to the lens is greater than the focal length.

fracted rays are parallel. In addition, if point X_1 is located arbitrarily far away so that its arriving rays are parallel, then $s_2 = f$.

The above equations can be rearranged to obtain:

$$1/s_1 + 1/s_2 = 1/f$$

$$f = s_1 s_2/(s_1 + s_2)$$

If $s_1 < f$, then s_2 becomes negative; the light from X_1 will diverge from a point X_2 located on the same side of the lens as X_1. This is shown in Fig. 5.11.

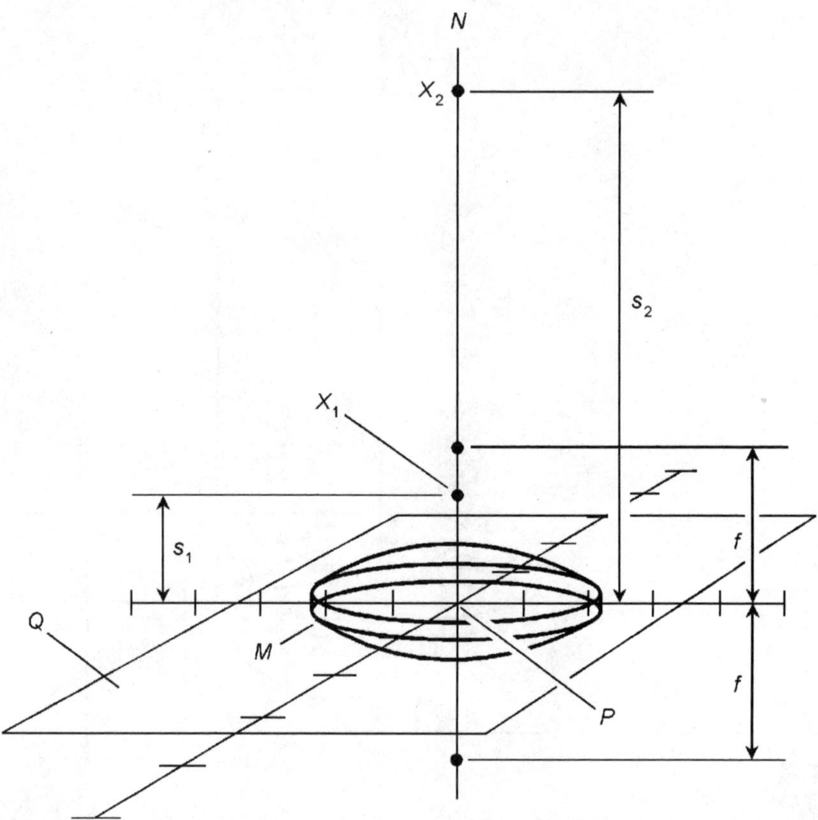

Figure 5.11 Refraction through a symmetrical convex lens when the distance from the source to the lens is less than the focal length.

Refraction through concave lens

Refer to Fig. 5.12. Let M be a symmetrical concave lens whose virtual focal length is f (in meters), and is defined as a negative length. Let N be a line passing through the center of M, such that N is normal to the plane Q defined by M. Let X_1 be a point source of light located on line N, at a distance s_1 (in meters) from the center point P of M. The light refracted through M will diverge from a point X_2 that is also located on line N. Let s_2 be the distance (in meters) of X_2 from P, and let s_2 be defined as a negative distance. Then the following equations hold:

$$1/s_1 + 1/s_2 = 1/f$$

$$f = s_1 s_2/(s_1 + s_2)$$

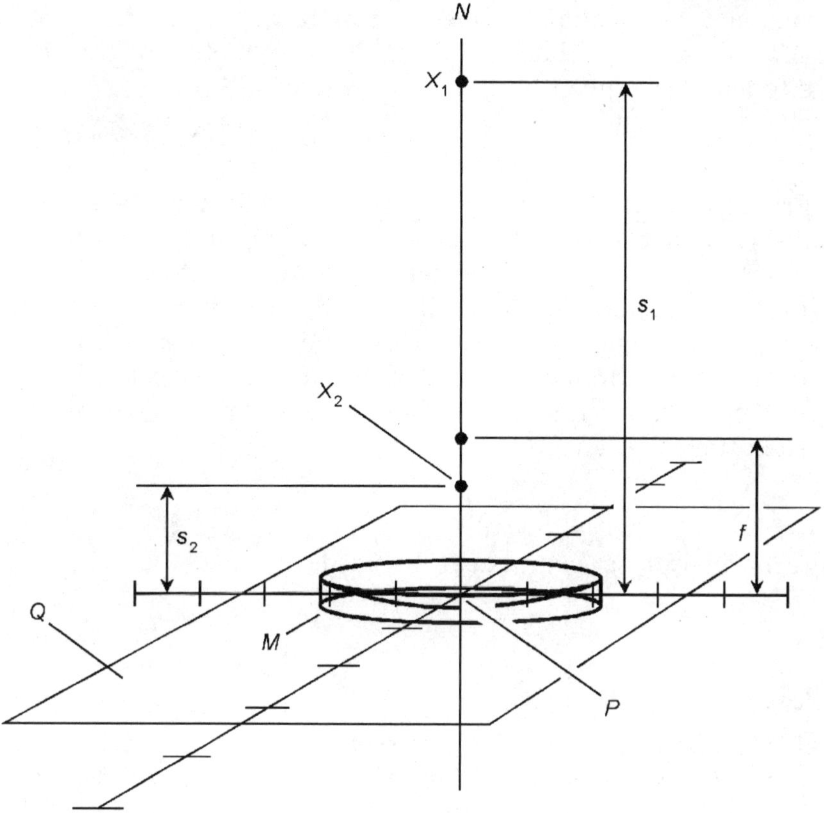

Figure 5.12 Refraction through a symmetrical concave lens.

Diopter of lens

Let M be a lens whose focal length (in meters) is f. The *diopter*, also called the *power*, of the lens (denoted δ and expressed per meter) is equal to the reciprocal of this length:

$$\delta = 1/f = f^{-1}$$

The diopter of a convex lens (also called magnifying or converging) is considered positive; the diopter of a concave (also called minimizing or diverging) lens is considered negative.

Composite lenses

Let M_1 and M_2 be symmetrical lenses in close proximity with their principal axes aligned. Either lens can be convex or concave. Let δ_1 be the diopter of M_1 (per meter); let δ_2 be the diopter of M_2 (per meter). Then the diopter δ (per meter) of the composite lens M is given by:

$$\delta = \delta_1 + \delta_2$$

From this formula, the analogous formula for focal length can be derived. Recall that diopter and focal length are inverses of each other when the displacement units are specified in meters. Let M_1 and M_2 be symmetrical lenses in close proximity with their principal axes aligned. Either lens can be convex or concave. Let f_1 be the focal length of M_1 (in meters); let f_2 be the focal length of M_2 (in meters). Then the focal length f (in meters) of the composite lens M is given by:

$$f = (f_1^{-1} + f_2^{-1})^{-1}$$

Alternatively, this formula can be used:

$$f = f_1 f_2/(f_1 + f_2)$$

Telescopic magnification

Refer to Fig. 5.13. Let T be a refracting telescope consisting of two convex lenses: an objective (M_1) and an eyepiece (M_2). Let f_1 be the focal length (in meters) of the objective lens; let f_2 be the focal length (in meters) of the eyepiece. In a magnifying telescope, $f_2 < f_1$. Assume that M_1 and M_2 are placed along a common principal axis, that the distance between M_1 and M_2 is adjusted for proper focus, and that the object(s) being observed

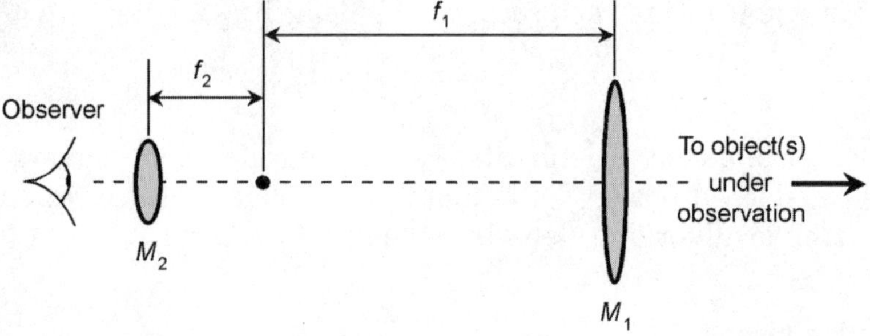

Figure 5.13 Determination of the magnification of a refracting telescope.

are arbitrarily distant. The *telescopic magnification* (a dimensionless quantity, denoted T_{mag} in this context) is given by:

$$T_{mag} = f_1/f_2$$

Alternatively, M_1 can be a concave mirror; in that case T is a reflecting telescope. Telescopic magnification is sometimes erroneously called "power."

Microscopic magnification

Refer to Fig. 5.14. Let U be a microscope consisting of two convex lenses: an objective (M_1) and an eyepiece (M_2). Let f_1 be the focal length (in meters) of the objective lens; let f_2 be the focal length (in meters) of the eyepiece. Assume that M_1 and M_2 are placed along a common principal axis, and that the distance between M_1 and M_2 is adjusted for proper focus. Let s_1 represent the distance (in meters) from the objective to the real image it

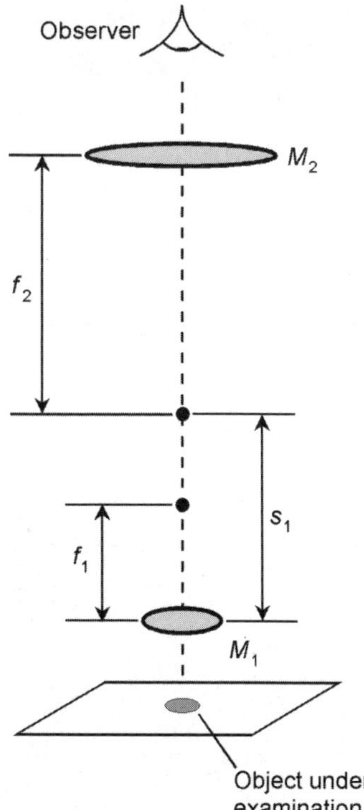

Figure 5.14 Determination of the magnification of a microscope.

forms of the object under examination. The *microscopic magnification* (a dimensionless quantity, denoted U_{mag} in this context) is given by:

$$U_{mag} = ((s_1 - f_1)/f_1)) ((f_2 + 0.25)/f_2)$$

The quantity 0.25 represents the nominal near point of the human eye, which is the closest distance over which the eye can focus on an object: approximately 0.25 meters. Microscopic magnification is sometimes erroneously called "power."

Diffraction through slit

Refer to Fig. 5.15. Suppose a beam of monochromatic light of wavelength λ (in nanometers) shines toward a barrier B at a right angle, and that the beam encounters a slit S of width w (in nanometers) in the barrier. The light diffracts through S. Let L represent the line of the direct beam, normal to B and passing through S. The waves in the diffracted beam cancel in phase, resulting in minimum brilliance, at angles θ_n relative to L, where n is a positive integer. These angles are given by:

$$\theta_n = \sin^{-1}(n\lambda/w)$$

Diffraction through grating

Refer to Fig. 5.16. Suppose a beam of monochromatic light of wavelength λ (in nanometers) shines toward a grating G at a

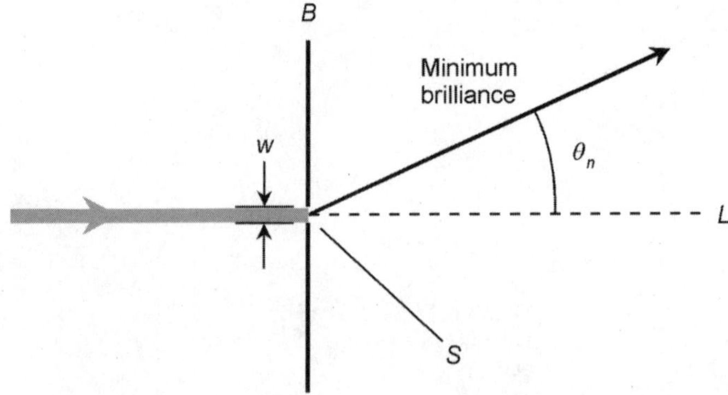

Figure 5.15 Diffraction of monochromatic light through a single slit.

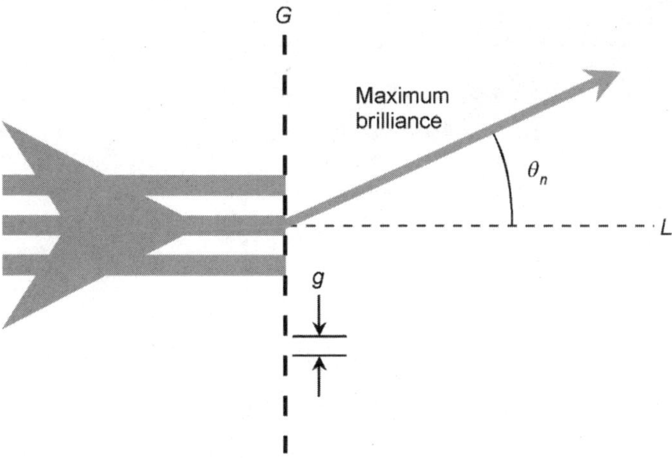

Figure 5.16 Diffraction of monochromatic light through a grating.

right angle. Suppose G consists of slits of uniform width and spacing, and that adjacent slits are separated by distance g (in nanometers). The light will diffract through G, forming a pattern of light and dark beams. Let L represent the line of the direct beam, normal to G. The waves in the diffracted beam will add in phase, resulting in maximum brilliance, at certain angles θ_n relative to L, where n is a positive integer. These angles can be determined according to the following equation:

$$\theta_n = \sin^{-1}(n\lambda/g)$$

Relativistic and Atomic Physics

This section deals with the special theory of relativity, quantum physics, electron behavior, nuclear physics, and radioactivity.

Relativistic addition of speeds

Refer to Fig. 5.17. Suppose two objects X and Y are moving in the same direction along a straight line L with respect to an observer Z. Suppose X moves at constant speed v_X relative to Z; suppose Y moves at constant speed v_Y (in the same units as v_X) relative to X, in the same direction that X moves relative to Z. Let c be the speed of light in a vacuum, expressed in the same units as v_X and v_Y. The speed of Y as measured along L relative

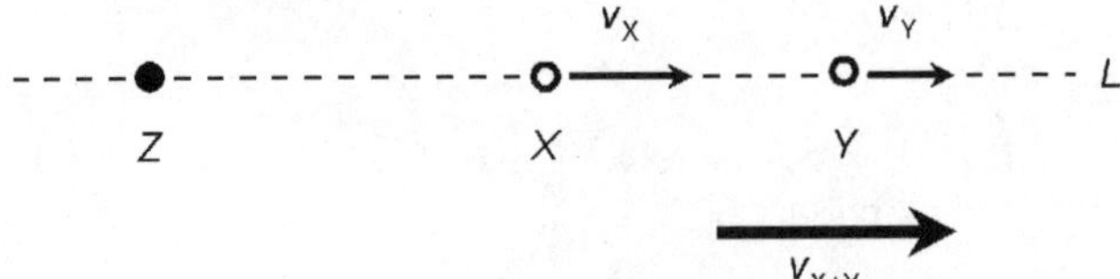

Figure 5.17 Relativistic addition of velocities. The composite speed, v_{X+Y}, can never exceed the speed of light in a vacuum.

to Z (denoted v_{X+Y} and expressed in the same units as v_X, v_Y, and c), is given by:

$$v_{X+Y} = (v_X + v_Y)/(1 + v_X v_Y/c^2)$$

If the speeds are all expressed in kilometers per second, then $c \approx 2.99792 \times 10^5$, and $c^2 \approx 8.98752 \times 10^{10}$.

Relativistic dilation of time

Refer to Fig. 5.18. Suppose an object X moves at constant velocity v along a line L relative to an observer Z. Suppose X carries a precision clock K_X that Z can constantly monitor. Suppose Z has a clock K_Z identical to K_X, and that K_X and K_Z are synchronized at a specific time so both clocks read zero: $t_X = 0$ and $t_Z = 0$. Let c be the speed of light in a vacuum, expressed

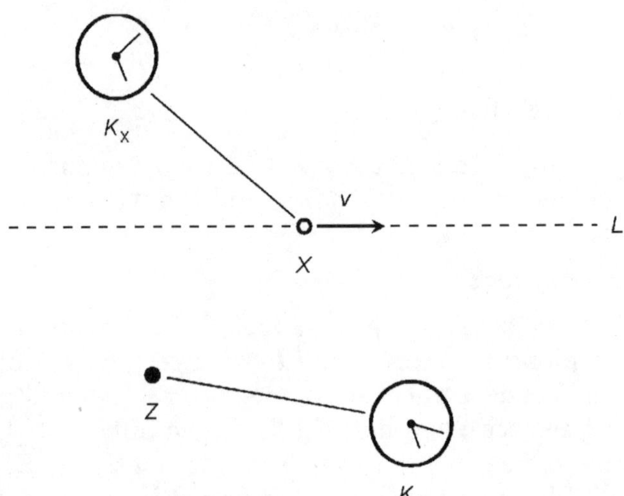

Figure 5.18 Relativistic time dilation. The "moving" clock appears to run more slowly than the "stationary" clock.

in the same units as v. Neglect the effects of variable observation-path delay between X and Z. Suppose Z observes the passage of t_Z seconds. At the time Z observes K_Z to read t_Z seconds, Z will observe K_X to read t_X seconds, where:

$$t_X = t_Z (1 - v^2/c^2)^{1/2}$$

If the speeds are all expressed in kilometers per second, then $c \approx 2.99792 \times 10^5$, and $c^2 \approx 8.98752 \times 10^{10}$.

Twin Paradox for relativistic time dilation

The formula in the preceding paragraph applies equally well when X and Z (the observer's and the object's reference frames) are transposed, even if all other considerations remain unchanged. This gives rise to the so-called *twin paradox*, in which clock K_X lags clock K_Z from the reference frame of Z, and K_Z lags K_X from the reference frame of X. This absurdity is overcome in the general theory of relativity, in which time dilation is considered to result from acceleration or the presence of a gravitational field, rather than from relative motion.

Relativistic spatial contraction

Refer to Fig. 5.19. Suppose an object X moves at constant velocity v (in meters per second) along a line L relative to an observer Z. Suppose X has length s_{X1} (in meters) at rest, measured in the dimension parallel to L. Let c be the speed of light in a vacuum (in meters per second). Suppose Z observes the length

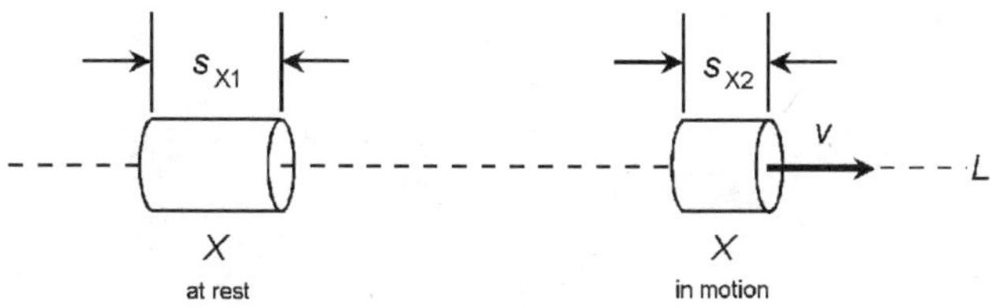

Figure 5.19 Relativistic spatial contraction occurs along an axis parallel to the velocity vector (instantaneous direction of motion).

of X to be s_{X2} (in meters, in the dimension of L) when X is in motion at velocity v. The following equation holds:

$$s_{X2} = s_{X1} (1 - v^2/c^2)^{1/2}$$

If the speeds are all expressed in meters per second, then $c \approx 2.99792 \times 10^8$, and $c^2 \approx 8.98752 \times 10^{16}$.

Relativistic mass increase

Suppose an object X moves at constant velocity v along a line L relative to an observer Z. Suppose X has mass m_{X1} at rest. Let c be the speed of light in a vacuum, expressed in the same units as v. Suppose Z observes the mass of X to be m_{X2} (in the same units as m_{X1}) when X is in motion at velocity v. The following equation holds:

$$m_{X2} = m_{X1}/(1 - v^2/c^2)^{1/2}$$

If the speeds are all expressed in kilometers per second, then $c \approx 2.99792 \times 10^5$, and $c^2 \approx 8.98752 \times 10^{10}$.

Relativistic linear momentum

Suppose an object X moves with constant velocity vector $\mathbf{v}$ (in meters per second in a specified direction) along a line L relative to an observer Z. Suppose X has mass m (in kilograms) at rest. Let c be the speed of light in a vacuum (in meters per second). Then Z will observe the *relativistic linear momentum* vector, $\mathbf{p}$ (whose magnitude is expressed in kilogram-meters per second, and whose direction is the same as that of $\mathbf{v}$), of X as follows:

$$\mathbf{p} = \mathbf{v}\ (m/(1 - |\mathbf{v}|^2/c^2)^{1/2})$$

If the speeds are all expressed in meters per second, then $c \approx 2.99792 \times 10^8$, and $c^2 \approx 8.98752 \times 10^{16}$.

Relativistic energy

Suppose an object X moves with constant velocity v (in meters per second) along a line L relative to an observer Z. Suppose X has mass m (in kilograms) at rest. Let c be the speed of light in a vacuum (in meters per second). Then Z will observe the *relativistic energy*, E (in joules), of X as follows:

$$E = mc^2/(1 - v^2/c^2)^{1/2}$$

If the speeds are all expressed in meters per second, then $c \approx 2.99792 \times 10^8$, and $c^2 \approx 8.98752 \times 10^{16}$.

Photon energy

Let f be the frequency of an electromagnetic wave disturbance (in hertz). Let λ be the wavelength (in meters); let c be the speed of propagation (in meters per second). The *photon energy* (denoted E_p and expressed in joules) contained in a single quantum of that wave is given by the following formulas:

$$E_p = hf$$

$$E_p = hc/\lambda$$

where h is *Planck's constant*, approximately equal to 6.6261×10^{-34} joule-seconds, and $c \approx 2.99792 \times 10^8$ meters per second.

Photon momentum

Let f be the frequency of an electromagnetic wave disturbance (in hertz). Let λ be the wavelength (in meters); let c be the speed of propagation (in meters per second). The *photon momentum* (denoted p_p and expressed in kilogram-meters per second) contained in a single quantum of that wave is given by the following formulas:

$$p_p = hf/c$$

$$p_p = h/\lambda$$

where $h \approx 6.6261 \times 10^{-34}$ joule-seconds, and $c \approx 2.99792 \times 10^8$ meters per second.

de Broglie wavelength

Let m be the mass (in kilograms) of a particle P moving at velocity v (in meters per second). Neglecting relativistic effects, the *de Broglie wavelength* (denoted λ_d and expressed in meters) of P is given by:

$$\lambda_d = h/mv$$

where $h \approx 6.6261 \times 10^{-34}$ joule-seconds. When relativistic effects are taken into account, the formula becomes:

$$\lambda_d = (h/(mv))(1 - v^2/c^2)^{1/2}$$

If the speeds are all expressed in meters per second, then $c \approx 2.99792 \times 10^8$, and $c^2 \approx 8.98752 \times 10^{16}$.

Compton effect

Suppose a photon P, having wavelength λ_1 (in meters), collides with an electron E. Suppose E is free and stationary, and has a rest mass m_e (in kilograms). Further suppose that when P strikes E, the direction of P is changed by an angle ϕ (in radians). Then the wavelength of P will change to a new value λ_2, such $\lambda_2 > \lambda_1$. The scattered-photon wavelength λ_2 (in meters) depends on the deflection angle ϕ, and is given by:

$$\lambda_2 = \lambda_1 + (h - h \cos \phi)/m_e c,$$

where $h \approx 6.6261 \times 10^{-34}$ joule-seconds and $c \approx 2.99792 \times 10^8$ meters per second. The rest mass of the electron, m_e, is approximately equal to 9.109×10^{-31} kilograms.

Balmer series

Suppose a sample of hydrogen gas is excited so its atoms emit photons. A set of emission lines known as the *Balmer series* occurs at wavelengths (in meters) defined by the following formula:

$$\lambda_n = (4n^2 + 16n + 16)/(n^2R + 4nR)$$

where $n \in \{1, 2, 3, \ldots\}$ and R is the *Rydberg constant,* approximately equal to 1.097×10^7 and is expressed per meter.

Lyman series

Suppose a sample of hydrogen gas is excited so its atoms emit photons. A set of emission lines known as the *Lyman series* occurs at wavelengths (in meters) defined by the following formula:

$$\lambda_n = (n^2 + 2n + 1)/(n^2R + 2nR)$$

where $n \in \{1, 2, 3, \ldots\}$ and R is the Rydberg constant, approximately equal to 1.097×10^7 and is expressed per meter.

Paschen series

Suppose a sample of hydrogen gas is excited so its atoms emit photons. A set of emission lines known as the *Paschen series* occurs at wavelengths (in meters) defined by the following formula:

$$\lambda_n = (9n^2 + 54n + 81)/(n^2R + 6nR)$$

where $n \in \{1, 2, 3, \ldots\}$ and R is the Rydberg constant, approximately equal to 1.097×10^7 and is expressed per meter.

Principal quantum numbers

In a hydrogen atom, the energy contained in the electron is defined in terms of *principal quantum numbers* (positive integers denoted n). The energy level e_k (in electronvolts) corresponding to a specific principal quantum number $n = k$ is given by the following formula:

$$e_k = -13.6/(k^2)$$

where $k \in \{1, 2, 3, \ldots\}$.

Orbital quantum numbers

In a hydrogen atom, the angular momentum contained in the electron is defined in terms of *orbital quantum numbers* (positive integers denoted l). The angular momentum p_k (in electronvolts) corresponding to a specific orbital quantum number $l = k$ and principal quantum number n is given by the following formula:

$$p_k = h\,(k^2 - k)^{1/2}/(2\pi)$$

where $k \in \{1, 2, 3, \ldots, (n - 1)\}$ and h is *Planck's constant*, approximately equal to 6.6261×10^{-34} joule-seconds.

Radioactive decay

Radioactive decay is measured in terms of *half life*, denoted t and expressed in time units that depend on the speed with which the substance in question decays (seconds, minutes, hours, days, or years). The half life is the length of time, as measured from a defined starting point t_0, after which exactly half (50 percent) of the radioactive atoms in the sample are unaltered. Half life is independent of t_0.

Suppose a radioactive sample has a half life of t time units. Suppose the substance is tested for radioactivity at a specific time, and then is tested again x time units later. The proportion of radioactive atoms $P_{\%}$ that are unaltered after the passage of x time units is given by the following formula:

$$P = 1/2^{(x/t)}$$

Radioactive decay can also be expressed in terms of the percentage of atoms that remain unaltered. Suppose a radioactive sample has a half life of t time units. Suppose the substance is tested for radioactivity at a specific time, and then is tested again x time units later. The percentage of radioactive atoms $P_{\%}$ that are unaltered after the passage of x time units is given by the following formula:

$$P_{\%} = 100/2^{(x/t)}$$

Ionizing-radiation dose

Suppose a sample of material has mass m (in kilograms) and absorbs an amount of ionizing-radiation energy e (in joules). The *ionizing-radiation dose* (denoted D and expressed in grays) is given by:

$$D = e/m$$

A dose of one gray is equivalent to one joule per kilogram of absorbed radiation.

Chemical Elements

The following paragraphs depict the chemical elements in order by atomic number, and describe some of their characteristics and practical uses.

1. Hydrogen

Symbol, H. A colorless, odorless gas at room temperature. The most common isotope has atomic weight 1.00794. The lightest and most abundant element in the universe. Used in the semiconductor manufacturing process, in glow-discharge lamps, in the refining processes of certain metals, and in some lighter-than-air balloons. Highly flammable; combines readily with oxygen to form water (H_2O). In recent years, this element has been the subject of research as a source of energy for use in internal combustion engines, furnaces, and nuclear fusion reactors. The isotope having one neutron in the nucleus is called *deuterium*. The isotope having two neutrons in the nucleus is called *tritium*.

- Electrons in first energy level: 1

2. Helium

Symbol, He. A colorless, odorless gas at room temperature. The most common isotope has atomic weight 4.0026. The second lightest and second most abundant element in the universe. Product of hydrogen fusion. Used in helium-neon (He-Ne) lasers, which produce visible red coherent light. Also used in place of nitrogen in self-contained underwater breathing apparatus (SCUBA) diving equipment. Used in medical equipment, especially in magnetic-resonance imaging (MRI). Liquid helium does not solidify, even at temperatures arbitrarily close to absolute zero.

- Electrons in first energy level: 2

3. Lithium

Symbol, Li. Classified as an alkali metal. In pure form it is silver-colored. The lightest elemental metal. The most common isotope has atomic weight 6.941. Used in certain types of lubricants, as an alloying agent, in ceramics, and in electrochemical cells known for long life in low-current applications. Also used in the synthesis of organic compounds. This element reacts rapidly with oxygen at standard atmospheric temperature and pressure. It is not found free in nature.

- Electrons in first energy level: 2
- Electrons in second energy level: 1

4. Beryllium

Symbol, Be. Classified as an alkaline earth. In pure form it has a grayish color similar to that of steel. Has a relatively high melting point. The most common isotope has atomic weight 9.01218. Found in various dielectrics, alloys, and phosphors. The oxide of this element is used as an electrical insulator. The metal is used in aircraft, spacecraft, and high-speed, high-altitude missiles. In crude form, the mineral containing this element is known as *beryl*.

- Electrons in first energy level: 2
- Electrons in second energy level: 2

5. Boron

Symbol, B. Classified as a metalloid. The most common isotope has atomic weight 10.82. Can exist as a powder or as a black, hard metalloid. The metalloid form is a poor electrical conductor, and is heat-resistant. Used in nuclear reactors, in the manufacture of electronic semiconductor devices, and in various industrial applications, especially insulating materials and bleach. In crude form, the mineral containing this element is known as *kernite*. Boron is not found free in nature.

- Electrons in first energy level: 2
- Electrons in second energy level: 3

6. Carbon

Symbol, C. A non-metallic element that is a solid at room temperature. Has a characteristic hexagonal crystal structure. Known as the basis of life on Earth. The most common isotope has atomic weight 12.011. Exists in three well-known forms: *graphite* (a black powder) which is common, *diamond* (a clear solid) which is rare, and *amorphous*. Occasionally, *white carbon* is observed along with graphite. Used in electrochemical cells, air-cleaning filters, thermocouples, and noninductive electrical resistors. Also used in medicine to absorb poisons and toxins in the stomach and intestines. Abundant in mineral rocks such as limestone. Carbon dioxide is found in the atmosphere of the earth, and plays a role in planetary heat retention.

- Electrons in first energy level: 2
- Electrons in second energy level: 4

7. Nitrogen

Symbol, N. A non-metallic element that is a colorless, odorless gas at room temperature. The most common isotope has atomic weight 14.007. The most abundant component of the earth's atmosphere (approximately 78 percent at the surface). Reacts to some extent with certain combinations of other elements. In its liquid form, this element is used in medicine to freeze skin lesions for removal. It is also important as a component of proteins, and is used to make ammonia. Nitrogen is used in the manufacture of semiconductors (it provides a stable environment) and also by the pharmaceutical industry.

- Electrons in first energy level: 2
- Electrons in second energy level: 5

8. Oxygen

Symbol, O. A non-metallic element that is a colorless, odorless gas at room temperature. The most common isotope has atomic weight 15.999. The second most abundant component of the earth's atmosphere (approximately 21 percent at the surface). Combines readily with many other elements, particularly metals. One of the oxides of iron, for example, is known as common rust. Normally, two atoms of oxygen combine to form a molecule (O_2). In this form, oxygen is essential for the sustenance of many forms of life on Earth. When three oxygen atoms form a molecule (O_3), the element is called *ozone*. This form of the element is beneficial in the upper atmosphere, because it reduces the amount of ultraviolet radiation reaching the earth's surface. Ozone is, ironically, also known as an irritant and pollutant in the surface air over heavily populated areas.

- Electrons in first energy level: 2
- Electrons in second energy level: 6

9. Fluorine

Symbol, F. The most common isotope has atomic weight 18.998. A gaseous element of the halogen family. Has a characteristic greenish or yellowish color. Reacts readily with many other elements. Fluorine compounds are used in some refrigerant and propellant chemicals. Human-synthesized chlorofluorocarbons have been implicated as a potential aggravating factor in the depletion of the ozone layer in the earth's upper atmosphere. Combined with certain other elements such as sodium, this element forms compounds that are known to retard or prevent tooth decay in humans. This element has been considered for use as a rocket propellant. It is used in the production of plastics.

- Electrons in first energy level: 2
- Electrons in second energy level: 7

10. Neon

Symbol, Ne. The most common isotope has atomic weight 20.179. A noble gas present in trace amounts in the atmosphere. Used in specialized low-frequency oscillators, flip-flops, voltage regulators, indicator lamps, and alphanumeric displays. Found in certain types of lamps used in special-effect lighting. Also used in lightning arrestors, cathode-ray-tube (CRT) displays, and radio-frequency-measuring devices (wavemeters). Neon lamps produce a characteristic red-orange glow. Neon is used in helium-neon (He-Ne) lasers, which produce a brilliant crimson visible-light output. Neon is obtained commercially by fractional distillation when air is liquefied.

- Electrons in first energy level: 2
- Electrons in second energy level: 8

11. Sodium

Symbol, Na. The most common isotope has atomic weight 22.9898. An element of the alkali-metal group. A solid at room temperature. Used in gas-discharge lamps. Has a characteristic candle-flame-colored glow when it fluoresces. Used in industrial chemicals; combined with chlorine, it forms common table salt. Various compounds of this element are used in the food industry for flavoring and for protection against spoilage. Reacts readily with various other elements. In its pure form, it must be protected from direct contact with the atmosphere, or it will corrode almost instantly.

- Electrons in first energy level: 2
- Electrons in second energy level: 8
- Electrons in third energy level: 1

12. Magnesium

Symbol, Mg. The most common isotope has atomic weight 24.305. A member of the alkaline earth group. At room temperature it is a whitish metal. Compounds of this element are used as phosphors for some types of CRT. Also mixed with aluminum to form strong alloys used in various structures such as antennas and antenna support towers. In medicine, oxides and sulfides of this element are used as antacids or laxatives. Reacts readily with various other elements. When heated in the atmosphere, it can burn with a brilliant white glow.

- Electrons in first energy level: 2
- Electrons in second energy level: 8
- Electrons in third energy level: 2

13. Aluminum

Symbol, Al. The most common isotope has atomic weight 26.98. A metallic element and a good electrical conductor. Has many of the same characteristics as magnesium, except it reacts less easily with oxygen in the atmosphere. Used as chassis in electronic equipment, in wireless-communications hardware, and in electrical wiring. Mixed with magnesium to form strong alloys used in various structures such as antennas and antenna support towers. In medicine, oxides of this element are used as antacids. Serves as the electrode material in electrolytic capacitors. Can be doped to form a semiconductor, or oxidized to form a dielectric.

- Electrons in first energy level: 2
- Electrons in second energy level: 8
- Electrons in third energy level: 3

14. Silicon

Symbol, Si. The most common isotope has atomic weight 28.086. A metalloid abundant in the earth's crust. Especially common in rocks such as granite, and in many types of sand. Used in the manufacture of glass. Processed to form semiconductor materials extensively used in electronic and computer systems. This element, and various compounds of it, are used in diodes, transistors, and integrated circuits (ICs). Can be oxidized to form a dielectric.

- Electrons in first energy level: 2
- Electrons in second energy level: 8
- Electrons in third energy level: 4

15. Phosphorus

Symbol, P. The most common isotope has atomic weight 30.974. A nonmetallic element of the nitrogen family. Found in certain types of rock. Used in the manufacture of various fertilizers. Also used in cleaning detergents. Commonly employed as a dopant in semiconductor manufacture, and as an alloy constituent in some electrical and electronic components.

- Electrons in first energy level: 2
- Electrons in second energy level: 8
- Electrons in third energy level: 5

16. Sulfur

Symbol, S. Also spelled *sulphur*. The most common isotope has atomic weight 32.06. A nonmetallic element. Reacts with some other elements. Compounds of sulfur are used in rechargeable electrochemical cells and in power transformers. Well known as a component in gunpowder; also used in the manufacture of matches and fireworks. Some compounds of this element are used in medicine; an example is magnesium sulfate, also known as Epsom salt.

- Electrons in first energy level: 2
- Electrons in second energy level: 8
- Electrons in third energy level: 6

17. Chlorine

Symbol, Cl. The most common isotope has atomic weight 35.453. A gas at room temperature, and a member of the halogen family. Reacts readily with various other elements. Potentially dangerous; displaces oxygen in the blood if inhaled. Liquefied compounds of this element are used industrially as bleaching agents. Because it kills bacteria, chlorine is sometimes used to render water safe to drink. Has a characteristic smell, similar to that of ozone. Chlorine is a component of common table salt (sodium chloride) and also of salt substitutes (potassium chloride).

- Electrons in first energy level: 2
- Electrons in second energy level: 8
- Electrons in third energy level: 7

18. Argon

Symbol, A or Ar. The most common isotope has atomic weight 39.94. A gas at room temperature; classified as a noble gas. Present in small amounts in the atmosphere. Used in some types of lasers and glow lamps. In general, this element does not react readily with others.

- Electrons in first energy level: 2
- Electrons in second energy level: 8
- Electrons in third energy level: 8

19. Potassium

Symbol, K. The most common isotope has atomic weight 39.098. A member of the alkali metal group. Compounds of this element are used in CRT phosphor coatings, in electroplating, and as ferroelectric substances. Also used in the manufacture of glass, and in certain detergents. Combines with chlorine to form a common table-salt substitute (potassium chloride). Obtained in nature from the mineral carnallite.

- Electrons in first energy level: 2
- Electrons in second energy level: 8
- Electrons in third energy level: 8
- Electrons in fourth energy level: 1

20. Calcium

Symbol, Ca. The most common isotope has atomic weight 40.08. A metallic element of the alkaline-earth group. Calcium carbonate, or calcite, is abundant in the earth's crust, especially in limestone. Calcium is also an important constituent of bones in vertebrate animals. Compounds of this element are used as phosphor coatings in CRTs; various display colors can be obtained by using different compounds of calcium. Calcium is also used in cement and in the glass-manufacturing industry.

- Electrons in first energy level: 2
- Electrons in second energy level: 8
- Electrons in third energy level: 8
- Electrons in fourth energy level: 2

21. Scandium

Symbol, Sc. The most common isotope has atomic weight 44.956. In pure form, it is a soft metal. Classified as a transition metal. This element is used along with aluminum in the manufacture of alloys for use in sporting equipment such as bicycle frames and baseball bats. Also used in the aerospace industry.

- Electrons in first energy level: 2
- Electrons in second energy level: 8
- Electrons in third energy level: 9
- Electrons in fourth energy level: 2

22. Titanium

Symbol, Ti. The most common isotope has atomic weight 47.88. Classified as a transition metal. Well known for its mechanical strength and resistance to corrosion. Used in a wide variety of applications, including aircraft frames, jet engines, antenna towers, and building supports. Certain compounds of this element, especially oxides, can be used as dielectric materials.

- Electrons in first energy level: 2
- Electrons in second energy level: 8
- Electrons in third energy level: 10
- Electrons in fourth energy level: 2

23. Vanadium

Symbol, V. The most common isotope has atomic weight 50.94. Classified as a transition metal. In its pure form it is whitish in color. Used in the manufacture of specialized industrial alloys. Also used in certain types of dyes. Can serve as a catalyst in various chemical reactions. Used in lithium-ion electrochemical cells and batteries, and in superconducting magnets. This element is toxic and must be handled with care.

- Electrons in first energy level: 2
- Electrons in second energy level: 8
- Electrons in third energy level: 11
- Electrons in fourth energy level: 2

24. Chromium

Symbol, Cr. The most common isotope has atomic weight 51.996. Classified as a transition metal. In pure form, it is grayish in color. Used as a plating for metals to improve resistance to corrosion. Also used in the manufacture of stainless steel. Compounds of this element are used in specialized recording tapes and in thermocouple devices.

- Electrons in first energy level: 2
- Electrons in second energy level: 8

- Electrons in third energy level: 13
- Electrons in fourth energy level: 1

25. Manganese

Symbol, Mn. The most common isotope has atomic weight 54.938. Classified as a transition metal. In pure form it is grayish. An alloy of manganese is used in the manufacture of permanent magnets. Some batteries contain compounds of this element. Also used in the manufacture of steel and certain ceramics.

- Electrons in first energy level: 2
- Electrons in second energy level: 8
- Electrons in third energy level: 13
- Electrons in fourth energy level: 2

26. Iron

Symbol, Fe. The most common isotope has atomic weight 55.847. In pure form, it is a dull gray metal. Well known for its magnetic properties. Found in abundance in the earth's crust. High-grade iron ore is known as hematite; a lower grade is taconite. Used in magnetic circuits and transformer cores. Also used in thermocouples. An alloy of this element (steel) is used extensively in buildings, bridges, and antenna support towers.

- Electrons in first energy level: 2
- Electrons in second energy level: 8
- Electrons in third energy level: 14
- Electrons in fourth energy level: 2

27. Cobalt

Symbol, Co. The most common isotope has atomic weight 58.94. Classified as a transition metal. In pure form it is silvery in color. Used in the manufacture of stainless steel. Also used in making certain types of ceramics and glass. The isotope Co-60 is radioactive and has medical applications in radiology. Cobalt chloride, a salt, is sometimes used to absorb moisture in packing materials.

- Electrons in first energy level: 2
- Electrons in second energy level: 8
- Electrons in third energy level: 15
- Electrons in fourth energy level: 2

28. Nickel

Symbol, Ni. The most common isotope has atomic weight 58.69. Classified as a transition metal. In its pure form it is light gray to white. Nickel-cadmium (NiCd) and nickel-metal-hydride (NiMH) are used in rechargeable electrochemical cells, especially the types used in portable electronic equipment and notebook computers. Nickel compounds are used in the manufacture of specialized semiconductor diodes. Alloys of this element are used as resistance wire. Elemental nickel is employed in some electron tubes. This element is also used in electroplating.

- Electrons in first energy level: 2
- Electrons in second energy level: 8
- Electrons in third energy level: 16
- Electrons in fourth energy level: 2

29. Copper

Symbol, Cu. The most common isotope has atomic weight 63.546. Classified as a transition metal. In pure form, it has a characteristic red or wine color. An excellent conductor of electricity and heat. Used in the manufacture of wires and cables. Oxides of this element are used in specialized diodes and photoelectric cells. Also used in some plumbing systems, and in the manufacture of coins.

- Electrons in first energy level: 2
- Electrons in second energy level: 8
- Electrons in third energy level: 18
- Electrons in fourth energy level: 1

30. Zinc

Symbol, Zn. The most common isotope has atomic weight 65.39. Classified as a transition metal. In pure form, it is a dull blue-gray color. Used as the negative-electrode material in electrochemical cells, and as a protective coating for metals, especially iron and steel. Zinc is a component of brass and bronze alloys. Certain zinc compounds are used as CRT phosphors. The oxide of this element is used to protect skin and help heal superficial skin injuries. Zinc can be immersed in strong acids to liberate hydrogen gas.

- Electrons in first energy level: 2
- Electrons in second energy level: 8
- Electrons in third energy level: 18
- Electrons in fourth energy level: 2

31. Gallium

Symbol, Ga. The most common isotope has atomic weight 69.72. A semiconducting metal. In pure form it is light gray to white. Forms a compound with arsenic (GaAs) that is used in low-noise, high-gain field-effect transistors (FETs), and also in specialized diodes. In nature, this element is commonly found with aluminum in the mineral *bauxite*.

- Electrons in first energy level: 2
- Electrons in second energy level: 8
- Electrons in third energy level: 18
- Electrons in fourth energy level: 3

32. Germanium

Symbol, Ge. The most common isotope has atomic weight 72.59. A semiconducting metalloid. Used in specialized diodes, transistors, rectifiers, and photoelectric cells. In semiconductor components, germanium has been largely replaced by silicon. This is mainly because silicon and its compounds are more resistant to heat than germanium and its compounds. Elemental germanium is obtained as a byproduct in the refining of other metals, particularly zinc and copper.

- Electrons in first energy level: 2
- Electrons in second energy level: 8
- Electrons in third energy level: 18
- Electrons in fourth energy level: 4

33. Arsenic

Symbol, As. The most common isotope has atomic weight 74.91. A metalloid used as a dopant in the manufacture of semiconductors. In its pure form, it is gray in color. Forms a compound with gallium (GaAs) known as gallium arsenide, which is used in low-noise, high-gain FETs, and also in specialized diodes. The element is noted for its toxicity.

- Electrons in first energy level: 2
- Electrons in second energy level: 8
- Electrons in third energy level: 18
- Electrons in fourth energy level: 5

34. Selenium

Symbol, Se. The most common isotope has atomic weight 78.96. Classified as a non-metal. In its pure form, it is gray in color. A semiconducting element used in diodes, rectifiers. Recognized for its photoconductive properties (the electrical conductivity varies depending on the intensity of visible light, infrared, and/or ultraviolet radiation that strikes the surface). Used in some types of photoelectric cells and video cameras.

- Electrons in first energy level: 2
- Electrons in second energy level: 8
- Electrons in third energy level: 18
- Electrons in fourth energy level: 6

35. Bromine

Symbol, Br. The most common isotope has atomic weight 79.90. A nonmetallic element of the halogen family. A reddish-brown liquid at room temperature. Has a characteristic unpleasant odor. Reacts readily with various other elements. Known as a poison. Obtained from sea water, where it occurs as a component of bromide salts. Was used at one time as a disinfectant in swimming pools; chlorine and ozone are favored for that purpose today.

- Electrons in first energy level: 2
- Electrons in second energy level: 8
- Electrons in third energy level: 18
- Electrons in fourth energy level: 7

36. Krypton

Symbol, Kr. The most common isotope has atomic weight 83.80. Classified as a noble gas. Colorless and odorless. Present in trace amounts in the earth's atmosphere. Some common isotopes of this element are radioactive. Used in specialized electric lamps; in this respect it is similar to neon and xenon gases. This element is liberated as a byproduct when air is liquefied.

- Electrons in first energy level: 2
- Electrons in second energy level: 8
- Electrons in third energy level: 18
- Electrons in fourth energy level: 8

37. Rubidium

Symbol, Rb. The most common isotope has atomic weight 85.468. Classified as an alkali metal. In pure form, it is silver-colored. Reacts easily with oxygen and chlorine. Used in certain types of photoelectric cells. Acts as a catalyst in some chemical reactions.

- Electrons in first energy level: 2
- Electrons in second energy level: 8
- Electrons in third energy level: 18
- Electrons in fourth energy level: 8
- Electrons in fifth energy level: 1

38. Strontium

Symbol, Sr. The most common isotope has atomic weight 87.62. A metallic element of the alkaline-earth group. In pure form it is gold-colored. Compounds of this element are used in the manufacture of specialized ceramic dielectrics. When used in fireworks, this element produces brilliant colors. One isotope of this element, Sr-90, is radioactive. Fallout containing Sr-90 was a cause for concern during above-ground nuclear-weapons tests carried out in the 1950s and 1960s.

- Electrons in first energy level: 2
- Electrons in second energy level: 8
- Electrons in third energy level: 18
- Electrons in fourth energy level: 8
- Electrons in fifth energy level: 2

39. Yttrium

Symbol, Y. The most common isotope has atomic weight 88.906. Classified as a transition metal. In its pure form it is silver-colored. This element, and compounds containing it, are used in electro-optical devices, particularly lamps and lasers. It is also used in radar devices, and as a red phosphor in color television picture tubes and computer monitors. Yttrium occurs naturally in lunar rock.

- Electrons in first energy level: 2
- Electrons in second energy level: 8
- Electrons in third energy level: 18
- Electrons in fourth energy level: 9
- Electrons in fifth energy level: 2

40. Zirconium

Symbol, Zr. The most common isotope has atomic weight 91.22. Classified as a transition metal. In its pure form it is grayish in color. The oxide of this element is used as a dielectric at high temperatures. It is also used in nuclear devices. Zinc-beryllium-zirconium silicate is employed as a CRT phosphor. Carbonates, phosphates, and salts of this element are used in a wide variety of products including paper coatings, paints, adhesives, fillers, ceramics, and textiles.

- Electrons in first energy level: 2
- Electrons in second energy level: 8
- Electrons in third energy level: 18
- Electrons in fourth energy level: 10
- Electrons in fifth energy level: 2

41. Niobium

Symbol, Nb. The most common isotope has atomic weight 92.91. Classified as a transition metal. In industry, this element is sometimes called *columbium*. In pure form it is shiny, and is light gray to white in color. In industry, it is used in specialized welding processes. Niobium is also used to produce equipment for the pharmaceutical, medical, aerospace, and electronic industries. It is especially favored for medical implants because it is well tolerated by the body. This element is known for its resistance to corrosion.

- Electrons in first energy level: 2
- Electrons in second energy level: 8
- Electrons in third energy level: 18
- Electrons in fourth energy level: 12
- Electrons in fifth energy level: 1

42. Molybdenum

Symbol, Mo. The most common isotope has atomic weight 95.94. Classified as a transition metal. In its pure form, it is hard and silver-white. Used in the grids and plates of certain vacuum tubes. Also used as a catalyst, as a component of hard alloys for the aeronautical and aerospace industries, and in steel-hardening processes. It is known for high thermal conductivity, low thermal-expansion coefficient, high melting point, and resistance to corrosion. Most molybdenum compounds are relatively non-toxic.

- Electrons in first energy level: 2
- Electrons in second energy level: 8
- Electrons in third energy level: 18
- Electrons in fourth energy level: 13
- Electrons in fifth energy level: 1

43. Technetium

Symbol, Tc. Formerly called *masurium*. The most common isotope has atomic weight 98. Classified as a transition metal. In its pure form, it is grayish in color. This element is not found in nature; it occurs when the uranium atom is split by nuclear fission. It also occurs when molybdenum is bombarded by high-speed deuterium nuclei (particles consisting of one proton and one neutron). This element is radioactive. It has applications in radiology, where it is used as a tracer, especially in the non-invasive diagnosis of heart disease.

- Electrons in first energy level: 2
- Electrons in second energy level: 8
- Electrons in third energy level: 18
- Electrons in fourth energy level: 14
- Electrons in fifth energy level: 1

44. Ruthenium

Symbol, Ru. The most common isotope has atomic weight 101.07. A rare element, classified as a transition metal. In pure form it is silver-colored. When mixed with platinum and/or palladium, this element can produce hard alloys that are useful in industrial bearings, the tips of pens, and in dental instruments. Ruthenium is highly resistant to corrosion, even by strong acids; it is added to some titanium alloys to improve durability. In nature, ruthenium is found along with platinum.

- Electrons in first energy level: 2
- Electrons in second energy level: 8
- Electrons in third energy level: 18

- Electrons in fourth energy level: 15
- Electrons in fifth energy level: 1

45. Rhodium

Symbol, Rh. The most common isotope has atomic weight 102.906. Classified as a transition metal. In its pure form it is silver-colored. Occurs in nature along with platinum and nickel. It is used in scientific work. In particular, it makes a good silvering for first-surface mirrors in optical devices and instruments. It is used as a plating material in jewelry manufacture.

- Electrons in first energy level: 2
- Electrons in second energy level: 8
- Electrons in third energy level: 18
- Electrons in fourth energy level: 16
- Electrons in fifth energy level: 1

46. Palladium

Symbol, Pd. The most common isotope has atomic weight 106.42. Classified as a transition metal. In its pure form it is light gray to white. In nature, palladium is found with copper ore. It is used in certain types of medical instruments, in jewelry, and in photographic printing.

- Electrons in first energy level: 2
- Electrons in second energy level: 8
- Electrons in third energy level: 18
- Electrons in fourth energy level: 18
- Electrons in fifth energy level: 0

47. Silver

Symbol, Ag. The most common isotope has atomic weight 107.87. Classified as a transition metal. In its pure form it is a bright, shiny, silverish-white color. An excellent conductor of electricity and heat. Resists corrosion. Used in circuits where low resistance and/or high Q factor (selectivity) are mandatory.

Also used for plating of electrical contacts. Certain silver compounds darken when exposed to infrared, visible light, or ultraviolet; this makes them useful in photographic film. Silver is considered a precious metal, and is used in the manufacture of jewelry and coins.

- Electrons in first energy level: 2
- Electrons in second energy level: 8
- Electrons in third energy level: 18
- Electrons in fourth energy level: 18
- Electrons in fifth energy level: 1

48. Cadmium

Symbol, Cd. The most common isotope has atomic weight 112.41. Classified as a transition metal. In its pure form it is silver-colored. Used with nickel in the manufacture of rechargeable electrochemical cells. Also employed as a protective plating. Helium-cadmium lasers are used in compact disc mastering, holography, data storage, stereolithography, Raman spectroscopy, interferometry, industrial defect detection, and particle counting. Cadmium occurs as a byproduct in the process of refining zinc ores. Highly toxic, similar to lead. The disposal process is regulated by law in some communities.

- Electrons in first energy level: 2
- Electrons in second energy level: 8
- Electrons in third energy level: 18
- Electrons in fourth energy level: 18
- Electrons in fifth energy level: 2

49. Indium

Symbol, In. The most common isotope has atomic weight 114.82. A metallic element used as a dopant in semiconductor processing. In pure form it is silver-colored. In nature, it is often found along with zinc. Certain compounds of this element are used as semiconductors; indium antimonide is an example. Because of its durability, indium is used to coat mechanical bearings. Industrial uses also include soldering, preforms, sputtering targets, and fusible alloys.

- Electrons in first energy level: 2
- Electrons in second energy level: 8
- Electrons in third energy level: 18
- Electrons in fourth energy level: 18
- Electrons in fifth energy level: 3

50. Tin

Symbol, Sn. The most common isotope has atomic weight 118.71. In pure form it is a white or grayish metal. It changes color (from white to gray) when it is cooled through a certain temperature range. It is ductile and malleable. Mixed with lead to manufacture solder. Tin foil is used to form the plates of some fixed capacitors. Compounds of tin can be used to manufacture resistors. Tin plating can protect metals against corrosion to a limited extent, although it is affected by strong acids. The element is toxic, in a manner similar to lead.

- Electrons in first energy level: 2
- Electrons in second energy level: 8
- Electrons in third energy level: 18
- Electrons in fourth energy level: 18
- Electrons in fifth energy level: 4

51. Antimony

Symbol, Sb. The most common isotope has atomic weight 121.76. Classified as a metalloid. In pure form, it is blue-white or blue-gray in color. Has a characteristic flakiness and brittleness. It is a comparatively poor electrical conductor. Burns with a bright glow. Used a dopant in the manufacture of N-type semiconductor material. The compound indium antimonide is used in some transistors, diodes, and integrated circuits. Industrially, antimony is used as a hardener for lead and also for certain plastics.

- Electrons in first energy level: 2
- Electrons in second energy level: 8
- Electrons in third energy level: 18
- Electrons in fourth energy level: 18
- Electrons in fifth energy level: 5

52. Tellurium

Symbol, Te. The most common isotope has atomic weight 127.60. A rare metalloid element related to selenium. In pure form, it is silverish-white and has high luster. In nature it is found along with other metals such as copper. It has a characteristic brittleness. Electrically, this element is a semiconductor that exhibits photoconductive characteristics. Used in the manufacture of solid-state electronic devices, especially for use in temperature-sensing and light-sensing apparatus. Also used to color glass. If this element is inhaled, even in minuscule amounts, it causes an onion-like or garlic-like odor to appear on the breath.

- Electrons in first energy level: 2
- Electrons in second energy level: 8
- Electrons in third energy level: 18
- Electrons in fourth energy level: 18
- Electrons in fifth energy level: 6

53. Iodine

Symbol, I. The most common isotope has atomic weight 126.905. A member of the halogen family. In pure form it has a black or purple-black color. Well known as a poison; solutions are used as a disinfectant for superficial skin wounds. Despite its toxicity in large amounts or high concentrations, iodine in trace amounts is an essential nutrient; the best sources are kelp and ocean shellfish. Radioactive isotopes of this element are used in medical radiology procedures. Solutions of the element can also be used as a dye. When dilute solutions of this element come into contact with starches, a characteristic dark purple stain is produced.

- Electrons in first energy level: 2
- Electrons in second energy level: 8
- Electrons in third energy level: 18
- Electrons in fourth energy level: 18
- Electrons in fifth energy level: 7

54. Xenon

Symbol, Xe. The most common isotope has atomic weight 131.29. Classified as a noble gas. Colorless and odorless; present in trace amounts in the earth's atmosphere. Also found in the atmosphere of Mars. Used in thyratrons, bubble chambers, electric lamps, flash tubes, and lasers. The xenon flash tube produces brilliant, blue-white visible output. Xenon gas can be isolated when air is liquefied. This element can combine with certain other elements, notably fluorine, hydrogen, and oxygen to form compounds. When put under extreme pressure, xenon becomes metallic.

- Electrons in first energy level: 2
- Electrons in second energy level: 8
- Electrons in third energy level: 18
- Electrons in fourth energy level: 18
- Electrons in fifth energy level: 8

55. Cesium

Symbol, Cs. Also spelled *caesium* (in Britain). The most common isotope has atomic weight 132.91. Classified as an alkali metal. In pure form, it is silver-white in color, is ductile, and is malleable. The oscillations of cesium atoms have been employed as an atomic time standard. The element can be used as the light-sensitive material in phototubes, and in arc lamps to produce infrared (IR) output. In vacuum tubes, cesium is used as a "getter" to remove residual traces of gas that remain after the evacuation process. The hydroxide of this element has extremely high pH, and is caustic.

- Electrons in first energy level: 2
- Electrons in second energy level: 8

- Electrons in third energy level: 18
- Electrons in fourth energy level: 18
- Electrons in fifth energy level: 8
- Electrons in sixth energy level: 1

56. Barium

Symbol, Ba. The most common isotope has atomic weight 137.36. Classified as an alkaline earth. In pure form it is silver-white in color, and is relatively soft; it is sometimes mistaken for lead. Various compounds of barium are used as dielectrics and ferroelectric materials. This element oxidizes readily. The oxides of barium and strontium are used as coatings of vacuum-tube cathodes to increase electron emission. Barium is used in certain types of paint. Isotopes of barium have been used in the radiological diagnoses of digestive-system diseases.

- Electrons in first energy level: 2
- Electrons in second energy level: 8
- Electrons in third energy level: 18
- Electrons in fourth energy level: 18
- Electrons in fifth energy level: 8
- Electrons in sixth energy level: 2

57. Lanthanum

Symbol, La. The most common isotope has atomic weight 138.906. Classified as a rare earth. In pure form it is white in color, malleable, and quite soft. Used in precision optical lenses. Lanthanum oxidizes easily, and reacts with a variety of other elements to form compounds. In nature, this element is found in various minerals. The process of isolating lanthanum from the other elements in these minerals is complicated and expensive.

- Electrons in first energy level: 2
- Electrons in second energy level: 8
- Electrons in third energy level: 18
- Electrons in fourth energy level: 18

- Electrons in fifth energy level: 9
- Electrons in sixth energy level: 2

58. Cerium

Symbol, Ce. The most common isotope has atomic weight 140.13. Classified as a rare earth. In pure form it is light silvery-gray. It reacts readily with various other elements, and is malleable and ductile. It oxidizes easily, and has been known to burn with a flame when exposed directly to air. There are numerous isotopes, some radioactive. Used in heat-resistant industrial alloys, despite its potential flammability in the presence of oxygen. It is commonly found in natural minerals.

- Electrons in first energy level: 2
- Electrons in second energy level: 8
- Electrons in third energy level: 18
- Electrons in fourth energy level: 20
- Electrons in fifth energy level: 8
- Electrons in sixth energy level: 2

59. Praseodymium

Symbol, Pr. The most common isotope has atomic weight 140.908. Classified as a rare earth. In pure form it is silver-gray, soft, malleable, and ductile. When exposed to air, it develops a characteristic green coating of oxide, although it is not as highly reactive as cerium. This element is a component of the flint material used in cigarette lighters. It is also added to certain types of industrial glass. Praseodymium salts are found in various natural minerals.

- Electrons in first energy level: 2
- Electrons in second energy level: 8
- Electrons in third energy level: 18
- Electrons in fourth energy level: 21
- Electrons in fifth energy level: 8
- Electrons in sixth energy level: 2

60. Neodymium

Symbol, Nd. The most common isotope has atomic weight 144.24. Classified as a rare earth. In pure form it is shiny, and is silvery in color. Neodymium oxidizes rapidly. It is used in low-to-medium-power lasers along with yttrium/aluminum/garnet (YAG) crystal. This *neodymium-YAG laser* is employed in jobs where high precision is required. Neodymium is also used to tint glass. It is used in the manufacture of infrared (IR) filters, and is also a component of the flint material used in cigarette lighters.

- Electrons in first energy level: 2
- Electrons in second energy level: 8
- Electrons in third energy level: 18
- Electrons in fourth energy level: 22
- Electrons in fifth energy level: 8
- Electrons in sixth energy level: 2

61. Promethium

Symbol, Pm. Formerly called *illinium*. The most common isotope has atomic weight 145. Classified as a rare earth. In pure form it is gray in color, and is highly radioactive. Promethium is dangerous to handle because of this radioactivity. Derived from the fission of uranium, thorium, and plutonium. This element has not been found in nature on the earth's surface. An isotope of this element is used in specialized photovoltaic cells and batteries.

- Electrons in first energy level: 2
- Electrons in second energy level: 8
- Electrons in third energy level: 18
- Electrons in fourth energy level: 23
- Electrons in fifth energy level: 8
- Electrons in sixth energy level: 2

62. Samarium

Symbol, Sm. The most common isotope has atomic weight 150.36. Classified as a rare earth. In pure form it is silvery-white in color with high luster. It is comparatively stable in open air. Used in the manufacture of permanent magnets, and in specialized alloys. Also employed in nuclear reactors and in various electronic devices. In nature, samarium is found in minerals along with other rare-earth elements.

- Electrons in first energy level: 2
- Electrons in second energy level: 8
- Electrons in third energy level: 18
- Electrons in fourth energy level: 24
- Electrons in fifth energy level: 8
- Electrons in sixth energy level: 2

63. Europium

Symbol, Eu. The most common isotope has atomic weight 151.96. Classified as a rare earth. In pure form it is silver-gray in color, and has ductility similar to that of lead. This element oxidizes rapidly when exposed to open air. It reacts with other elements in a manner similar to calcium. It is used in CRT displays to produce a red color. It is also used in lasers, and as a neutron absorber in nuclear reactors. Europium is one of the rarest and most expensive of the rare-earth metals.

- Electrons in first energy level: 2
- Electrons in second energy level: 8
- Electrons in third energy level: 18
- Electrons in fourth energy level: 25
- Electrons in fifth energy level: 8
- Electrons in sixth energy level: 2

64. Gadolinium

Symbol, Gd. The most common isotope has atomic weight 157.25. Classified as a rare earth. In pure form it is silver in color, is ductile, and is malleable. This element is fairly stable in dry air, but when the humidity is high it tends to oxidize. Compounds of this element are used in CRT phosphors. This element enhances the resistance of certain metallic alloys to oxidation, and also improves their ability to withstand heat.

The pure metal has ferromagnetic and superconductive properties. In nature, gadolinium is found along with other rare earths in various minerals.

- Electrons in first energy level: 2
- Electrons in second energy level: 8
- Electrons in third energy level: 18
- Electrons in fourth energy level: 25
- Electrons in fifth energy level: 9
- Electrons in sixth energy level: 2

65. Terbium

Symbol, Tb. The most common isotope has atomic weight 158.93. Classified as a rare earth. In pure form it is silver-gray, soft, malleable, and ductile. It is fairly stable when exposed to dry air. The oxide is used to provide green color in CRT displays. Sodium terbium borate is used in the semiconductor industry. In nature, it is found in minerals with other rare-earth elements.

- Electrons in first energy level: 2
- Electrons in second energy level: 8
- Electrons in third energy level: 18
- Electrons in fourth energy level: 27
- Electrons in fifth energy level: 8
- Electrons in sixth energy level: 2

66. Dysprosium

Symbol, Dy. The most common isotope has atomic weight 162.5. Classified as a rare earth. In pure form it is a bright, shiny silver color. It is soft and malleable, but it has a relatively high melting point. It is fairly stable when exposed to dry air. Dysprosium has ferromagnetic properties. It is used in certain types of lasers with vanadium. In nature, it is found in minerals with other rare-earth elements.

- Electrons in first energy level: 2
- Electrons in second energy level: 8

- Electrons in third energy level: 18
- Electrons in fourth energy level: 28
- Electrons in fifth energy level: 8
- Electrons in sixth energy level: 2

67. Holmium

Symbol, Ho. The most common isotope has atomic weight 164.93. Classified as a rare earth. In pure form it is silver in color. It is soft and malleable. It is fairly stable in dry air, but if the humidity or temperature are high, it oxidizes quickly. Forms ferromagnetic compounds. Used in nuclear reactors. In nature, it is found in minerals with other rare-earth elements.

- Electrons in first energy level: 2
- Electrons in second energy level: 8
- Electrons in third energy level: 18
- Electrons in fourth energy level: 29
- Electrons in fifth energy level: 8
- Electrons in sixth energy level: 2

68. Erbium

Symbol, Er. The most common isotope has atomic weight 167.26. Classified as a rare earth. In pure form it is silverish, soft, malleable, and ductile. This element is comparatively stable when exposed to air. Erbium can be added to vanadium to soften that element. Several of the isotopes are radioactive. This element is used in the manufacture of certain ceramics; erbium oxide has a pink tint and is sometimes used to color glass. In nature, it is found in minerals with other rare-earth elements.

- Electrons in first energy level: 2
- Electrons in second energy level: 8
- Electrons in third energy level: 18
- Electrons in fourth energy level: 30
- Electrons in fifth energy level: 8
- Electrons in sixth energy level: 2

69. Thulium

Symbol, Tm. The most common isotope has atomic weight 168.93. Classified as a rare earth. In pure form, this element is grayish in color, soft, malleable, and ductile. The natural isotope is stable. Used in power supplies for X-ray generating equipment. Compounds of this element are used in ferromagnetic materials that maintain low loss at ultra-high and microwave frequencies. In nature, it is found in minerals with other rare-earth elements, although it is comparatively rare.

- Electrons in first energy level: 2
- Electrons in second energy level: 8
- Electrons in third energy level: 18
- Electrons in fourth energy level: 31
- Electrons in fifth energy level: 8
- Electrons in sixth energy level: 2

70. Ytterbium

Symbol, Yb. The most common isotope has atomic weight 173.04. Classified as a rare earth. In pure form it is silver-white in color, soft, malleable, and ductile. It is somewhat susceptible to oxidation if exposed to air, especially if the humidity is high. Ytterbium is used in the manufacture of some stainless steel alloys. Has been used in portable X-ray generating equipment. One form of this element behaves as a pressure-sensitive semiconductor under high pressure. Derived from various ores including gadolinite and monazite.

- Electrons in first energy level: 2
- Electrons in second energy level: 8
- Electrons in third energy level: 18
- Electrons in fourth energy level: 32
- Electrons in fifth energy level: 8
- Electrons in sixth energy level: 2

71. Lutetium

Symbol, Lu. The most common isotope has atomic weight 174.967. Classified as a rare earth. In its pure form, it is silver-white and radioactive, with a half-life on the order of thousands of millions of years. It is fairly resistant to corrosion. This element has been used as a catalyst in certain industrial processes. In nature, it is found in minerals along with other rare-earth elements, but is difficult to separate and purify.

- Electrons in first energy level: 2
- Electrons in second energy level: 8
- Electrons in third energy level: 18
- Electrons in fourth energy level: 32
- Electrons in fifth energy level: 9
- Electrons in sixth energy level: 2

72. Hafnium

Symbol, Hf. The most common isotope has atomic weight 178.49. Classified as a transition metal. In pure form, it is silver-colored, shiny, and ductile. This element is highly corrosion-resistant. It readily emits electrons and absorbs neutrons. Used in nuclear reactors, incandescent lamps, and gas-filled lamps. Also used as a "getter" to remove residual gases from vacuum tubes. In nature, it is found in minerals with zirconium. It is difficult to separate pure hafnium from these minerals.

- Electrons in first energy level: 2
- Electrons in second energy level: 8
- Electrons in third energy level: 18
- Electrons in fourth energy level: 32
- Electrons in fifth energy level: 10
- Electrons in sixth energy level: 2

73. Tantalum

Symbol, Ta. The most common isotope has atomic weight 180.95. Classified as a transition metal; an element of the vanadium family. In pure form it is grayish-silver in color, ductile, and hard, with a high melting point. This element is highly resistant to corrosion at moderate temperatures. It is used in the manufacture of high-capacitance, close-tolerance electrolytic capacitors. Also used in the elements of vacuum tubes, in some camera lenses, and in the manufacture of diodes and resistors. The chemical and pharmaceutical industries use tantalum, which is not affected by body fluids, to manufacture various products. This element is resistant to corrosion and is used in hostile industrial environments.

- Electrons in first energy level: 2
- Electrons in second energy level: 8
- Electrons in third energy level: 18
- Electrons in fourth energy level: 32
- Electrons in fifth energy level: 11
- Electrons in sixth energy level: 2

74. Tungsten

Symbol, W. Also known as *wolfram*. The most common isotope has atomic weight 183.85. Classified as a transition metal. In pure form it is silver-colored. It has an extremely high melting point, is known for its high tensile strength, and is relatively resistant to corrosion. Used in switch and relay contacts, in the filaments of electron tubes, and as the filaments in incandescent lamps. Also employed as the anodes of X-ray tubes, and in sealing between glass and metal. Alloys containing tungsten are used in cutting tools. Compounds of tungsten are used in paints, leather tanning, and fluorescent lamps.

- Electrons in first energy level: 2
- Electrons in second energy level: 8
- Electrons in third energy level: 18
- Electrons in fourth energy level: 32
- Electrons in fifth energy level: 12
- Electrons in sixth energy level: 2

75. Rhenium

Symbol, Re. The most common isotope has atomic weight 186.207. Classified as a transition metal. In pure form it is silver-white, has high density, and has a high melting point. Annealed rhenium is ductile, and can be drawn easily into wire. Used in thermocouples, mass spectrography equipment, flash lamps, and durable high-voltage electrical contacts. Alloyed with molybdenum, this element becomes superconductive at approximately −263 degrees Celsius. In nature, it is found with copper-sulfide ore.

- Electrons in first energy level: 2
- Electrons in second energy level: 8
- Electrons in third energy level: 18
- Electrons in fourth energy level: 32
- Electrons in fifth energy level: 13
- Electrons in sixth energy level: 2

76. Osmium

Symbol, Os. The most common isotope has atomic weight 190.2. A transition metal of the platinum group. In pure form, it is bluish-silver in color, dense, hard, and brittle. Known for durability. Alloys containing osmium are used in lamp filaments, electrical contacts, the tips of writing instruments, and the bearings of precision analog electromechanical meters. This element is dangerous if it gets into the air, even in low concentrations. In nature it is found in mineral deposits with platinum and nickel.

- Electrons in first energy level: 2
- Electrons in second energy level: 8
- Electrons in third energy level: 18
- Electrons in fourth energy level: 32
- Electrons in fifth energy level: 14
- Electrons in sixth energy level: 2

77. Iridium

Symbol, Ir. The most common isotope has atomic weight 192.22. A transition metal of the platinum group. In pure form, it is yellowish-white in color with high luster; it is hard, brittle, and has high density. It is extremely resistant to corrosion. It is used as a hardening agent for platinum. Because of its durability, it is used in bearings, writing-instrument tips, and electrical contacts that must open and close frequently. Salts of this element have various bright colors. It is somewhat famous as a component of the standard meter bar in Paris, France. Occurs naturally along with platinum and nickel.

- Electrons in first energy level: 2
- Electrons in second energy level: 8
- Electrons in third energy level: 18
- Electrons in fourth energy level: 32
- Electrons in fifth energy level: 15
- Electrons in sixth energy level: 2

78. Platinum

Symbol, Pt. The most common isotope has atomic weight 195.08. Classified as a transition metal. In pure form, it has a brilliant, shiny, white luster. It is malleable and ductile. This element resists corrosion; used as a coating for electrodes and switch contacts in electronic and computer systems. Also used in specialized vacuum tubes, and in thermocouple-type meters. An alloy of platinum and cobalt is used in the manufacture of strong permanent magnets. In some applications it acts as a catalyst. Platinum expands and contracts with temperature changes in a manner similar to lime-silica glass, so it is used as the electrode material in evacuated tubes made from this type of glass. In nature, platinum is found along with other metals in the same group.

- Electrons in first energy level: 2
- Electrons in second energy level: 8
- Electrons in third energy level: 18
- Electrons in fourth energy level: 32
- Electrons in fifth energy level: 17
- Electrons in sixth energy level: 1

79. Gold

Symbol, Au. The most common isotope has atomic weight 196.967. A transition metal. In pure form it is shiny, yellowish, ductile, malleable, and comparatively soft. Gold resists corrosion, and is used as a coating for electrodes and switch contacts in electronic and computer systems. Also used in specialized semiconductor devices as electrodes or as a dopant. It is most well known for its use in jewelry. A radioactive isotope of this element is used for treating certain types of cancer. A compound of gold and sodium is used for treating arthritis. In nature, gold is found free, often appearing as shiny nuggets or grains recognizable to the trained eye.

- Electrons in first energy level: 2
- Electrons in second energy level: 8
- Electrons in third energy level: 18
- Electrons in fourth energy level: 32
- Electrons in fifth energy level: 18
- Electrons in sixth energy level: 1

80. Mercury

Symbol, Hg. The most common isotope has atomic weight 200.59. Classified as a transition metal. In pure form, it is silver-colored and liquid at room temperature. It is a relatively poor conductor of heat, but a good conductor of electric current. Used in switches, relays, high-voltage rectifiers, high-vacuum pumps, lamps, barometers, thermometers, and electrochemical cells. It is also used in the manufacture of certain pesticides. It forms compounds with many other elements. Mercury salts are used in certain explosives devices and paints. The element is highly toxic. It has been identified as a major pollutant in some lakes and streams. It is not often found pure in nature. It is usually obtained by heating the mineral *cinnabar*.

- Electrons in first energy level: 2
- Electrons in second energy level: 8
- Electrons in third energy level: 18
- Electrons in fourth energy level: 32
- Electrons in fifth energy level: 18
- Electrons in sixth energy level: 2

81. Thallium

Symbol, Tl. The most common isotope has atomic weight 204.38. A metallic element. In pure form it is bluish-gray or dull gray, soft, malleable, and ductile. Compounds of this element exhibit photoconductivity, and are used in photoelectric cells at infrared (IR) wavelengths. The element has been tested in the treatment of certain skin infections, but because of its toxicity and the fact that it may be carcinogenic, it has not gained widespread medical acceptance. Compounds of thallium are used in the manufacture of pesticides, special glass, and in high-refraction optics. Thallium occurs in various natural minerals, and is commonly found in ores containing zinc and lead.

- Electrons in first energy level: 2
- Electrons in second energy level: 8
- Electrons in third energy level: 18
- Electrons in fourth energy level: 32
- Electrons in fifth energy level: 18
- Electrons in sixth energy level: 3

82. Lead

Symbol, Pb. The most common isotope has atomic weight 207.2. A metallic element. In pure form it is dull gray or blue-gray, soft, and malleable. It is a relatively poor conductor of electric current. Exhibits relatively low melting temperature. Lead is corrosion-resistant and has historically been used to contain caustic substances. It was used for thousands of years in plumbing, but recently has been replaced by copper or plastic because lead is toxic. Used in rechargeable cells and batteries, and as fuse elements. Lead is alloyed with tin to make solder for use in electronic equipment. It is also employed as a shield against ionizing radiation, as a sound absorber, and in the manufacture of certain types of glass. This element is rare in nature; it is usually obtained from the mineral *galena*.

- Electrons in first energy level: 2
- Electrons in second energy level: 8

- Electrons in third energy level: 18
- Electrons in fourth energy level: 32
- Electrons in fifth energy level: 18
- Electrons in sixth energy level: 4

83. Bismuth

Symbol, Bi. The most common isotope has atomic weight 208.98. A metallic element. In pure form it is pinkish-white and brittle. Exhibits magnetoresistive properties. Bismuth is an extremely poor conductor of heat and electricity. When a sample of the element is subjected to a magnetic field, its electrical conductivity decreases. Used in fuses, thermocouples, thermocouple type meters, and nuclear reactors. Also used in skin cosmetics. Compounds of this element are used in medicine for the relief of mild gastrointestinal upset. Alloys containing bismuth have low melting points, and are used in fire-detection systems. The element is found pure in nature, but industrially it is obtained from the refining of metals.

- Electrons in first energy level: 2
- Electrons in second energy level: 8
- Electrons in third energy level: 18
- Electrons in fourth energy level: 32
- Electrons in fifth energy level: 18
- Electrons in sixth energy level: 5

84. Polonium

Symbol, Po. The most common isotope has atomic weight 209. Classified as a metalloid. It is produced from the decay of radium and is sometimes called *radium-F*. Polonium is radioactive; it emits primarily alpha particles. The half-life depends on the isotope and can range from a few weeks to more than 100 years. Polonium is dangerous to handle and can be deadly when ingested because of the ionizing radiation it emits. It is rare in nature, and occurs in *pitchblende* ore along with uranium. Polonium can be obtained in the laboratory by subjecting bismuth to high-speed neutron bombardment.

- Electrons in first energy level: 2
- Electrons in second energy level: 8
- Electrons in third energy level: 18
- Electrons in fourth energy level: 32
- Electrons in fifth energy level: 18
- Electrons in sixth energy level: 6

85. Astatine

Symbol, At. The most common isotope has atomic weight 210. Formerly called *alabamine*. Classified as a halogen. The element is radioactive, and is believed to accumulate in the thyroid gland in a manner similar to iodine. It is produced from radioactive decay; it can also be obtained in the laboratory by subjecting bismuth to alpha-particle bombardment. This element is extremely rare in nature.

- Electrons in first energy level: 2
- Electrons in second energy level: 8
- Electrons in third energy level: 18
- Electrons in fourth energy level: 32
- Electrons in fifth energy level: 18
- Electrons in sixth energy level: 7

86. Radon

Symbol, Rn. The most common isotope has atomic weight 222. Classified as a noble gas. It is radioactive, emitting primarily alpha particles, and has a short half-life. Radon is a colorless gas that results from the disintegration of radium. This gas has been found in basements, subterranean mines, caverns, and other enclosed underground spaces. It has also been observed dissolved in some mineral springs. Used in medical radiation therapy. Known to contribute to lung cancer if regularly inhaled over a long period of time.

- Electrons in first energy level: 2
- Electrons in second energy level: 8

- Electrons in third energy level: 18
- Electrons in fourth energy level: 32
- Electrons in fifth energy level: 18
- Electrons in sixth energy level: 8

87. Francium

Symbol, Fr. The most common isotope has atomic weight 223. Classified as an alkali metal. This element is radioactive, and all isotopes decay rapidly. Produced as a result of the radioactive disintegration of actinium. It can also be obtained in the laboratory by subjecting thorium to high-speed proton bombardment. But because of its unstable nature, it is impractical to isolate pure francium.

- Electrons in first energy level: 2
- Electrons in second energy level: 8
- Electrons in third energy level: 18
- Electrons in fourth energy level: 32
- Electrons in fifth energy level: 18
- Electrons in sixth energy level: 8
- Electrons in seventh energy level: 1

88. Radium

Symbol, Ra. The most common isotope has atomic weight 226. Classified as an alkaline earth. In pure form it is silver-gray, but darkens quickly when exposed to air. This element is radioactive, emitting alpha particles, beta particles, and gamma rays. It has a moderately long half-life. Radium is used in the treatment of certain cancers. The element is luminescent and was used at one time in "glow-in-the-dark" wristwatches. Radium releases radon gas slowly over time. It occurs naturally with uranium in pitchblende ore.

- Electrons in first energy level: 2
- Electrons in second energy level: 8
- Electrons in third energy level: 18

- Electrons in fourth energy level: 32
- Electrons in fifth energy level: 18
- Electrons in sixth energy level: 8
- Electrons in seventh energy level: 2

89. Actinium

Symbol, Ac. The most common isotope has atomic weight 227. Classified as a rare earth. In pure form it is silver-gray in color. This element is radioactive, emitting beta particles. The most common isotope has a half-life of 21.6 years. It behaves in a manner similar to lanthanum. In nature, it is found along with uranium.

- Electrons in first energy level: 2
- Electrons in second energy level: 8
- Electrons in third energy level: 18
- Electrons in fourth energy level: 32
- Electrons in fifth energy level: 18
- Electrons in sixth energy level: 9
- Electrons in seventh energy level: 2

90. Thorium

Symbol, Th. The most common isotope has atomic weight 232.038. Classified as a rare earth, and a member of the actinide series. In pure form it is silver-colored, soft, ductile, and malleable. It is fairly stable in air, but gradually darkens. The naturally-occurring isotope is radioactive, emitting alpha particles, and has an extremely long half-life, on the order of several billion years. Some isotopes are used in specialized alloys and compounds, in the manufacture of photoelectric cells, and to coat the tungsten in electron-tube and incandescent-lamp filaments. When heated sufficiently, thorium burns with a brilliant white light. This element is fairly abundant in nature, existing in *thorite* and *monazite* ores. It holds promise as a source of nuclear energy for the future. The energy released by the radioactive decay of thorium and uranium is believed to be responsible for much of the internal heating of the earth.

- Electrons in first energy level: 2
- Electrons in second energy level: 8
- Electrons in third energy level: 18
- Electrons in fourth energy level: 32
- Electrons in fifth energy level: 18
- Electrons in sixth energy level: 10
- Electrons in seventh energy level: 2

91. Protactinium

Symbol, Pa. Formerly called *protoactinium*. The most common isotope has atomic weight 231.036. Classified as a rare earth. In pure form it is silver-colored. It is stable in air, and is dangerously radioactive, emitting alpha particles. The most common isotope has a half-life of about 33,000 years. This element is produced from the fission of uranium, plutonium, and thorium. It occurs naturally along with uranium in pitchblende ore, but is much less abundant than uranium.

- Electrons in first energy level: 2
- Electrons in second energy level: 8
- Electrons in third energy level: 18
- Electrons in fourth energy level: 32
- Electrons in fifth energy level: 20
- Electrons in sixth energy level: 9
- Electrons in seventh energy level: 2

92. Uranium

Symbol, U. The most common isotope has atomic weight 238.029. Classified as a rare earth. In pure form it is silver-colored, malleable, and ductile. In oxidizes readily when exposed to air. Uranium is toxic as well as dangerously radioactive. The naturally occurring isotope emits neutrons and gamma rays, and has a half life of 4.5 billion years. It is probably best known as fuel for nuclear fission reactors. The element is also employed in a variety of industrial and aerospace applications. Uranium occurs naturally in *carnotite*, pitchblende, and various other ores. The energy released by the radioactive decay of uranium and thorium is believed to be responsible for much of the internal heating of the earth.

- Electrons in first energy level: 2
- Electrons in second energy level: 8
- Electrons in third energy level: 18
- Electrons in fourth energy level: 32
- Electrons in fifth energy level: 21
- Electrons in sixth energy level: 9
- Electrons in seventh energy level: 2

93. Neptunium

Symbol, Np. The most common isotope has atomic weight 237. Classified as a rare earth. In pure form it is silver-colored, and reacts with various other elements to form compounds. It is used in some neutron-detection devices. Neptunium is primarily a human-made element; it occurs naturally in minute amounts.

- Electrons in first energy level: 2
- Electrons in second energy level: 8
- Electrons in third energy level: 18
- Electrons in fourth energy level: 32
- Electrons in fifth energy level: 23
- Electrons in sixth energy level: 8
- Electrons in seventh energy level: 2

94. Plutonium

Symbol, Pu. The most common isotope has atomic weight 244. Classified as a rare earth. In pure form it is silver-colored; when it is exposed to air, a yellow oxide layer forms. Plutonium reacts with various other elements to form compounds. It is used in nuclear reactors, and in the manufacture of nuclear bombs. It is dangerous because of the high level of ionizing radiation it emits, and because it can explode if a quantity reaches the critical mass. Plutonium is primarily a human-made element; it occurs naturally in minute amounts.

- Electrons in first energy level: 2
- Electrons in second energy level: 8
- Electrons in third energy level: 18
- Electrons in fourth energy level: 32
- Electrons in fifth energy level: 24
- Electrons in sixth energy level: 8
- Electrons in seventh energy level: 2

95. Americium

Symbol, Am. The most common isotope has atomic weight 243. Classified as a rare earth. In pure form it is silver-white and malleable; it oxidizes slowly when exposed to air. Used in high-tech smoke detectors. This element, like most transuranic elements, is dangerously radioactive. Americium is a human-made element, not known to occur in nature.

- Electrons in first energy level: 2
- Electrons in second energy level: 8
- Electrons in third energy level: 18
- Electrons in fourth energy level: 32
- Electrons in fifth energy level: 25
- Electrons in sixth energy level: 8
- Electrons in seventh energy level: 2

96. Curium

Symbol, Cm. The most common isotope has atomic weight 247. Classified as a rare earth. In pure form it is silvery in color, and it reacts readily with various other elements. This element, like most transuranic elements, is dangerously radioactive. When humans are exposed for long periods, the element can accumulate in the bones and interfere with blood production. It holds some promise as a power source. Curium is a human-made element, not known to occur in nature, although minute amounts might exist in some minerals.

- Electrons in first energy level: 2
- Electrons in second energy level: 8
- Electrons in third energy level: 18
- Electrons in fourth energy level: 32
- Electrons in fifth energy level: 25
- Electrons in sixth energy level: 9
- Electrons in seventh energy level: 2

97. Berkelium

Symbol, Bk. The most common isotope has atomic weight 247. Classified as a rare earth. It is radioactive with a short half-life. Berkelium is a human-made element, and is not known to occur in nature.

- Electrons in first energy level: 2
- Electrons in second energy level: 8
- Electrons in third energy level: 18
- Electrons in fourth energy level: 32
- Electrons in fifth energy level: 26
- Electrons in sixth energy level: 9
- Electrons in seventh energy level: 2

98. Californium

Symbol, Cf. The most common isotope has atomic weight 251. Classified as a rare earth. It is radioactive, emitting neutrons in large quantities. Californium is used as a portable source of neutrons, and in the process of locating precious-metal deposits. It has a short half-life. It is a human-made element, not known to occur in nature.

- Electrons in first energy level: 2
- Electrons in second energy level: 8
- Electrons in third energy level: 18

- Electrons in fourth energy level: 32
- Electrons in fifth energy level: 28
- Electrons in sixth energy level: 8
- Electrons in seventh energy level: 2

99. Einsteinium

Symbol, E or Es. The most common isotope has atomic weight 252. Classified as a rare earth. It is radioactive with a short half-life. Einsteinium is a human-made element, and is not known to occur in nature.

- Electrons in first energy level: 2
- Electrons in second energy level: 8
- Electrons in third energy level: 18
- Electrons in fourth energy level: 32
- Electrons in fifth energy level: 29
- Electrons in sixth energy level: 8
- Electrons in seventh energy level: 2

100. Fermium

Symbol, Fm. The most common isotope has atomic weight 257. Classified as a rare earth. It has a short half-life, is human-made, and is not known to occur in nature.

- Electrons in first energy level: 2
- Electrons in second energy level: 8
- Electrons in third energy level: 18
- Electrons in fourth energy level: 32
- Electrons in fifth energy level: 30
- Electrons in sixth energy level: 8
- Electrons in seventh energy level: 2

101. Mendelevium

Symbol, Md or Mv. The most common isotope has atomic weight 258. Classified as a rare earth. It has a short half-life, is human-made, and is not known to occur in nature.

- Electrons in first energy level: 2
- Electrons in second energy level: 8
- Electrons in third energy level: 18
- Electrons in fourth energy level: 32
- Electrons in fifth energy level: 31
- Electrons in sixth energy level: 8
- Electrons in seventh energy level: 2

102. Nobelium

Symbol, No. The most common isotope has atomic weight 259. Classified as a rare earth. It has a short half-life (seconds or minutes, depending on the isotope), is human-made, and is not known to occur in nature.

- Electrons in first energy level: 2
- Electrons in second energy level: 8
- Electrons in third energy level: 18
- Electrons in fourth energy level: 32
- Electrons in fifth energy level: 32
- Electrons in sixth energy level: 8
- Electrons in seventh energy level: 2

103. Lawrencium

Symbol, Lr or Lw. The most common isotope has atomic weight 262. Classified as a rare earth. It has a half life less than one minute, is human-made, and is not known to occur in nature.

- Electrons in first energy level: 2
- Electrons in second energy level: 8
- Electrons in third energy level: 18
- Electrons in fourth energy level: 32
- Electrons in fifth energy level: 32
- Electrons in sixth energy level: 9
- Electrons in seventh energy level: 2

104. Rutherfordium

Symbol, Rf. Also called *unnilquadium* (Unq) and *Kurchatovium* (Ku). The most common isotope has atomic weight 261. Classified as a transition metal. It has a half-life on the order of a few seconds to a few tenths of a second (depending on the isotope), is human-made, and is not known to occur in nature.

- Electrons in first energy level: 2
- Electrons in second energy level: 8
- Electrons in third energy level: 18
- Electrons in fourth energy level: 32
- Electrons in fifth energy level: 32
- Electrons in sixth energy level: 10
- Electrons in seventh energy level: 2

105. Dubnium

Symbol, Db. Also called *unnilpentium* (Unp) and *Hahnium* (Ha). The most common isotope has atomic weight 262. Classified as a transition metal. It has a half-life on the order of a few seconds to a few tenths of a second (depending on the isotope), is human-made, and is not known to occur in nature.

- Electrons in first energy level: 2
- Electrons in second energy level: 8
- Electrons in third energy level: 18
- Electrons in fourth energy level: 32
- Electrons in fifth energy level: 32
- Electrons in sixth energy level: 11
- Electrons in seventh energy level: 2

106. Seaborgium

Symbol, Sg. Also called *unnilhexium* (Unh). The most common isotope has atomic weight 263. Classified as a transition metal. It has a half-life on the order of one second or less, is human-made, and is not known to occur in nature.

- Electrons in first energy level: 2
- Electrons in second energy level: 8
- Electrons in third energy level: 18
- Electrons in fourth energy level: 32
- Electrons in fifth energy level: 32
- Electrons in sixth energy level: 12
- Electrons in seventh energy level: 2

107. Bohrium

Symbol, Bh. Also called *unnilseptium* (Uns). The most common isotope has atomic weight 262. Classified as a transition metal. It is human-made, and is not known to occur in nature.

- Electrons in first energy level: 2
- Electrons in second energy level: 8
- Electrons in third energy level: 18
- Electrons in fourth energy level: 32
- Electrons in fifth energy level: 32
- Electrons in sixth energy level: 13
- Electrons in seventh energy level: 2

108. Hassium

Symbol, Hs. Also called *unniloctium* (Uno). The most common isotope has atomic weight 265. Classified as a transition metal. It is human-made, and not known to occur in nature.

- Electrons in first energy level: 2
- Electrons in second energy level: 8
- Electrons in third energy level: 18
- Electrons in fourth energy level: 32
- Electrons in fifth energy level: 32
- Electrons in sixth energy level: 14
- Electrons in seventh energy level: 2

109. Meitnerium

Symbol, Mt. Also called *unnilenium* (Une). The most common isotope has atomic weight 266. Classified as a transition metal. It is human-made, and not known to occur in nature.

- Electrons in first energy level: 2
- Electrons in second energy level: 8
- Electrons in third energy level: 18
- Electrons in fourth energy level: 32
- Electrons in fifth energy level: 32
- Electrons in sixth energy level: 15
- Electrons in seventh energy level: 2

110. Ununnilium

Symbol, Uun. The most common isotope has atomic weight 269. Classified as a transition metal. It is human-made, and not known to occur in nature.

- Electrons in first energy level: 2
- Electrons in second energy level: 8
- Electrons in third energy level: 18
- Electrons in fourth energy level: 32
- Electrons in fifth energy level: 32
- Electrons in sixth energy level: 17
- Electrons in seventh energy level: 1

111. Unununium

Symbol, Uuu. The most common isotope has atomic weight 272. Classified as a transition metal. It is human-made, and not known to occur in nature.

- Electrons in first energy level: 2
- Electrons in second energy level: 8
- Electrons in third energy level: 18
- Electrons in fourth energy level: 32

- Electrons in fifth energy level: 32
- Electrons in sixth energy level: 18
- Electrons in seventh energy level: 1

112. Ununbium

Symbol, Uub. The most common isotope has atomic weight 277. Classified as a transition metal. It is human-made, and not known to occur in nature.

- Electrons in first energy level: 2
- Electrons in second energy level: 8
- Electrons in third energy level: 18
- Electrons in fourth energy level: 32
- Electrons in fifth energy level: 32
- Electrons in sixth energy level: 18
- Electrons in seventh energy level: 2

113.

As of this writing, no identifiable atoms of an element with atomic number 113 have been reported. The synthesis of or appearance of such an atom is believed possible because of the observation of ununqadium (Uuq, element 114) in the laboratory.

114. Ununquadium

Symbol, Uuq. The most common isotope has atomic weight 285. First reported in January 1999. It is human-made, and not known to occur in nature.

115.

As of this writing, no identifiable atoms of an element with atomic number 115 have been reported. The synthesis or appearance of such an atom is believed possible, because of the observation of ununhexium (Uuh, element 116) in the laboratory.

116. Ununhexium

Symbol, Uuh. The most common isotope has atomic weight 289. First reported in January 1999. It is a decomposition product of ununoctium, and it in turn decomposes into ununquadium. It is not known to occur in nature.

117.

As of this writing, no identifiable atoms of an element with atomic number 117 have been reported. The synthesis or appearance of such an atom is believed possible, because of the observation of ununoctium (Uuo, element 118) in the laboratory.

118. Ununoctium

Symbol, Uuo. The most common isotope has atomic weight 293. It is the result of the fusion of krypton and lead, and decomposes into ununhexium. It is not known to occur in nature.

Chemical Compounds and Mixtures

The following is a list of some chemical combinations used in scientific and industrial components, devices, and systems.

Alnico

Trade name for an alloy used in strong permanent magnets. Contains aluminum, nickel, and cobalt. Sometimes also contains copper and/or titanium.

Alumel

Trade name for an alloy used in thermocouple devices. Composed of nickel (three parts) and aluminum (one part).

Alumina

Trade name for an aluminum-oxide ceramic used in electron tube insulators and as a substrate in the fabrication of thin-film circuits.

Aluminum antimonide

Formula, AlSb. A crystalline compound useful as a semiconductor dopant.

Barium-strontium oxides

The combined oxides of barium and strontium, employed as coatings of vacuum-tube cathodes to increase electron emission at relatively low temperatures.

Barium-strontium titanate

A compound of barium, strontium, oxygen, and titanium, used as a ceramic dielectric material. Exhibits ferroelectric properties and a high dielectric constant.

Barium titanate

Formula, $BaTi0_2$. A ceramic employed as the dielectric in capacitors. Has a high dielectric constant and some ferroelectric properties.

Beryllia

Formula, BeO. Trade name for beryllium oxide, used in various forms as an insulator and structural element (as in resistor cores).

Cadmium borate

Formula, $(CdO + B_2O_3)$: Mn. Used as a phosphor coating in CRT screens; characteristic fluorescence is green or orange.

Cadmium selenide

A compound consisting of cadmium and selenium. Exhibits photoconductive properties. Useful as a semiconductor in photoelectric cells.

Cadmium silicate

Formula, $CdO + SiO_2$. Used as a phosphor coating in CRT screens; characteristic fluorescence is orange-yellow.

Cadmium sulfide

A compound consisting of cadmium and sulfur. Exhibits photoconductive properties. Useful as a semiconductor in photoelectric cells.

Cadmium tungstate

Formula, $CdO + WO_3$. Used as a phosphor coating in CRT screens; characteristic fluorescence is blue-white.

Calcium phosphate

Formula, $Ca_3(PO_4)_2$. Used as a phosphor coating in CRT screens; characteristic fluorescence is white.

Calcium silicate

Formula, $(CaO + SiO_2)$: Mn. Used as a phosphor coating in CRT screens; characteristic fluorescence ranges from orange to green.

Calcium tungstate

Formula, $CaWO_4$. Used as a phosphor coating in CRT screens; characteristic fluorescence is blue.

Chromel

Trade name for a nickel/chromium/iron alloy that is used in the manufacture of thermocouples.

Chromium dioxide

Formula, CrO_2. Used in the manufacture of specialized thermocouples and recording tape.

Constantan

Trade name for an alloy of copper and nickel, used in thermocouples and standard resistors.

Copper oxides and sulfides

Compounds with semiconducting properties, occasionally used in the manufacture of rectifiers, meters, modulators, and photocells. These compounds have been largely replaced by silicon in recent years.

Ferrite

Trade name for a ferromagnetic material consisting of iron oxide and one or more other metals. Used as core material for inductors and switching elements. Also used in CRT deflection coils, and in loopstick receiving antennas at very low, low, medium, and high radio frequencies.

Gallium arsenide

Formula, GaAs. A compound of gallium and arsenic, used as a semiconductor material in low-noise diodes, varactors, and FETs.

Gallium phosphide

A compound of gallium and phosphorus, used as a semiconductor material in light-emitting diodes (LEDs).

Garnet

A mineral containing silicon and various other elements, forming a hard crystalline material. Mixed with aluminum and yttrium, garnet is used in solid-state lasers.

Germanium dioxide

Formula, GeO_2. A gray or white powder obtainable from various sources; it is reduced in an atmosphere of hydrogen or helium to yield elemental germanium.

Helium/neon

Abbreviation, He-Ne. A mixture of these two gases is used in low-cost lasers for various applications. The output is in the red part of the visible spectrum.

Indium antimonide

A combination of indium and antimony, used in the manufacture of semiconductor components.

Iron oxides

Compounds consisting of iron and oxygen. The most familiar example is common rust. Used in specialized rechargeable cells and batteries.

Lead peroxide

A compound used as a constituent of the positive electrodes in lead-acid electrochemical storage cells and batteries.

Magnesium fluoride

Used as a phosphor coating on the screens of long-persistence CRTs. The fluorescence is orange.

Magnesium silicate

Used as a phosphor coating on the screens of CRTs. Fluorescence is orange-red.

Magnesium tungstate

Used as a phosphor coating on the screens of CRTs. Fluorescence is blue-white.

Magnet steel

A high-retentivity alloy of chromium, cobalt, manganese, steel, and tungsten, employed in the manufacture of permanent magnets.

Manganese dioxide

Formula, MnO_2. Mixed with powdered carbon and used as a depolarizing agent in electrochemical dry cells.

Manganin

Trade name for a low-temperature-coefficient alloy used in making wire for precision resistors. Consists of copper 84 percent, manganese 12 percent, and nickel 4 percent.

Monel

Trade name for an alloy primarily consisting of nickel, copper, iron, manganese, and trace amounts of various other metals.

Mercuric iodide

Formula, HgI_2. A compound whose crystals are used as detectors in high-resolution gamma-ray spectroscopy.

Mercuric oxide

A compound used in the cathodes of electrochemical mercury cells and batteries.

Mercury cadmium telluride

Formula HgCdTe. An alloy used as a semiconductor in transistors, integrated circuits, and IR detectors.

Neodymium/yttrium/aluminum/garnet

Abbreviation, neodymium-YAG. A mixture used in low-power, solid-state lasers. Employed in medical applications and other jobs where high precision is required.

Nichrome

Trade name for a nickel-chromium alloy used in the form of a wire or strip for resistors and heater elements.

Nickel/cadmium

Abbreviation, NiCd or NICAD. A mixture used in rechargeable electrochemical cells and batteries.

Nickel hydroxide

A compound used in rechargeable electrochemical power supplies. Examples are nickel/cadmium (NiCd or NICAD) and nickel/metal hydride (NiMH) batteries, used in older notebook computers.

Nickel/iron

A mixture used as a specialized rechargeable electrochemical cell in which the active positive plate material consists of nickel hydroxide, the active negative plate material is powdered iron oxide mixed with cadmium, and the electrolyte is potassium hydroxide.

Nickel oxide

A compound of nickel and oxygen, used in specialized semiconductor components, particularly diodes.

Nickel silver

An alloy of copper, nickel, and zinc, sometimes used for making resistance wire. Also called *German silver*.

Platinum/tellurium

These two metals, when placed in direct contact, form a thermocouple used in specialized metering devices.

Potassium chloride

Formula, KCl. A compound used as a phosphor coating on the screen of long-persistence CRTs. Fluorescence is magenta or white. Also used as a salt substitute for people on prescribed low-sodium diets.

Potassium cyanide

Formula, KCN. A highly toxic salt used as an electrolyte in electroplating.

Potassium hydroxide

A compound used in rechargeable electrochemical cells and batteries, along with various other compounds and mixtures. An example is the nickel/cadmium (NiCd or NICAD) cell.

Proustite

Trade name for crystalline silver arsenide trisulfide. Artificial crystals of this compound are used in tunable IR emitting devices.

Silicon carbide

Formula, SiC. A compound of silicon and carbon, used as a semiconductor, an abrasive material, and a refractory substance. In industrial applications, this compound is sometimes called by its trade name, *Carborundum*.

Silicon dioxide

Formula, SiO_2. Also called *silica*. Used in IR emitting devices. In the passivation of transistors and integrated circuits, a thin layer of silicon dioxide is grown on the surface of the wafer to protect the otherwise exposed junctions.

Silicon oxides

A mixture of silicon monoxide (SiO) and silicon dioxide (SiO_2) that exhibits dielectric properties. Used in the manufacture of metal-oxide-semiconductor (MOS) devices.

Silicon steel

A high-permeability, high-resistance steel containing 2 to 3 percent silicon. Used as core material in transformers and other electromagnetic devices.

Silver solder

A solder consisting of an alloy of silver, copper, and zinc. Has a comparatively high melting temperature.

Sodium iodide

A crystalline compound that sparkles when exposed to high-speed subatomic particles or radioactivity. Useful as a detector or counter of ionizing radiation.

Sodium silicate

Also called *water glass*. A compound used as a fireproofing agent and protective coating.

Steel

An alloy of iron, carbon, and other metals, used in the construction of antenna support towers, in permanent magnets and electromagnets, and as the core material for high-tensile-strength wire.

Sulfur hexafluoride

A gas employed as a coolant and insulant in some power transformers.

Sulfuric acid

Formula, H_2SO_4. An acid consisting of hydrogen, sulfur, and oxygen. Used in a dilute solution or paste as the electrolyte in rechargeable lead-acid cells and batteries.

Tantalum nitride

A compound used in the manufacture of specialized, close-tolerance, thin-film resistors.

Thallium oxysulfide

A compound of thallium, oxygen, and sulfur, used as the light-sensitive material in photoelectric cells.

Thorium oxide

A compound mixed with tungsten to increase electron emissivity in the filaments and cathodes of electron tubes.

Tin/lead

These two elements are commonly alloyed to make solder. Usually combined in a tin-to-lead ratio of 50:50 or 60:40.

Tin oxide

A combination of tin and oxygen, useful as resistive material in the manufacture of thin-film resistors.

Titanium dioxide

Formula, TiO_2. A compound consisting of titanium and oxygen. Useful as a dielectric material.

Yttrium/aluminum/garnet

Abbreviation, YAG. A crystalline mixture used along with various elements, such as neodymium, in low-power, solid-state lasers.

Yttrium/iron/garnet

Abbreviation, YIG. A crystalline mixture used in acoustic delay lines, parametric amplifiers, and filters.

Zinc aluminate

Either of two similar compounds used as phosphor coatings in CRT screens. One form glows blue; the other form glows red.

Zinc beryllium silicate

A compound used as a phosphor coating in CRT screens. Fluorescence is yellow.

Zinc beryllium zirconium silicate

A compound used as a phosphor coating in CRT screens. Fluorescence is white.

Zinc borate

A compound used as a phosphor coating in CRT screens. Fluorescence is yellow-orange.

Zinc cadmium sulfide

Either of two similar compounds used as phosphor coatings in CRT screens. One form glows blue; the other form glows red.

Zinc germanate

A compound used as a phosphor coating in CRT screens. Fluorescence is yellow-green.

Zinc magnesium fluoride

A compound used as a phosphor coating in CRT screens. Fluorescence is orange.

Zinc orthoscilicate

Also called by the trade name *Willemite*. A compound used as a phosphor coating in CRT screens. Fluorescence is yellow-green.

Zinc oxide

A compound used as a phosphor coating in CRT screens. Fluorescence is blue-green. Also used in the manufacture of certain electronic components, such as voltage-dependent resistors (varistors). A cream containing this compound is used for relief of skin irriration.

Zinc silicate

A compound used as a phosphor coating in CRT screens. Fluorescence is blue.

Zinc sulfide

A compound used as a phosphor coating in CRT screens. Fluorescence is blue-green or yellow-green.

Zirconia

Any of various compounds containing zirconium, especially its oxide (ZrO_2), valued for high-temperature dielectric properties.

Chapter

6

Data Tables

This chapter contains information in tabular form of of interest to engineers, mathematicians, and general scientists.

Prefix multipliers Table 6.1 Page 478	This table lists names and multiplication factors for decimal (power-of-10) prefix multipliers and binary (power-of-2) prefix multipliers.
SI unit conversions Table 6.2 Page 479	This is a conversion database for basic SI units to and from various other units. The first column lists units to be converted; the second column lists units to be derived. The third column lists factors by which units in the first column must be multiplied to obtain units in the second column. The fourth column lists factors by which units in the second column must be multiplied to obtain units in the first column.

Electrical unit conversions
Table 6.3
Page 480

This is a conversion database for electrical SI units to and from various other units. The first column lists units to be converted; the second column lists units to be derived. The third column lists numbers by which units in the first column must be multiplied to obtain units in the second column. The fourth column lists numbers by which units in the second column must be multiplied to obtain units in the first column.

Magnetic unit conversions
Table 6.4
Page 482

This is a conversion database for magnetic SI units to and from various other units. The first column lists units to be converted; the second column lists units to be derived. The third column lists factors by which units in the first column must be multiplied to obtain units in the second column. The fourth column lists factors by which units in the second column must be multiplied to obtain units in the first column.

Miscellaneous unit conversions
Table 6.5
Page 483

This is a conversion database for miscellaneous SI units to and from various other units. The first column lists units to be converted; the second column lists units to be derived. The third column lists factors by which units in the first column must be multiplied to obtain units in the second column. The fourth column lists factors by which units in the second column must be multiplied to obtain units in the first column.

Constants
Table 6.6
Page 485

This table lists common physical, electrical, and chemical constants. Expressed units can be converted to other units by using Tables 6.2 through 6.5.

Chemical symbols and atomic numbers
Table 6.7
Page 486

This table lists the known chemical elements in alphabetical order, with their chemical symbols and atomic numbers. For further information about these elements, refer to the section "Chemical Elements" in Chapter 5.

Derivatives
Table 6.8
Page 489

This table lists some common first derivatives. Functions are in the first column; their first derivatives are in the second column. For further information about differentiation, refer to the section "Scalar Differentiation" in Chapter 3.

Indefinite integrals
Table 6.9
Page 490

This table lists some common indefinite integrals (antiderivatives). Functions are in the first column; their antiderivatives are in the second column. For further information about integration, refer to the section "Scalar Integration" in Chapter 3.

Fourier series
Table 6.10
Page 496

This table lists some common Fourier series. Descriptions are in the first column; the first few terms of the expansions are in the second column.

Fourier transforms
Table 6.11
Page 497

This table lists some common Fourier transforms. Functions are in the first column; transforms are in the second column.

Orthogonal polynomials
Table 6.12
Page 499

This table depicts expansions of Tschebshev, Hermite, and Laguerre polynomials. Symbols are in the first column; expansions are in the second column.

Laplace transforms
Table 6.13
Page 500

This table depicts one-dimensional Laplace transforms. Image functions are in the first column, original functions are in the second column.

Lowercase Greek alphabet Table 6.14 Page 503	This table lists lowercase Greek letters, character names (as written in English), and usages. These symbols are often italicized in printed texts.
Uppercase Greek alphabet Table 6.15 Page 505	This table lists uppercase Greek letters, character names (as written in English), and usages. These symbols are often italicized in printed texts.
General mathematical symbols Table 6.16 Page 506	This table lists symbols used to depict operations, relations, and specifications in mathematics. Some of these symbols may appear italicized in printed texts.
Number conversion Table 6.17 Page 511	This table compares decimal, binary, octal, and hexadecimal numbers for decimal values 0 through 256.
Flip-flops Table 6.18 Page 516	Table A denotes the states for flip-flops of the R-S type; table B denotes the states for flip-flops of the J-K type.
Logic gates Table 6.19 Page 517	This table denotes the states for logic gates of NOT, OR, AND, NOR, NAND, XOR, and XNOR types. The NOT gate (inverter) is specified as having one input and one output. The other gates are all specified as having two inputs and one output.
Wire gauge Table 6.20 Page 518	Table A shows designators and diameters for American Wire Gauge (AWG). Table B shows designators and diameters for British Standard Wire Gauge (NBS SWG). Table C shows designators and diameters for Birmingham Wire Gauge (BWG).

Current-carrying capacity Table 6.21 Page 520	This table denotes the maximum safe continuous DC carrying capacities, in amperes, for American Wire Gauges (AWG) 8 through 20 in open air at room temperature.
Resistivity Table 6.22 Page 521	This table shows resistivity values, in micro-ohms per meter, for solid copper wire of sizes AWG No. 2 through No. 30 carrying DC at room temperature.
Permeability Table 6.23 Page 521	This table shows the permeability factors of some materials, in terms of the extent to which a substance concentrates magnetic lines of flux. Free space is assumed to have permeability 1.
Solder data Table 6.24 Page 522	This table denotes the most common types of solder used in industry, with melting points in degrees Fahrenheit and Celsius.
Radio spectrum Table 6.25 Page 522	This table denotes bands in the radio-frequency (RF) spectrum, with frequency and wavelength ranges defined in the common units for each band.
Schematic symbols Table 6.26 Page 523	This table shows common discrete-component symbols used in electronic circuit diagrams.
TV broadcast channels Table 6.27 Page 536	Table A lists frequencies for very-high-frequency (VHF) television broadcast channels. Table B lists frequencies for ultra-high-frequency (UHF) television channels.
Q signals Table 6.28 Page 538	This table lists signals commonly used by Amateur Radio and some military communications personnel.

Ten-code signals Table 6.29 Page 541	This table lists signals commonly used by Citizens Radio and law-enforcement communications personnel.
Morse code Table 6.30 Page 549	This table lists "dot/dash" symbols for International Morse code, occasionally used by Amateur Radio and military communications personnel.
Phonetic alphabet Table 6.31 Page 550	This table lists standard words to define letters of the alphabet in voice communications modes. These designators are used by Amateur Radio, military, and law-enforcement personnel.
Time conversion Table 6.32 Page 551	This is a conversion database for time designations between Coordinated Universal Time (UTC) and various time zones in the United States.

TABLE 6.1 Prefix Multipliers and Their Abbreviations

Designator	Symbol	Decimal	Binary
yocto-	y	10^{-24}	2^{-80}
zepto-	z	10^{-21}	2^{-70}
atto-	a	10^{-18}	2^{-60}
femto-	f	10^{-15}	2^{-50}
pico-	p	10^{-12}	2^{-40}
nano-	n	10^{-9}	2^{-30}
micro-	μ or mm	10^{-6}	2^{-20}
milli-	m	10^{-3}	2^{-10}
centi-	c	10^{-2}	—
deci-	d	10^{-1}	—
(none)	—	10^{0}	2^{0}
deka-	da or D	10^{1}	—
hecto-	h	10^{2}	—
kilo-	K or k	10^{3}	2^{10}
mega-	M	10^{6}	2^{20}
giga-	G	10^{9}	2^{30}
tera-	T	10^{12}	2^{40}
peta-	P	10^{15}	2^{50}
exa-	E	10^{18}	2^{60}
zetta-	Z	10^{21}	2^{70}
yotta-	Y	10^{24}	2^{80}

TABLE 6.2 SI Unit Conversions*

To convert:	To:	Multiply by:	Conversely, multiply by:
meters (m)	Angstroms	10^{10}	10^{-10}
meters (m)	nanometers (nm)	10^{9}	10^{-9}
meters (m)	microns (μ)	10^{6}	10^{-6}
meters (m)	millimeters (mm)	10^{3}	10^{-3}
meters (m)	centimeters (cm)	10^{2}	10^{-2}
meters (m)	inches (in)	39.37	0.02540
meters (m)	feet (ft)	3.281	0.3048
meters (m)	yards (yd)	1.094	0.9144
meters (m)	kilometers (km)	10^{-3}	10^{3}
meters (m)	statute miles (mi)	6.214×10^{-4}	1.609×10^{3}
meters (m)	nautical miles	5.397×10^{-4}	1.853×10^{3}
meters (m)	light seconds	3.336×10^{-9}	2.998×10^{8}
meters (m)	astronomical units (AU)	6.685×10^{-12}	1.496×10^{11}
meters (m)	light years	1.057×10^{-16}	9.461×10^{15}
meters (m)	parsecs (pc)	3.241×10^{-17}	3.085×10^{16}
kilograms (kg)	atomic mass units (amu)	6.022×10^{26}	1.661×10^{-27}
kilograms (kg)	nanograms (ng)	10^{12}	10^{-12}
kilograms (kg)	micrograms (μg)	10^{9}	10^{-9}
kilograms (kg)	milligrams (mg)	10^{6}	10^{-6}
kilograms (kg)	grams (g)	10^{3}	10^{-3}
kilograms (kg)	ounces (oz)	35.28	0.02834
kilograms (kg)	pounds (lb)	2.205	0.4535
kilograms (kg)	English tons	1.103×10^{-3}	907.0
seconds (s)	minutes (min)	0.01667	60.00
seconds (s)	hours (h)	2.778×10^{-4}	3.600×10^{3}
seconds (s)	days (dy)	1.157×10^{-5}	8.640×10^{4}
seconds (s)	years (yr)	3.169×10^{-8}	3.156×10^{7}
seconds (s)	centuries	3.169×10^{-10}	3.156×10^{9}
seconds (s)	millenia	3.169×10^{-11}	3.156×10^{10}
degrees Kelvin (°K)	degrees Celsius (°C)	Subtract 273	Add 273
degrees Kelvin (°K)	degrees Fahrenheit (°F)	Multiply by 1.80, then subtract 459	Multiply by 0.556, then add 255
degrees Kelvin (°K)	degrees Rankine (°R)	1.80	0.556
amperes (A)	carriers per second	6.24×10^{18}	1.60×10^{-19}
amperes (A)	statamperes (statA)	2.998×10^{9}	3.336×10^{-10}
amperes (A)	nanoamperes (nA)	10^{9}	10^{-9}
amperes (A)	microamperes (μA)	10^{6}	10^{-6}
amperes (A)	abamperes (abA)	0.10000	10.000
amperes (A)	milliamperes (mA)	10^{3}	10^{-3}
candela (cd)	microwatts per steradian (μW/sr)	1.464×10^{3}	6.831×10^{-4}

TABLE 6.2 SI Unit Conversions* (*Continued*)

To convert:	To:	Multiply by:	Conversely, multiply by:
candela (cd)	milliwatts per steradian (mW/sr)	1.464	0.6831
candela (cd)	lumens per steradian (lum/sr)	identical; no conversion	identical; no conversion
candela (cd)	watts per steradian (W/sr)	1.464×10^{-3}	683.1
moles (mol)	coulombs (C)	9.65×10^{4}	1.04×10^{-5}

*When no coefficient is given, the coefficient is meant to be precisely equal to 1.

TABLE 6.3 Electrical Unit Conversions*

To convert:	To:	Multiply by:	Conversely, multiply by:
unit electric charges	coulombs (C)	1.60×10^{-19}	6.24×10^{18}
unit electric charges	abcoulombs (abC)	1.60×10^{-20}	6.24×10^{19}
unit electric charges	statcoulombs (statC)	4.80×10^{-10}	2.08×10^{9}
coulombs (C)	unit electric charges	6.24×10^{18}	1.60×10^{-19}
coulombs (C)	statcoulombs (statC)	2.998×10^{9}	3.336×10^{-10}
coulombs (C)	abcoulombs (abC)	0.1000	10.000
joules (J)	electronvolts (eV)	6.242×10^{18}	1.602×10^{-19}
joules (J)	ergs (erg)	10^{7}	10^{-7}
joules (J)	calories (cal)	0.2389	4.1859
joules (J)	British thermal units (Btu)	9.478×10^{-4}	1.055×10^{3}
joules (J)	watt-hours (Wh)	2.778×10^{-4}	3.600×10^{3}
joules (J)	kilowatt-hours (kWh)	2.778×10^{-7}	3.600×10^{6}
volts (V)	abvolts (abV)	10^{8}	10^{-8}
volts (V)	microvolts (μV)	10^{6}	10^{-6}
volts (V)	millivolts (mV)	10^{3}	10^{-3}
volts (V)	statvolts (statV)	3.336×10^{-3}	299.8
volts (V)	kilovolts (kV)	10^{-3}	10^{3}
volts (V)	megavolts (MV)	10^{-6}	10^{6}
ohms (Ω)	abohms (abΩ)	10^{9}	10^{-9}
ohms (Ω)	megohms (MΩ)	10^{-6}	10^{6}
ohms (Ω)	kilohms (kΩ)	10^{-3}	10^{3}
ohms (Ω)	statohms (statΩ)	1.113×10^{-12}	8.988×10^{11}
siemens (S)	statsiemens (statS)	8.988×10^{11}	1.113×10^{-12}
siemens (S)	microsiemens (μS)	10^{6}	10^{-6}
siemens (S)	millisiemens (mS)	10^{3}	10^{-3}
siemens (S)	absiemens (abS)	10^{-9}	10^{9}
watts (W)	picowatts (pW)	10^{12}	10^{-12}
watts (W)	nanowatts (nW)	10^{9}	10^{-9}

TABLE 6.3 Electrical Unit Conversions* (*Continued*)

To convert:	To:	Multiply by:	Conversely, multiply by:
watts (W)	microwatts (μW)	10^6	10^{-6}
watts (W)	milliwatts (mW)	10^3	10^{-3}
watts (W)	British thermal units per hour (Btu/hr)	3.412	0.2931
watts (W)	horsepower (hp)	1.341×10^{-3}	745.7
watts (W)	kilowatts (kW)	10^{-3}	10^3
watts (W)	megawatts (MW)	10^{-6}	10^6
watts (W)	gigawatts (GW)	10^{-9}	10^9
hertz (Hz)	degrees per second (deg/s)	360.0	0.002778
hertz (Hz)	radians per second (rad/s)	6.283	0.1592
hertz (Hz)	kilohertz (kHz)	10^{-3}	10^3
hertz (Hz)	megahertz (MHz)	10^{-6}	10^6
hertz (Hz)	gigahertz (GHz)	10^{-9}	10^9
hertz (Hz)	terahertz (THz)	10^{-12}	10^{12}
farads (F)	picofarads (pF)	10^{12}	10^{-12}
farads (F)	statfarads (statF)	8.898×10^{11}	1.113×10^{-12}
farads (F)	nanofarads (nF)	10^9	10^{-9}
farads (F)	microfarads (μF)	10^6	10^{-6}
farads (F)	abfarads (abF)	10^{-9}	10^9
henrys (H)	nanohenrys (nH)	10^9	10^{-9}
henrys (H)	abhenrys (abH)	10^9	10^{-9}
henrys (H)	microhenrys (μH)	10^6	10^{-6}
henrys (H)	millihenrys (mH)	10^3	10^{-3}
henrys (H)	stathenrys (statH)	1.113×10^{-12}	8.898×10^{11}
volts per meter (V/m)	picovolts per meter (pV/m)	10^{12}	10^{-12}
volts per meter (V/m)	nanovolts per meter (nV/m)	10^9	10^{-9}
volts per meter (V/m)	microvolts per meter (μV/m)	10^6	10^{-6}
volts per meter (V/m)	millivolts per meter (mV/m)	10^3	10^{-3}
volts per meter (V/m)	volts per foot (v/ft)	3.281	0.3048
watts per square meter (W/m^2)	picowatts per square meter (pW/m^2)	10^{12}	10^{-12}
watts per square meter (W/m^2)	nanowatts per square meter (pW/m^2)	10^9	10^{-9}
watts per square meter (W/m^2)	microwatts per square meter (μW/m^2)	10^6	10^{-6}
watts per square meter (W/m^2)	milliwatts per square meter (mW/m^2)	10^3	10^{-3}

TABLE 6.3 Electrical Unit Conversions* (*Continued*)

To convert:	To:	Multiply by:	Conversely, multiply by:
watts per square meter (W/m^2)	watts per square foot (W/ft^2)	0.09294	10.76
watts per square meter (W/m^2)	watts per square inch (W/in^2)	6.452×10^{-4}	1.550×10^3
watts per square meter (W/m^2)	watts per square centimeter (W/cm^2)	10^{-4}	10^4
watts per square meter (W/m^2)	watts per square millimeter (W/mm^2)	10^{-6}	10^6

*When no coefficient is given, the coefficient is meant to be precisely equal to 1.

TABLE 6.4 Magnetic Unit Conversions*

To convert:	To:	Multiply by:	Conversely, multiply by:
webers (Wb)	maxwells (Mx)	10^8	10^{-8}
webers (Wb)	ampere-microhenrys ($A\mu H$)	10^6	10^{-6}
webers (Wb)	ampere-millihenrys (AmH)	10^3	10^{-3}
webers (Wb)	unit poles	1.257×10^{-7}	7.956×10^6
teslas (T)	maxwells per square meter (Mx/m^2)	10^8	10^{-8}
teslas (T)	gauss (G)	10^4	10^{-4}
teslas (T)	maxwells per square centimeter (Mx/cm^2)	10^4	10^{-4}
teslas (T)	maxwells per square millimeter (Mx/mm^2)	10^2	10^{-2}
teslas (T)	webers per square centimeter (W/cm^2)	10^4	10^4
teslas (T)	webers per square millimeter (W/mm^2)	10^{-6}	10^6
oersteds (Oe)	microamperes per meter ($\mu A/m$)	7.956×10^7	1.257×10^{-8}
oersteds (Oe)	milliamperes per meter (mA/m)	7.956×10^4	1.257×10^{-5}
oersteds (Oe)	amperes per meter (A/m)	79.56	0.01257
ampere-turns (AT)	microampere-turns (μAT)	10^6	10^{-6}
ampere-turns (AT)	milliampere-turns (mAT)	10^3	10^{-3}
ampere-turns (AT)	gilberts (G)	1.256	0.7956

*When no coefficient is given, the coefficient is meant to be precisely equal to 1.

TABLE 6.5 Miscellaneous Unit Conversions*

To convert:	To:	Multiply by:	Conversely, multiply by:
square meters (m^2)	square Angstroms	10^{20}	10^{-20}
square meters (m^2)	square nanometers (nm^2)	10^{18}	10^{-18}
square meters (m^2)	square microns (μ^2)	10^{12}	10^{-12}
square meters (m^2)	square millimeters (mm^2)	10^{6}	10^{-6}
square meters (m^2)	square centimeters (cm^2)	10^{4}	10^{-4}
square meters (m^2)	square inches (in^2)	1.550×10^{3}	6.452×10^{-4}
square meters (m^2)	square feet (ft^2)	10.76	0.09294
square meters (m^2)	acres	2.471×10^{-4}	4.047×10^{3}
square meters (m^2)	hectares	10^{-4}	10^{4}
square meters (m^2)	square kilometers (km^2)	10^{-6}	10^{6}
square meters (m^2)	square statute miles (mi^2)	3.863×10^{-7}	2.589×10^{6}
square meters (m^2)	square nautical miles	2.910×10^{-7}	3.434×10^{6}
square meters (m^2)	square light years	1.117×10^{-17}	8.951×10^{16}
square meters (m^2)	square parsecs (pc^2)	1.051×10^{-33}	9.517×10^{32}
cubic meters (m^3)	cubic Angstroms	10^{30}	10^{-30}
cubic meters (m^3)	cubic nanometers (nm^3)	10^{27}	10^{-27}
cubic meters (m^3)	cubic microns (μ^3)	10^{18}	10^{-18}
cubic meters (m^3)	cubic millimeters (mm^3)	10^{9}	10^{-9}
cubic meters (m^3)	cubic centimeters (cm^3)	10^{6}	10^{-6}
cubic meters (m^3)	milliliters (ml)	10^{6}	10^{-6}
cubic meters (m^3)	liters (l)	10^{3}	10^{-3}
cubic meters (m^3)	U.S. gallons (gal)	264.2	3.785×10^{-3}
cubic meters (m^3)	cubic inches (in^3)	6.102×10^{4}	1.639×10^{-5}
cubic meters (m^3)	cubic feet (ft^3)	35.32	0.02831
cubic meters (m^3)	cubic kilometers (km^3)	10^{-9}	10^{9}
cubic meters (m^3)	cubic statute miles (mi^3)	2.399×10^{-10}	4.166×10^{9}
cubic meters (m^3)	cubic nautical miles	1.572×10^{-10}	6.362×10^{9}
cubic meters (m^3)	cubic light seconds	3.711×10^{-26}	2.695×10^{25}

TABLE 6.5 Miscellaneous Unit Conversions* (*Continued*)

To convert:	To:	Multiply by:	Conversely, multiply by:
cubic meters (m^3)	cubic astronomical units (AU^3)	2.987×10^{-34}	3.348×10^{33}
cubic meters (m^3)	cubic light years	1.181×10^{-48}	8.469×10^{47}
cubic meters (m^3)	cubic parsecs (pc^3)	3.406×10^{-50}	2.936×10^{49}
radians (rad)	degrees (° or deg)	57.30	0.01745
meters per second (m/s)	inches per second (in/s)	39.37	0.02540
meters per second (m/s)	kilometers per hour (km/hr)	3.600	0.2778
meters per second (m/s)	feet per second (ft/s)	3.281	0.3048
meters per second (m/s)	statute miles per hour (mi/hr)	2.237	0.4470
meters per second (m/s)	knots (kt)	1.942	0.5149
meters per second (m/s)	kilometers per minute (km/min)	0.06000	16.67
meters per second (m/s)	kilometers per second (km/s)	10^{-3}	10^{3}
radians per second (rad/s)	degrees per second (°/s or deg/s)	57.30	0.01745
radians per second (rad/s)	revolutions per second (rev/s or rps)	0.1592	6.283
radians per second (rad/s)	revolutions per minute (rev/min or rpm)	2.653×10^{-3}	377.0
meters per second per second (m/s^2)	inches per second per second (in/s^2)	39.37	0.02540
meters per second per second (m/s^2)	feet per second per second (ft/s^2)	3.281	0.3048
radians per second per second (rad/s^2)	degrees per second per second ($°/s^2$ or deg/s^2)	57.30	0.01745
radians per second per second (rad/s^2)	revolutions per second per second (rev/s^2 or rps/s)	0.1592	6.283
radians per second per second (rad/s^2)	revolutions per minute per second (rev/min/s or rpm/s)	2.653×10^{-3}	377.0
newtons (N)	dynes	10^{5}	10^{-5}
newtons (N)	ounces (oz)	3.597	0.2780
newtons (N)	pounds (lb)	0.2248	4.448

*When no coefficient is given, the coefficient is meant to be precisely equal to 1.

TABLE 6.6 Physical, Electrical, and Chemical Constants

Quantity or phenomenon	Value	Symbol
Mass of sun	1.989×10^{30} kg	m_{sun}
Mass of earth	5.974×10^{24} kg	m_{earth}
Avogadro's number	6.022169×10^{23} mol^{-1}	N or N_A
Mass of moon	7.348×10^{22} kg	m_{moon}
Mean radius of sun	6.970×10^{8} m	r_{sun}
Speed of electromagnetic-field propagation in free space	2.99792×10^{8} m/s	c
Faraday constant	9.64867×10^{4} C/mol	F
Mean radius of earth	6.371×10^{6} m	r_{earth}
Mean orbital speed of earth	2.978×10^{4} m/s	
Base of natural logarithms	2.718282	e or ε
Ratio of circle circumference to radius	3.14159	π
Mean radius of moon	1.738×10^{6} m	r_{moon}
Characteristic impedance of free space	376.7 Ω	Z_0
Speed of sound in dry air at standard atmospheric temperature and pressure	344 m/s	
Gravitational acceleration at sea level	9.8067 m/s^2	g
Gas constant	8.31434 J/K/mol	R or R_0
Fine structure constant	7.2974×10^{-3}	α
Wien's constant	0.0029 m · K	σ_W
Second radiation constant	0.0143883 m · K	c_2
Permeability of free space	1.257×10^{-6} H/m	μ_0
Stefan-Boltzmann constant	5.66961×10^{-8} W/m^2/K^4	σ
Gravitational constant	6.6732×10^{-11} N · m^2/kg^2	G
Permittivity of free space	8.85×10^{-12} F/m	ε_0
Boltzmann's constant	1.380622×10^{-23} J/K	k
First radiation constant	4.99258×10^{-24} J · m	c_1
Atomic mass unit (AMU)	1.66053×10^{-27} kg	u
Bohr magneton	9.2741×10^{-24} J/T	μ_B
Bohr radius	5.2918×10^{-11} m	α_0
Nuclear magneton	5.0510×10^{-27} J/T	μ_n
Mass of alpha particle	6.64×10^{-27} kg	m_α
Mass of neutron at rest	1.67492×10^{-27} kg	m_n
Mass of proton at rest	1.67261×10^{-27} kg	m_p
Compton wavelength of proton	1.3214×10^{-15} m	λ_{cp}
Mass of electron at rest	9.10956×10^{-31} kg	m_e
Radius of electron	2.81794×10^{-15} m	r_e
Elementary charge	1.60219×10^{-19} C	e
Charge-to-mass ratio of electron	1.7588×10^{11} C/kg	e/m_e
Compton wavelength of electron	2.4263×10^{-12} m	λ_c
Planck's constant	6.6262×10^{-34} J · s	h
Quantum-charge ratio	4.1357×10^{-5} J · s/C	h/e
Rydberg constant	1.0974×10^{7} m^{-1}	R_∞
Euler's constant	0.577216	γ

TABLE 6.7 The Chemical Elements in Alphabetical Order by Name, Including Chemical Symbols and Atomic Numbers 1 Through 118*

Element name	Chemical symbol	Atomic number
Actinium	Ac	89
Aluminum	Al	13
Americium	Am	95
Antimony	Sb	51
Argon	Ar	18
Arsenic	As	33
Astatine	At	85
Barium	Ba	56
Berkelium	Bk	97
Beryllium	Be	4
Bismuth	Bi	83
Bohrium	Bh	107
Boron	B	5
Bromine	Br	35
Cadmium	Cd	48
Calcium	Ca	20
Californium	Cf	98
Carbon	C	6
Cerium	Ce	58
Cesium	Cs	55
Chlorine	Cl	17
Chromium	Cr	24
Cobalt	Co	27
Copper	Cu	29
Curium	Cm	96
Dubnium	Db	105
Dysprosium	Dy	66
Einsteinium	Es	99
Erbium	Er	68
Europium	Eu	63
Fermium	Fm	100
Fluorine	F	9
Francium	Fr	87
Gadolinium	Gd	64
Gallium	Ga	31
Germanium	Ge	32
Gold	Au	79
Hafnium	Hf	72
Hassium	Hs	108
Helium	He	2
Holmium	Ho	67
Hydrogen	H	1

TABLE 6.7 The Chemical Elements in Alphabetical Order by Name, Including Chemical Symbols and Atomic Numbers 1 Through 118* (*Continued*)

Element name	Chemical symbol	Atomic number
Indium	In	49
Iodine	I	53
Iridium	Ir	77
Iron	Fe	26
Krypton	Kr	36
Lanthanum	La	57
Lawrencium	Lr or Lw	103
Lead	Pb	82
Lithium	Li	3
Lutetium	Lu	71
Magnesium	Mg	12
Manganese	Mn	25
Meitnerium	Mt	109
Mendelevium	Md	101
Mercury	Hg	80
Molybdenum	Mo	42
Neodymium	Nd	60
Neon	Ne	10
Neptunium	Np	93
Nickel	Ni	28
Niobium	Nb	41
Nitrogen	N	7
Nobelium	No	102
Osmium	Os	76
Oxygen	O	8
Palladium	Pd	46
Phosphorus	P	15
Platinum	Pt	78
Plutonium	Pu	94
Polonium	Po	84
Potassium	K	19
Praseodymium	Pr	59
Promethium	Pm	61
Protactinium	Pa	91
Radium	Ra	88
Radon	Rn	86
Rhenium	Re	75
Rhodium	Rh	45
Rubidium	Rb	37
Ruthenium	Ru	44

TABLE 6.7 The Chemical Elements in Alphabetical Order by Name, Including Chemical Symbols and Atomic Numbers 1 Through 118* (*Continued*)

Element name	Chemical symbol	Atomic number
Rutherfordium	Rf	104
Samarium	Sm	62
Scandium	Sc	21
Seaborgium	Sg	106
Selenium	Se	34
Silicon	Si	14
Silver	Ag	47
Sodium	Na	11
Strontium	Sr	38
Sulfur	S	16
Tantalum	Ta	73
Technetium	Tc	43
Tellurium	Te	52
Terbium	Tb	65
Thallium	Tl	81
Thorium	Th	90
Thulium	Tm	69
Tin	Sn	50
Titanium	Ti	22
Tungsten	W	74
Ununbium	Uub	112
Ununhexium	Uuh	116
Ununnilium	Uun	110
Ununoctium	Uuo	118
Ununquadium	Uuq	114
Unununium	Uuu	111
Uranium	U	92
Vanadium	V	23
Xenon	Xe	54
Ytterbium	Yb	70
Yttrium	Y	39
Zinc	Zn	30
Zirconium	Zr	40

*As of the time of writing, there were no known elements with atomic numbers 113, 115, or 117. For further details about each element, please refer to the section "Chemical Elements" in Chapter 5.

TABLE 6.8 Derivatives*

Function	Derivative
$f(x) = a$	$f'(x) = 0$
$f(x) = ax$	$f'(x) = a$
$f(x) = ax^n$	$f'(x) = nax^{n-1}$
$f(x) = 1/x$	$f'(x) = \ln \|x\|$
$f(x) = \ln x$	$f'(x) = 1/x$
$f(x) = \ln g(x)$	$f'(x) = g^{-1}(x)\, g'(x)$
$f(x) = 1/x^a$	$f'(x) = -a/(x^{a+1})$
$f(x) = e^x$	$f'(x) = e^x$
$f(x) = a^x$	$f'(x) = a^x \ln a$
$f(x) = a^{g(x)}$	$f'(x) = (a^{g(x)})\,(\ln a)\,(g'(x))$
$f(x) = e^{ax}$	$f'(x) = ae^x$
$f(x) = e^{g(x)}$	$f'(x) = e^{g(x)}\, g'(x)$
$f(x) = \sin x$	$f'(x) = \cos x$
$f(x) = \cos x$	$f'(x) = -\sin x$
$f(x) = \tan x$	$f'(x) = \sec^2 x$
$f(x) = \csc x$	$f'(x) = -\csc x \cot x$
$f(x) = \sec x$	$f'(x) = \sec x \tan x$
$f(x) = \cot x$	$f'(x) = -\csc^2 x$
$f(x) = \arcsin x = \sin^{-1} x$	$f'(x) = 1/(1 - x^2)^{1/2}$
$f(x) = \arccos x = \cos^{-1} x$	$f'(x) = -1/(1 - x^2)^{1/2}$
$f(x) = \arctan x = \tan^{-1} x$	$f'(x) = 1/(1 + x^2)$
$f(x) = \text{arccsc}\, x = \csc^{-1} x$	$f'(x) = -1/[x\,(x^2 - 1)^{1/2}]$
$f(x) = \text{arcsec}\, x = \sec^{-1} x$	$f'(x) = 1/[x\,(x^2 - 1)^{1/2}]$
$f(x) = \text{arccot}\, x = \cot^{-1} x$	$f'(x) = -1/(1 + x^2)$
$f(x) = \sinh x$	$f'(x) = \cosh x$
$f(x) = \cosh x$	$f'(x) = \sinh x$
$f(x) = \tanh x$	$f'(x) = \text{sech}^2\, x$
$f(x) = \text{csch}\, x$	$f'(x) = -\text{csch}\, x \coth x$
$f(x) = \text{sech}\, x$	$f'(x) = \text{sech}\, x \tanh x$
$f(x) = \coth x$	$f'(x) = -\text{csch}^2\, x$
$f(x) = \text{arcsinh}\, x = \sinh^{-1} x$	$f'(x) = 1/(x^2 + 1)^{1/2}$
$f(x) = \text{arccosh}\, x = \cosh^{-1} x$	$f'(x) = 1/(x^2 - 1)^{1/2}$
$f(x) = \text{arctanh}\, x = \tanh^{-1} x$	$f'(x) = 1/(1 - x^2)$
$f(x) = \text{arccsch}\, x = \text{csch}^{-1} x$	$f'(x) = -1/[x(1 + x^2)^{1/2}]$ for $x > 0$
	$f'(x) = 1/[x(1 + x^2)^{1/2}]$ for $x < 0$
$f(x) = \text{arcsech}\, x = \text{sech}^{-1} x$	$f'(x) = -1/[x(1 - x^2)^{1/2}]$ for $x > 0$
	$f'(x) = 1/[x(1 - x^2)^{1/2}]$ for $x < 0$
$f(x) = \text{arccoth}\, x = \coth^{-1} x$	$f'(x) = 1/(1 - x^2)$

*Letters a, b, and c denote constants. Letters f, g, and h denote functions; m, n, and p denote integers; w, x, y, and z denote variables. The letter e represents the exponential constant (approximately 2.71828).

TABLE 6.9 Indefinite Integrals*

Function	Indefinite integral
$f(x) = 0$	$\int f(x)\,dx = c$
$f(x) = 1$	$\int f(x)\,dx = 1 + c$
$f(x) = a$	$\int f(x)\,dx = a + c$
$f(x) = x$	$\int f(x)\,dx = 0.5x^2 + c$
$f(x) = ax$	$\int f(x)\,dx = 0.5ax^2 + c$
$f(x) = ax^2$	$\int f(x)\,dx = (1/3)ax^3 + c$
$f(x) = ax^3$	$\int f(x)\,dx = (1/4)ax^4 + c$
$f(x) = ax^4$	$\int f(x)\,dx = (1/5)ax^5 + c$
$f(x) = ax^{-1}$	$\int f(x)\,dx = a \ln \lvert x \rvert + c$
$f(x) = ax^{-2}$	$\int f(x)\,dx = -ax^{-1} + c$
$f(x) = ax^{-3}$	$\int f(x)\,dx = -0.5ax^{-2} + c$
$f(x) = ax^{-4}$	$\int f(x)\,dx = (-1/3)ax^{-3} + c$
$f(x) = (ax + b)^{1/2}$	$\int f(x)\,dx = (2/3)(ax + b)^{3/2}\, a^{-1} + c$
$f(x) = (ax + b)^{-1/2}$	$\int f(x)\,dx = 2(ax + b)^{1/2}\, a^{-1} + c$
$f(x) = (ax + b)^{-1}$	$\int f(x)\,dx = a^{-1}(\ln (ax + b)) + c$
$f(x) = (ax + b)^{-2}$	$\int f(x)\,dx = -a^{-1}(ax + b)^{-1} + c$
$f(x) = (ax + b)^{-3}$	$\int f(x)\,dx = -0.5a^{-1}(ax + b)^{-2} + c$
$f(x) = (ax + b)^n$ where $n \neq -1$	$\int f(x)\,dx = (ax + b)^{n+1}(an + a)^{-1} + c$
$f(x) = x(ax + b)^{1/2}$	$\int f(x)\,dx = (1/15)a^{-2}(6ax - 4b)(ax + b)^{3/2} + c$
$f(x) = x(ax + b)^{-1/2}$	$\int f(x)\,dx = (1/3)a^{-2}(4ax - 4b)(ax + b)^{1/2} + c$
$f(x) = x(ax + b)^{-1}$	$\int f(x)\,dx = (a^{-1}x) - a^{-2}b \ln (ax + b) + c$
$f(x) = x(ax + b)^{-2}$	$\int f(x)\,dx = b(a^3x + a^2b)^{-1} + a^{-2} \ln (ax + b) + c$
$f(x) = x(ax + b)^{-3}$	$\int f(x)\,dx = (b - (a^3x + a^2b)^{-1})(2a^4x^2 + 4a^3\,bx + 2a^2b^2)^{-1} + c$
$f(x) = x^2(ax + b)^{1/2}$	$\int f(x)\,dx = (1/105)a^{-3}(30a^2x^2 - 24abx + 16b^2)(ax + b)^{3/2} + c$
$f(x) = x^2(ax + b)^{-1/2}$	$\int f(x)\,dx = (1/15)a^{-3}(6a^2x^2 - 8abx + 16b^2)(ax + b)^{1/2} + c$
$f(x) = x^2(ax + b)^{-1}$	$\int f(x)\,dx = ((ax + b)^2/2a^3) - a^{-3}(2abx + 2b^2) + (a^{-3}b^2) \ln (ax + b) + c$
$f(x) = x^2(ax + b)^{-2}$	$\int f(x)\,dx = a^{-2}x + a^{-3}b - a^{-3}b^2\,(ax + b)^{-1} - 2a^{-3}b \ln (ax + b) + c$
$f(x) = x^2(ax + b)^{-3}$	$\int f(x)\,dx = 2a^{-3}b(ax + b)^{-1} - b^2(2a^3(ax + b)^2)^{-1} + a^{-3} \ln (ax + b) + c$
$f(x) = (ax^2 + bx)^{-1}$	$\int f(x)\,dx = (b^{-2}a)(\ln ((ax + b)/x)) - (bx)^{-1} + c$
$f(x) = (ax^3 + bx^2)^{-1}$	$\int f(x)\,dx = (\ln (x(ax + b)^{-1}))/b$
$f(x) = (ax + b)(rx + s)$	$\int f(x)\,dx = (br - as)^{-1} \ln ((ax + b)^{-1}\,(rx + s)) + c$
$f(x) = (ax + b)(rx + s)^{-1}$	$\int f(x)\,dx = ar^{-1}x + r^{-2}(br - as) \ln (rx + s) + c$
$f(x) = (x^2 + a^2)^{1/2}$	$\int f(x)\,dx = 0.5x(x^2 + a^2)^{1/2} + 0.5a^2 \ln (x + (x^2 + a^2)^{1/2}) + c$

TABLE 6.9 Indefinite Integrals* (*Continued*)

Function	Indefinite integral
$f(x) = (x^2 - a^2)^{1/2}$	$\int f(x)\,dx = 0.5x(x^2 - a^2)^{1/2} + 0.5a^2 \ln (x + (x^2 - a^2)^{1/2}) + c$
$f(x) = (a^2 - x^2)^{1/2}$	$\int f(x)\,dx = 0.5x(a^2 - x^2)^{1/2} + 0.5a^2 \sin^{-1} (a^{-1}x) + c$
$f(x) = (x^2 + a^2)^{-1/2}$	$\int f(x)\,dx = \ln (x + (x^2 + a^2)^{1/2}) + c$
$f(x) = (x^2 - a^2)^{-1/2}$	$\int f(x)\,dx = \ln (x + (x^2 - a^2)^{1/2}) + c$
$f(x) = (a^2 - x^2)^{-1/2}$	$\int f(x)\,dx = \sin^{-1} (a^{-1}x) + c$
$f(x) = (x^2 + a^2)^{-1}$	$\int f(x)\,dx = a^{-1} \tan^{-1} (a^{-1}x) + c$
$f(x) = (x^2 - a^2)^{-1}$	$\int f(x)\,dx = 0.5a^{-1} \ln ((x + a)^{-1} (x - a)) + c$
$f(x) = (a^2 - x^2)^{-1}$ where $\lvert a\rvert > \lvert x\rvert$	$\int f(x)\,dx = 0.5a^{-1} \ln ((a - x)^{-1} (a + x)) + c$
$f(x) = (x^2 + a^2)^{-2}$	$\int f(x)\,dx = (2a^2x^2 + 2a^4)^{-1} x + 0.5a^{-3} \tan^{-1} (a^{-1}x) + c$
$f(x) = (x^2 - a^2)^{-2}$	$\int f(x)\,dx = (-x)(2a^2x^2 - 2a^4)^{-1} - 0.25a^{-3} \ln ((x + a)^{-1} (x - a)) + c$
$f(x) = (a^2 - x^2)^{-2}$ where $\lvert a\rvert > \lvert x\rvert$	$\int f(x)\,dx = -x(2a^4 - 2a^2x^2)^{-1} + 0.25a^{-3} \ln ((a - X)^{-1} (a + x)) + c$
$f(x) = x(x^2 + a^2)^{1/2}$	$\int f(x)\,dx = (1/3)(x^2 + a^2)^{3/2} + c$
$f(x) = x(x^2 - a^2)^{1/2}$	$\int f(x)\,dx = (1/3)(x^2 - a^2)^{3/2} + c$
$f(x) = x(a^2 - x^2)^{1/2}$	$\int f(x)\,dx = (-1/3)(a^2 - x^2)^{3/2} + c$
$f(x) = x(x^2 + a^2)^{-1/2}$	$\int f(x)\,dx = (x^2 + a^2)^{1/2} + c$
$f(x) = x(x^2 - a^2)^{-1/2}$	$\int f(x)\,dx = (x^2 - a^2)^{1/2} + c$
$f(x) = x(a^2 - x^2)^{-1/2}$	$\int f(x)\,dx = -(a^2 - x^2)^{1/2} + c$
$f(x) = x(x^2 + a^2)^{-1}$	$\int f(x)\,dx = 0.5 \ln (x^2 + a^2) + c$
$f(x) = x(x^2 - a^2)^{-1}$	$\int f(x)\,dx = 0.5 \ln (x^2 - a^2) + c$
$f(x) = x(a^2 - x^2)^{-1}$ where $\lvert a\rvert > \lvert x\rvert$	$\int f(x)\,dx = -0.5 \ln (a^2 - x^2) + c$
$f(x) = x(x^2 + a^2)^{-2}$	$\int f(x)\,dx = (-2x^2 - 2a^2)^{-1} + c$
$f(x) = x(x^2 - a^2)^{-2}$	$\int f(x)\,dx = -0.5(x^2 - a^2)^{-1} + c$
$f(x) = x(a^2 - x^2)^{-2}$ where $\lvert a\rvert > \lvert x\rvert$	$\int f(x)\,dx = 0.5(a^2 - x^2)^{-1} + c$
$f(x) = x^2(x^2 + a^2)^{1/2}$	$\int f(x)\,dx = 0.25x(x^2 + a^2)^{3/2} - (1/8)a^2x(x^2 + a^2)^{1/2} - (1/8)a^4 \ln (x + (x^2 + a^2)^{1/2}) + c$
$f(x) = x^2(x^2 - a^2)^{1/2}$	$\int f(x)\,dx = 0.25x(x^2 - a^2)^{3/2} + (1/8)a^2x(x^2 - a^2)^{1/2} - (1/8)a^4 \ln (x + (x^2 - a^2)^{1/2}) + c$
$f(x) = x^2(a^2 - x^2)^{1/2}$	$\int f(x)\,dx = -0.25x(a^2 - x^2)^{3/2} + (1/8)a^2x(a^2 - x^2)^{1/2} + (1/8)a^4 \sin^{-1} (a^{-1}x) + c$
$f(x) = x^2(x^2 + a^2)^{-1/2}$	$\int f(x)\,dx = 0.5(x^2 + a^2)^{1/2} - 0.5a^2 \ln (x + (x^2 + a^2)^{1/2}) + c$
$f(x) = x^2(x^2 - a^2)^{-1/2}$	$\int f(x)\,dx = 0.5x(x^2 - a^2)^{1/2} + 0.5a^2 \ln (x + (x^2 - a^2)^{1/2}) + c$
$f(x) = x^2(a^2 - x^2)^{-1/2}$	$\int f(x)\,dx = -0.5x(a^2 - x^2)^{1/2} + 0.5a^2 \sin^{-1} (a^{-1}x) + c$
$f(x) = x^2(x^2 + a^2)^{-1}$	$\int f(x)\,dx = x - a \tan^{-1} (a^{-1}x) + c$

TABLE 6.9 Indefinite Integrals* (*Continued*)

Function	Indefinite integral
$f(x) = x^2(x^2 - a^2)^{-1}$	$\int f(x)\,dx = x + 0.5a \ln((x + a)^{-1}(x - a)) + c$
$f(x) = x^2(a^2 - x^2)^{-1}$ where $\|a\| > \|x\|$	$\int f(x)\,dx = -x + 0.5a \ln((a - x)^{-1}(a + x)) + c$
$f(x) = x^2(x^2 + a^2)^{-2}$	$\int f(x)\,dx = - x(2x^2 + 2a^2)^{-1} + (2a)^{-1}\tan^{-1}(a^{-1}x) + c$
$f(x) = x^2(x^2 - a^2)^{-2}$	$\int f(x)\,dx = (-x)(2x^2 - 2a^2)^{-1} + 0.25\,a^{-1}\ln((x + a)^{-1}(x - a)) + c$
$f(x) = x^2(a^2 - x^2)^{-2}$ where $\|a\| > \|x\|$	$\int f(x)\,dx = x(2a^2 - 2x^2)^{-1} - 0.25a^{-1}\ln((a - x)^{-1}(a + x)) + c$
$f(x) = ax^n$	$\int f(x)\,dx = ax^{n+1}(n + 1)^{-1} + c$ provided that $n \neq -1$
$f(x) = a\,g(x)$	$\int f(x)\,dx = a\int g(x)\,dx + c$
$f(x) = g(x) + h(x)$	$\int f(x)\,dx = \int g(x)\,dx + \int h(x)\,dx + c$
$f(x) = h(x)\,g'(x)$	$\int f(x)\,dx = g(x)h(x) - \int g(x)h'(x) + c$
$f(x) = e^x$	$\int f(x)\,dx = e^x + c$
$f(x) = a\,e^{bx}$	$\int f(x)\,dx = a\,e^{bx}/b + c$
$f(x) = x^{-1}e^{bx}$	$\int f(x)\,dx = \ln x + c + bx + (2! \times 2)^{-1}b^2x^2 + (3! \times 3)^{-1}b^3x^3 + (4! \times 4)^{-1}b^4x^4 + \ldots$
$f(x) = x\,e^{bx}$	$\int f(x)\,dx = b^{-1}x\,e^{bx} - b^{-2}e^{bx} + c$
$f(x) = x^2e^{bx}$	$\int f(x)\,dx = b^{-1}x^2e^{bx} - 2b^{-2}x\,e^{bx} + 2b^{-3}e^{bx} + c$
$f(x) = \ln^{-1}x$	$\int f(x)\,dx = \ln(\ln x) + \ln x + c + (2! \times 2)^{-1}\ln^2 x + (3! \times 3)^{-1}\ln^3 x + \ldots$
$f(x) = x^{-2}\ln x$	$\int f(x)\,dx = - x^{-1}\ln x - x^{-1} + c$
$f(x) = x^{-1}\ln x$	$\int f(x)\,dx = 0.5\ln^2 x + c$
$f(x) = \ln x$	$\int f(x)\,dx = x\ln x - x + c$
$f(x) = x\ln x$	$\int f(x)\,dx = (1/2)x^2\ln x - (1/4)x^2 + c$
$f(x) = x^2\ln x$	$\int f(x)\,dx = (1/3)x^3\ln x - (1/9)x^3 + c$
$f(x) = \ln^2 x$	$\int f(x)\,dx = x\ln^2 x - 2x\ln x + 2x + c$
$f(x) = \sin x$	$\int f(x)\,dx = -\cos x + c$
$f(x) = \cos x$	$\int f(x)\,dx = \sin x + c$
$f(x) = \tan x$	$\int f(x)\,dx = \ln\|\sec x\| + c$
$f(x) = \csc x$	$\int f(x)\,dx = \ln\|\tan(0.5x)\| + c$
$f(x) = \sec x$	$\int f(x)\,dx = \ln\|\sec x + \tan x\| + c$
$f(x) = \cot x$	$\int f(x)\,dx = \ln\|\sin x\| + c$
$f(x) = \sin ax$	$\int f(x)\,dx = -a^{-1}\cos ax + c$
$f(x) = \cos ax$	$\int f(x)\,dx = a^{-1}\sin ax + c$
$f(x) = \tan ax$	$\int f(x)\,dx = a^{-1}\ln(\sec ax) + c$
$f(x) = \csc ax$	$\int f(x)\,dx = a^{-1}\ln(\tan(0.5ax)) + c$
$f(x) = \sec ax$	$\int f(x)\,dx = a^{-1}\ln(\tan(0.25\pi + 0.5ax)) + c$
$f(x) = \cot ax$	$\int f(x)\,dx = a^{-1}\ln(\sin ax) + c$
$f(x) = \sin^2 x$	$\int f(x)\,dx = 0.5(x - (0.5\sin(2x))) + c$
$f(x) = \cos^2 x$	$\int f(x)\,dx = 0.5(x + (0.5\sin(2x))) + c$
$f(x) = \tan^2 x$	$\int f(x)\,dx = \tan x - x + c$
$f(x) = \csc^2 x$	$\int f(x)\,dx = -\cot x + c$
$f(x) = \sec^2 x$	$\int f(x)\,dx = \tan x + c$

TABLE 6.9 Indefinite Integrals* (*Continued*)

Function	Indefinite integral
$f(x) = \cot^2 x$	$\int f(x)\,dx = -\cot x - x + c$
$f(x) = \sin^2 ax$	$\int f(x)\,dx = 0.5x - 0.25a^{-1}(\sin 2ax) + c$
$f(x) = \cos^2 ax$	$\int f(x)\,dx = 0.5x + 0.25a^{-1}(\sin 2ax) + c$
$f(x) = \tan^2 ax$	$\int f(x)\,dx = a^{-1}\tan ax - x + c$
$f(x) = \csc^2 ax$	$\int f(x)\,dx = -a^{-1}\cot ax + c$
$f(x) = \sec^2 ax$	$\int f(x)\,dx = a^{-1}\tan ax + c$
$f(x) = \cot^2 ax$	$\int f(x)\,dx = -a^{-1}\cot ax - x + c$
$f(x) = x \sin ax$	$\int f(x)\,dx = a^{-2}\sin ax - a^{-1}x\cos ax + c$
$f(x) = x \cos ax$	$\int f(x)\,dx = a^{-2}\cos ax + a^{-1}x\sin ax + c$
$f(x) = x^2 \sin ax$	$\int f(x)\,dx = 2a^{-2}x\sin ax$ $+ (2a^{-3} - a^{-1}x^2)\cos ax + c$
$f(x) = x^2 \cos ax$	$\int f(x)\,dx = 2a^{-2}x\cos ax$ $+ (a^{-1}x^2 - 2a^{-3})\sin ax + c$
$f(x) = (\sin x \cos x)^{-2}$	$\int f(x)\,dx = 2\cot 2x + c$
$f(x) = (\sin x \cos x)^{-1}$	$\int f(x)\,dx = \ln(\tan x) + c$
$f(x) = \sin x \cos x$	$\int f(x)\,dx = 0.5\sin^2 x + c$
$f(x) = \sin^2 x \cos^2 x$	$\int f(x)\,dx = (1/8)x - (1/32)\sin 4x + c$
$f(x) = (\sin ax \cos ax)^{-2}$	$\int f(x)\,dx = 2a^{-1}\cot 2ax + c$
$f(x) = (\sin ax \cos ax)^{-1}$	$\int f(x)\,dx = a^{-1}\ln(\tan ax) + c$
$f(x) = \sin ax \cos ax$	$\int f(x)\,dx = 0.5\,a^{-1}\sin^2 ax + c$
$f(x) = \sin^2 ax \cos^2 ax$	$\int f(x)\,dx = (1/8)x - (1/32)(a^{-1})\sin 4ax + c$
$f(x) = \sec x \tan x$	$\int f(x)\,dx = \sec x + c$
$f(x) = \sin^{-1} x$	$\int f(x)\,dx = x\sin^{-1} x + (1 - x^2)^{1/2} + c$
$f(x) = \cos^{-1} x$	$\int f(x)\,dx = x\cos^{-1} x - (1 - x^2)^{1/2} + c$
$f(x) = \tan^{-1} x$	$\int f(x)\,dx = x\tan^{-1} x - 0.5\ln(1 + x^2) + c$
$f(x) = \csc^{-1} x$	$\int f(x)\,dx = x\csc^{-1} x - \ln(x + (x^2 - 1)^{1/2}) + c$ when $-\pi/2 < \csc^{-1} x < 0$
	$\int f(x)\,dx = x\csc^{-1} x + \ln(x + (x^2 - 1)^{1/2}) + c$ when $0 < \csc^{-1} x < \pi/2$
$f(x) = \sec^{-1} x$	$\int f(x)\,dx = x\sec^{-1} x - \ln(x + (x^2 - 1)^{1/2}) + c$ when $0 < \sec^{-1} x < \pi/2$
	$\int f(x)\,dx = x\sec^{-1} x + \ln(x + (x^2 - 1)^{1/2}) + c$ when $\pi/2 < \sec^{-1} x < \pi$
$f(x) = \cot^{-1} x$	$\int f(x)\,dx = x\cot^{-1} x + 0.5\ln(1 + x^2) + c$
$f(x) = \sinh x$	$\int f(x)\,dx = \cosh x + c$
$f(x) = \cosh x$	$\int f(x)\,dx = \sinh x + c$
$f(x) = \tanh x$	$\int f(x)\,dx = \ln\lvert\cosh x\rvert + c$
$f(x) = \operatorname{csch} x$	$\int f(x)\,dx = \ln\lvert\tanh(0.5x)\rvert + c$
$f(x) = \operatorname{sech} x$	$\int f(x)\,dx = 2\tan^{-1}(e^x) + c$
$f(x) = \coth x$	$\int f(x)\,dx = \ln\lvert\sinh x\rvert + c$
$f(x) = \sinh ax$	$\int f(x)\,dx = a^{-1}\cosh ax + c$
$f(x) = \cosh ax$	$\int f(x)\,dx = a^{-1}\sinh ax + c$
$f(x) = \tanh ax$	$\int f(x)\,dx = a^{-1}\ln\lvert\cosh ax\rvert + c$
$f(x) = \operatorname{csch} ax$	$\int f(x)\,dx = a^{-1}\ln\lvert\tanh(0.5\,ax)\rvert + c$
$f(x) = \operatorname{sech} ax$	$\int f(x)\,dx = 2a^{-1}\tan^{-1}(e^{ax}) + c$
$f(x) = \coth ax$	$\int f(x)\,dx = a^{-1}\ln\lvert\sinh ax\rvert + c$
$f(x) = \sinh^2 x$	$\int f(x)\,dx = 0.5\sinh x\cosh x - 0.5x + c$

TABLE 6.9 Indefinite Integrals* (*Continued*)

Function	Indefinite integral
$f(x) = \cosh^2 x$	$\int f(x)\,dx = 0.5 \sinh x \cosh x + 0.5x + c$
$f(x) = \tanh^2 x$	$\int f(x)\,dx = x - \tanh x + c$
$f(x) = \operatorname{csch}^2 x$	$\int f(x)\,dx = -\coth x + c$
$f(x) = \operatorname{sech}^2 x$	$\int f(x)\,dx = \tanh x + c$
$f(x) = \coth^2 x$	$\int f(x)\,dx = x - \coth x + c$
$f(x) = \sinh^2 ax$	$\int f(x)\,dx = 0.5a^{-1} \sinh ax \cosh ax - 0.5x + c$
$f(x) = \cosh^2 ax$	$\int f(x)\,dx = 0.5a^{-1} \sinh ax \cosh ax + 0.5x + c$
$f(x) = \tanh^2 ax$	$\int f(x)\,dx = x - a^{-1} \tanh ax + c$
$f(x) = \operatorname{csch}^2 ax$	$\int f(x)\,dx = -a^{-1} \coth ax + c$
$f(x) = \operatorname{sech}^2 ax$	$\int f(x)\,dx = a^{-1} \tanh ax + c$
$f(x) = \coth^2 ax$	$\int f(x)\,dx = x - a^{-1} \coth ax + c$
$f(x) = (\sinh x)^1$	$\int f(x)\,dx = \ln \lvert\tanh 0.5x\rvert + c$
$f(x) = (\cosh x)^{-1}$	$\int f(x)\,dx = 2 \tan^{-1} e^x + c$
$f(x) = (\sinh ax)^{-1}$	$\int f(x)\,dx = a^{-1} \ln \lvert\tanh 0.5ax\rvert + c$
$f(x) = (\cosh ax)^{-1}$	$\int f(x)\,dx = 2a^{-1} \tan^{-1} e^{ax} + c$
$f(x) = (\sinh x)^{-2}$	$\int f(x)\,dx = \coth x + c$
$f(x) = (\cosh x)^{-2}$	$\int f(x)\,dx = \tanh x + c$
$f(x) = (\sinh ax)^{-2}$	$\int f(x)\,dx = a^{-1} \coth ax + c$
$f(x) = (\cosh ax)^{-2}$	$\int f(x)\,dx = a^{-1} \tanh ax + c$
$f(x) = (\sinh x \cosh x)^{-2}$	$\int f(x)\,dx = -2 \coth 2x + c$
$f(x) = (\sinh x \cosh x)^{-1}$	$\int f(x)\,dx = \ln \lvert\tanh x\rvert + c$
$f(x) = \sinh x \cosh x$	$\int f(x)\,dx = 0.5 \sinh^2 x + c$
$f(x) = \sinh^2 x \cosh^2 x$	$\int f(x)\,dx = (1/32) \sinh 4x - (1/8)x + c$
$f(x) = (\sinh ax \cosh ax)^{-2}$	$\int f(x)\,dx = -2a^{-1} \coth 2ax + c$
$f(x) = (\sinh ax \cosh ax)^{-1}$	$\int f(x)\,dx = a^{-1} \ln \lvert\tanh ax\rvert + c$
$f(x) = \sinh ax \cosh ax$	$\int f(x)\,dx = 0.5a^{-1} \sinh^2 ax + c$
$f(x) = \sinh^2 ax \cosh^2 ax$	$\int f(x)\,dx = (1/32)\, a^{-1} \sinh 4ax - (1/8)x + c$
$f(x) = \sinh^{-1} x$	$\int f(x)\,dx = x \sinh^{-1} x - (x^2 + 1)^{1/2} + c$
$f(x) = \cosh^{-1} x$	$\int f(x)\,dx = x \cosh^{-1} x + (x^2 - 1)^{1/2} + c$ when $\cosh^{-1} x < 0$
	$\int f(x)\,dx = x \cosh^{-1} x - (x^2 - 1)^{1/2} + c$ when $\cosh^{-1} x > 0$
$f(x) = \tanh^{-1} x$	$f(x)\,dx = x \tanh^{-1} x + 0.5 \ln (1 - x^2) + c$
$f(x) = \operatorname{csch}^{-1} x$	$\int f(x)\,dx = x \operatorname{csch}^{-1} x - \sinh^{-1} x + c$ when $x < 0$
	$\int f(x)\,dx = x \operatorname{csch}^{-1} x + \sinh^{-1} x + c$ when $x > 0$
$f(x) = \operatorname{sech}^{-1} x$	$\int f(x)\,dx = x \operatorname{sech}^{-1} x - \sin^{-1} x + c$ when $\operatorname{sech}^{-1} x < 0$
	$\int f(x)\,dx = x \operatorname{sech}^{-1} x + \sin^{-1} x + c$ when $\operatorname{sech}^{-1} x > 0$

TABLE 6.9 Indefinite Integrals* (*Continued*)

Function	Indefinite integral
$f(x) = \coth^{-1} x$	$\int f(x)\,dx = x \coth^{-1} x + 0.5 \ln (x^2 - 1) + c$
$f(x) = \sinh^{-1} ax$	$\int f(x)\,dx = x \sinh^{-1} ax - (x^2 + a^{-2})^{1/2} + c$
$f(x) = \cosh^{-1} ax$	$\int f(x)\,dx = x \cosh^{-1} ax + (x^2 - a^{-2})^{1/2} + c$ when $\cosh^{-1} ax < 0$
	$\int f(x)\,dx = x \cosh^{-1} ax - (x^2 - a^{-2})^{1/2} + c$ when $\cosh^{-1} ax > 0$
$f(x) = \tanh^{-1} ax$	$\int f(x)\,dx = x \tanh^{-1} ax + 0.5a^{-1} \ln (a^{-2} - x^2) + c$
$f(x) = \operatorname{csch}^{-1} ax$	$\int f(x)\,dx = x \operatorname{csch}^{-1} ax - a^{-1} \sinh^{-1} ax + c$ when $x < 0$
	$\int f(x)\,dx = x \operatorname{csch}^{-1} ax + a^{-1} \sinh^{-1} ax + c$ when $x > 0$
$f(x) = \operatorname{sech}^{-1} ax$	$\int f(x)\,dx = x \operatorname{sech}^{-1} ax - a^{-1} \sin^{-1} ax + c$ when $\operatorname{sech}^{-1} ax < 0$
	$\int f(x)\,dx = x \operatorname{sech}^{-1} ax + a^{-1} \sin^{-1} ax + c$ when $\operatorname{sech}^{-1} ax > 0$
$f(x) = \coth^{-1} ax$	$\int f(x)\,dx = x \coth^{-1} ax + 0.5a^{-1} \ln (x^2 - a^{-2}) + c$

*Letters a, b, r, and s denote general constants; c denotes the constant of integration; f, g, and h denote functions; w, x, y, and z denote variables. The letter e represents the exponential constant (approximately 2.71828)

TABLE 6.10 Fourier Series*

Description	Expansion
General Fourier series	$F(x) = a_0/2 + a_1 \cos(\pi x/L) + b_1 \sin(\pi x/L)$ $+ a_2 \cos(2\pi x/L) + b_2 \sin(2\pi x/L)$ $+ a_3 \cos(3\pi x/L) + b_3 \sin(3\pi x/L) + \ldots$ $+ a_n \cos(n\pi x/L) + b_n \sin(n\pi x/L) + \ldots$
Square wave $f(\theta) = -1$ for $-\pi < \theta < 0$ $f(\theta) = 1$ for $0 < \theta < \pi$	$F(\theta) = \sin\theta + (\sin 3\theta)/3 + (\sin 5\theta)/5$ $+ (\sin 7\theta)/7 + (\sin 9\theta)/9 + \ldots$
Ramp wave $f(\theta) = \theta$ for $-\pi < \theta < \pi$	$F(\theta) = \sin\theta - (\sin 2\theta)/2 + (\sin 3\theta/3)$ $- (\sin 4\theta)/4 + (\sin 5\theta)/5 - \ldots$
Inverted full-rectified sine wave $f(\theta) = \theta^2$ for $-\pi < \theta < \pi$	$F(\theta) = \pi/3 - 4\cos\theta + (4\cos 2\theta)/2^2$ $- (4\cos 3\theta)/3^2 + (4\cos 4\theta)/4^2$ $- (4\cos 5\theta)/5^2 - \ldots$
Sawtooth wave $f(\theta) = \|\theta\|$	$F(\theta) = \cos\theta + (\cos 3\theta)/3^2 + (\cos 5\theta)/5^2$ $+ (\cos 7\theta)/7^2 + (\cos 9\theta)/9^2 + \ldots$
Half-wave-rectified sine wave $f(\theta) = \sin\theta$ for $0 < \theta < \pi$ $f(\theta) = 0$ for $\pi \le \theta \le 2\pi$	$F(\theta) = \frac{1}{2} + (\pi/4)\sin\theta - (\cos 2\theta)/3$ $- (\cos 4\theta)/(3 \times 5) - (\cos 6\theta)/(5 \times 7) - \ldots$
Full-wave-rectified sine wave $f(\theta) =$ $\|\sin\theta\|$ for $-\pi < \theta < \pi$	$F(\theta) = \frac{1}{2} - (\cos 2\theta)/3 - (\cos 4\theta)/(3 \times 5)$ $- (\cos 6\theta)/(5 \times 7) - (\cos 8\theta)/(7 \times 9) - \ldots$
2π	$8 - \frac{8}{3} + \frac{8}{5} - \frac{8}{7} + \frac{8}{9} - \frac{8}{11} + \ldots$
π	$4 - \frac{4}{3} + \frac{4}{5} - \frac{4}{7} + \frac{4}{9} - \frac{4}{11} + \ldots$
$\pi/2$	$2 - \frac{2}{3} + \frac{2}{5} - \frac{2}{7} + \frac{2}{9} - \frac{2}{11} + \ldots$
$\pi/4$	$1 - \frac{1}{3} + \frac{1}{5} - \frac{1}{7} + \frac{1}{9} - \frac{1}{11} + \ldots$
$\pi/8$	$\frac{1}{2} - \frac{1}{6} + \frac{1}{10} - \frac{1}{14} + \frac{1}{18} - \frac{1}{22} + \ldots$
π^2	$6 + \frac{6}{4} + \frac{6}{9} + \frac{6}{16} + \ldots + 6/n^2 + \ldots$
$\pi^2/2$	$3 + \frac{3}{4} + \frac{3}{9} + \frac{3}{16} + \ldots + 3/n^2 + \ldots$
$\pi^2/3$	$2 + \frac{2}{4} + \frac{2}{9} + \frac{2}{16} + \ldots + 2/n^2 + \ldots$
$\pi^2/6$	$1 + \frac{1}{4} + \frac{1}{9} + \frac{1}{16} + \ldots + 1/n^2 + \ldots$
$\pi^2/8$	$1 + \frac{1}{9} + \frac{1}{25} + \ldots + 1/(2n - 1)^2 + \ldots$
$\pi^2/12$	$1 - \frac{1}{4} + \frac{1}{9} - \frac{1}{16} + \frac{1}{25} - \frac{1}{36} \pm \ldots$ $+ 1/(n^2)$ if n is odd ... $- 1/(n^2)$ if n is even ... $\pm \ldots$

*Functions are in the first column; initial terms are in the second column. Letters a and b (with or without subscripts) denote constants; n denotes a positive integer; f denotes a function; F denotes a Fourier series corresponding to the function f. Characters x and θ denote variables.

TABLE 6.11 Fourier Transforms*

Description	Transform
General Fourier transform of $f(x)$	$F(\omega) = \int_{-\infty}^{\infty} f(x)\, e^{-i\omega x}\, dx$
General Fourier sine transform of $f(x)$	$S(\omega) = \int_{0}^{\infty} f(x) \sin \omega x\, dx$
General Fourier cosine transform of $f(x)$	$C(\omega) = \int_{0}^{\infty} f(x) \cos \omega x\, dx$
$f(x) = x^{-1}$	$S(\omega) = 0.5\pi$
$f(x) = x^{-1/2}$	$S(\omega) = (0.5\pi/\omega)^{1/2}$
$f(x) = x^{-1/2}$	$C(\omega) = (0.5\pi/\omega)^{1/2}$
$f(x) = (x^2 + 1)^{-1}$	$F(\omega) = \pi e^{-\omega}$
$f(x) = (x^2 + 1)^{-1}$	$C(\omega) = 0.5\pi e^{-\omega}$
$f(x) = (x^2 + a^2)^{-1}$	$F(\omega) = a^{-1}\pi e^{-a\omega}$
$f(x) = (x^2 + a^2)^{-1}$	$C(\omega) = 0.5a^{-1}\pi e^{-a\omega}$
$f(x) = x(x^2 + 1)^{-1}$	$F(\omega) = -i\pi e^{-\omega}$
$f(x) = x(x^2 + 1)^{-1}$	$S(\omega) = 0.5\pi e^{-\omega}$
$f(x) = x(x^2 + a^2)^{-1}$	$F(\omega) = -i\pi e^{-a\omega}$
$f(x) = x(x^2 + a^2)^{-1}$	$S(\omega) = 0.5\pi e^{-a\omega}$
$f(x) = e^{-\lvert x\rvert}$	$F(\omega) = -(2\pi^{-1})^{1/2}(1 + \omega^2)^{-1}$
$f(x) = e^{-\lvert x\rvert}$	$S(\omega) = (2\pi^{-1})^{1/2}\omega(1 + \omega^2)^{-1}$
$f(x) = e^{-\lvert x\rvert}$	$C(\omega) = (2\pi^{-1})^{1/2}(1 + \omega^2)^{-1}$
$f(x) = e^{-a\lvert x\rvert}$	$F(\omega) = -a(2\pi^{-1})^{1/2}(a^2 + \omega^2)^{-1}$
$f(x) = e^{-a\lvert x\rvert}$	$S(\omega) = (2\pi^{-1})^{1/2}\omega(a^2 + \omega^2)^{-1}$
$f(x) = e^{-a\lvert x\rvert}$	$C(\omega) = (2\pi^{-1})^{1/2}a(a^2 + \omega^2)^{-1}$
$f(x) = \lvert x\rvert^{-1/2} e^{-\lvert x\rvert}$	$F(\omega) = (1 + \omega^2)^{-1/2}(1 + (1 + \omega^2)^{1/2})^{1/2}$
$f(x) = \lvert x\rvert^{-1/2} e^{-a\lvert x\rvert}$	$F(\omega) = (a^2 + \omega^2)^{-1/2}(a + (a^2 + \omega^2)^{1/2})^{1/2}$
$f(x) = x^{-1} \sin x$	$F(\omega) = (0.5\pi)^{1/2}$ when $\lvert\omega\rvert < 1$ $F(\omega) = 0$ when $\lvert\omega\rvert > 1$
$f(x) = x^{-1} \sin x$	$S(\omega) = 0.5 \ln ((\omega - 1)^{-1}(\omega + 1))$
$f(x) = x^{-1} \sin x$	$C(\omega) = 0.5\pi$ when $\omega < 1$ $C(\omega) = 0.25\pi$ when $\omega = 1$ $C(\omega) = 0$ when $\omega > 1$
$f(x) = x^{-1} \sin ax$	$F(\omega) = (0.5\pi)^{1/2}$ when $\lvert\omega\rvert < a$ $F(\omega) = 0$ when $\lvert\omega\rvert > a$
$f(x) = x^{-1} \sin ax$	$S(\omega) = 0.5 \ln ((\omega - a)^{-1}(\omega + a))$
$f(x) = x^{-1} \sin ax$	$C(\omega) = 0.5\pi$ when $\omega < a$ $C(\omega) = 0.25\pi$ when $\omega = a$ $C(\omega) = 0$ when $\omega > a$
$f(x) = x^{-2} \sin x$	$S(\omega) = 0.5\omega\pi$ when $\omega < 1$ $S(\omega) = 0.5\pi$ when $\omega > 1$
$f(x) = x^{-2} \sin ax$	$S(\omega) = 0.5\omega\pi$ when $\omega < a$ $S(\omega) = 0.5a\pi$ when $\omega > a$

TABLE 6.11 Fourier Transforms* (*Continued*)

Description	Transform
$f(x) = \sin x^2$	$F(\omega) = (2)^{-1/2}(\cos(0.25\omega^2) + 0.25\pi)$
$f(x) = \sin x^2$	$C(\omega) = (\pi/8)^{1/2}(\cos(0.25\omega^2) - \sin(0.25\omega^2))$
$f(x) = \sin ax^2$	$F(\omega) = (2a)^{-1/2}(\cos(0.25a^{-1}\omega^2) + 0.25\pi)$
$f(x) = \sin ax^2$	$C(\omega) = (a^{-1}\pi/8)^{1/2}(\cos(0.25a^{-1}\omega^2) - \sin(0.25a^{-1}\omega^2))$
$f(x) = x^{-1}\cos x$	$S(\omega) = 0$ when $\omega < 1$ $S(\omega) = 0.25\pi$ when $\omega = 1$ $S(\omega) = 0.5\pi$ when $\omega > 1$
$f(x) = x^{-1}\cos ax$	$S(\omega) = 0$ when $\omega < a$ $S(\omega) = 0.25\pi$ when $\omega = a$ $S(\omega) = 0.5\pi$ when $\omega > a$
$f(x) = \cos x^2$	$F(\omega) = (2)^{-1/2}(\cos(0.25\omega^2) - 0.25\pi)$
$f(x) = \cos x^2$	$C(\omega) = (0.125\pi)^{1/2}(\cos(0.25\omega^2) + \sin(0.25\omega^2))$
$f(x) = \cos ax^2$	$F(\omega) = (2a)^{-1/2}(\cos(0.25a^{-1}\omega^2) - 0.25\pi)$
$f(x) = \cos ax^2$	$C(\omega) = (a^{-1}\pi/8)^{1/2}(\cos(0.25a^{-1}\omega^2) + \sin(0.25a^{-1}\omega^2))$
$f(x) = \tan^{-1} x$	$S(\omega) = 0.5\omega^{-1}\pi e^{-\omega}$
$f(x) = \tan^{-1} ax$	$S(\omega) = 0.5\omega^{-1}\pi e^{-\omega/a}$
$f(x) = (\sinh x)/(\sinh \pi x)$	$F(\omega) \approx 0.1339(\cos a + \cosh\omega)^{-1}$
$f(x) = (\sinh ax)/(\sinh \pi x)$ when $-\pi < a < \pi$	$F(\omega) = (2\pi)^{-1/2}(\cos a + \cosh\omega)^{-1}\sin a$
$f(x) = (\cosh x)/(\cosh \pi x)$	$F(\omega) \approx 0.7979(0.8776 + \cosh\omega)^{-1}\cosh 0.5\omega$
$f(x) = (\cosh ax)/(\cosh \pi x)$ when $-\pi < a < \pi$	$F(\omega) = (2\pi^{-1})^{1/2}(\cos a + \cosh\omega)^{-1}\cos 0.5a \cosh 0.5\omega$
$f(x) = \csc x$	$S(\omega) = 0.5\pi \tanh(0.5\pi\omega)$
$f(x) = \csc ax$	$S(\omega) = 0.5a^{-1}\pi \tanh(0.5a^{-1}\pi\omega)$
$f(x) = \operatorname{sech} x$	$C(\omega) = 0.5\pi \operatorname{sech}(0.5\pi\omega)$
$f(x) = \operatorname{sech} ax$	$C(\omega) = 0.5a^{-1}\pi \operatorname{sech}(0.5a^{-1}\pi\omega)$

*Functions are in the first column; transform functions are in the second column. Letters a and b denote constants; n denotes a positive integer; i denotes the unit imaginary number $(-1)^{1/2}$; f denotes a function; F denotes a Fourier transform; S denotes a Fourier sine transform; C denotes a Fourier cosine transform. Characters x and ω denote variables.

TABLE 6.12 Orthogonal Polynomials*

Symbol	Polynomial expansion
$T_0\ (x)$	1
$T_1\ (x)$	x
$T_2\ (x)$	$2x^2 - 1$
$T_3\ (x)$	$4x^3 - 3x$
$T_4\ (x)$	$8x^4 - 8x^2 + 1$
$T_5\ (x)$	$16x^5 - 20x^3 + 5x$
$T_6\ (x)$	$32x^6 - 48x^4 + 18x^2 - 1$
$T_7\ (x)$	$64x^7 - 112x^5 + 56x^3 - 7x$
$T_8\ (x)$	$128x^8 - 256x^6 + 160x^4 - 32x^2 + 1$
$T_9\ (x)$	$256x^9 - 576x^7 + 432x^5 - 120x^3 + 9x$
$T_{10}\ (x)$	$512x^{10} - 1280x^8 + 1120x^6 - 400x^4 + 50x^2 - 1$
$U_0\ (x)$	1
$U_1\ (x)$	$2x$
$U_2\ (x)$	$4x^2 - 1$
$U_3\ (x)$	$8x^3 - 4x$
$U_4\ (x)$	$16x^4 - 12x^2 + 1$
$U_5\ (x)$	$32x^5 - 32x^3 + 6x$
$U_6\ (x)$	$64x^6 - 80x^4 + 24x^2 - 1$
$U_7\ (x)$	$128x^7 - 192x^5 + 80x^3 - 8x$
$U_8\ (x)$	$256x^8 - 448x^6 + 240x^4 - 40x^2 + 1$
$U_9\ (x)$	$512x^9 - 1024x^7 + 672x^5 - 160x^3 + 10x$
$U_{10}\ (x)$	$1024x^{10} - 2304x^8 + 1792x^6 - 560x^4 + 60x^2 - 1$
$H_0\ (x)$	1
$H_1\ (x)$	$2x$
$H_2\ (x)$	$4x^2 - 2$
$H_3\ (x)$	$8x3 - 12x$
$H_4\ (x)$	$16x^4 - 48x^2 + 12$
$H_5\ (x)$	$32x^5 - 160x^3 + 120x$
$H_6\ (x)$	$64x^6 - 480x^4 + 720x^2 - 120$
$H_7\ (x)$	$128x^7 - 1344x^5 + 3360x^3 - 1680x$
$H_8\ (x)$	$256x^8 - 3584x^6 + 13440x^4 - 13440x^2 + 1680$
$H_9\ (x)$	$512x^9 - 9216x^7 + 48384x^5 - 80640x^3 + 30240x$
$H_{10}\ (x)$	$1024x^{10} - 23040x^8 + 161280x^6 - 403200x^4 + 302400x^2 - 30240$
$L_0\ (x)$	1
$L_1\ (x)$	$-x + 1$
$L_2\ (x)$	$x^2 - 4x + 2$
$L_3\ (x)$	$-x^3 + 9x^2 - 18x + 6$
$L_4\ (x)$	$x^4 - 16x^3 + 72x^2 - 96x + 24$
$L_5\ (x)$	$-x^5 + 25x^4 - 200x^3 + 600x^2 - 600x + 120$
$L_6\ (x)$	$x^6 - 36x^5 + 450x^4 - 2400x^3 + 5400x^2 - 4320x + 720$
$L_7\ (x)$	$-x^7 + 49x^6 - 882x^5 + 7350x^4 - 29400x^3 + 52920x^2 - 35280x + 5040$
$L_8\ (x)$	$x^8 - 64x^7 + 1568x^6 - 18816x^5 + 117600x^4 - 376320x^3 + 564480x^2 - 322560x + 40320$

TABLE 6.12 Orthogonal Polynomials* (*Continued*)

Symbol	Polynomial expansion
$L_9(x)$	$-x^9 + 81x^8 - 2592x^7 + 42336x^6 - 381024x^5 + 1905120x^4 - 5080320x^3 + 6531840x^2 - 3265930x + 362880$
$L_{10}(x)$	$x^{10} - 100x^9 + 4050x^8 - 86400x^7 + 1058400x^6 - 7620480x^5 + 31752000x^4 - 72576000x^3 + 81648000x^2 - 36288000x + 3628800$

*Symbols are in the first column; expansions are in the second column. T_n represent Tschebyshev polynomials of the first kind; U_n represent Tschebyshev polynomials of the second kind; H_n represent Hermite polynomials; L_n represent Laguerre polynomials.

TABLE 6.13 One-dimensional Laplace Transforms*

Image function	Original function
$f(s) = s^{-1}$	$F(t) = 1$
$f(s) = s^{-2}$	$F(t) = t$
$f(s) = s^{-3}$	$F(t) = 0.5\, t^2$
$f(s) = s^{-4}$	$F(t) = (1/6)\, t^3$
$f(s) = s^{-5}$	$F(t) = (1/24)\, t^4$
$f(s) = s^{-n}$	$F(t) = (1/(n-1)!)\, t^{(n-1)}$
$f(s) = s^{-1/2}$	$F(t) = \pi^{-1/2}\, t^{-1/2}$
$f(s) = s^{-3/2}$	$F(t) = 2\pi^{-1/2}\, t^{1/2}$
$f(s) = s^{-5/2}$	$F(t) = (4/3)\, \pi^{-1/2}\, t^{3/2}$
$f(s) = s^{-7/2}$	$F(t) = (8/15)\, \pi^{-1/2}\, t^{5/2}$
$f(s) = s^{-9/2}$	$F(t) = (16/105)\, \pi^{-1/2}\, t^{7/2}$
$f(s) = s^{(-n-1/2)}$	$F(t) = ((1 \times 3 \times 5 \times \cdots \times (2n-1)) \times 2^n\, \pi^{-1/2}\, t^{(n-1/2)}$
$f(s) = (s + a)^{-1}$	$F(t) = e^{-at}$
$f(s) = (s - a)^{-1}$	$F(t) = e^{at}$
$f(s) = (s + a)^{-2}$	$F(t) = t\, e^{-at}$
$f(s) = (s - a)^{-2}$	$F(t) = t\, e^{at}$
$f(s) = (s + a)^{-3}$	$F(t) = 0.5\, t^2\, e^{-at}$
$f(s) = (s - a)^{-3}$	$F(t) = 0.5\, t^2\, e^{at}$
$f(s) = (s + a)^{-4}$	$F(t) = (1/6)\, t^3\, e^{-at}$
$f(s) = (s - a)^{-4}$	$F(t) = (1/6)\, t^3\, e^{at}$
$f(s) = (s + a)^{-5}$	$F(t) = (1/24)\, t^4\, e^{-at}$
$f(s) = (s - a)^{-5}$	$F(t) = (1/24)\, t^4\, e^{at}$
$f(s) = (s + a)^{-n}$	$F(t) = (1/(n-1)!)\, t^{(n-1)}\, e^{-at}$
$f(s) = (s - a)^{-n}$	$F(t) = (1/(n-1)!)\, t^{(n-1)}\, e^{at}$
$f(s) = s(s + a)^{-3/2}$	$F(t) = \pi^{-1/2}\, t^{-1/2}\, (e^{-at} - 2at\, e^{-at})$
$f(s) = (s^2 + a^2)^{-1}$	$F(t) = a^{-1} \sin at$
$f(s) = (s^2 - a^2)^{-1}$	$F(t) = a^{-1} \sinh at$
$f(s) = s(s^2 + a^2)^{-1}$	$F(t) = \cos at$
$f(s) = s(s^2 - a^2)^{-1}$	$F(t) = \cosh at$
$f(s) = (s^4 - a^4)^{-1}$	$F(t) = 0.5\, a^{-3} \sinh at - 0.5\, a^{-3} \sin at$
$f(s) = s(s^4 - a^4)^{-1}$	$F(t) = 0.5\, a^{-2} \cosh at - 0.5\, a^{-2} \cos at$
$f(s) = s^2\, (s^4 - a^4)^{-1}$	$F(t) = 0.5\, a^{-1} \sinh at + 0.5\, a^{-1} \sin at$

TABLE 6.13 One-dimensional Laplace Transforms* (*Continued*)

Image function	Original function
$f(s) = s^3 (s^4 - a^4)^{-1}$	$F(t) = 0.5 \cosh at + 0.5 \cos at$
$f(s) = (s^4 + 4a^4)^{-1}$	$F(t) = 0.25\, a^{-3} \sin at \cosh at$ $- 0.25\, a^{-3} \cos at \sinh at$
$f(s) = s(s^4 + 4a^4)^{-1}$	$F(t) = 0.5\, a^{-2} \sin at \sinh at$
$f(s) = s^2 (s^4 + 4a^4)^{-1}$	$F(t) = 0.5\, a^{-1} \sin at \cosh at$ $+ 0.5\, a^{-1} \cos at \sinh at$
$f(s) = s^3(s^4 + 4a^4)^{-1}$	$F(t) = \cosh at \cos at$
$f(s) = 4a^3(s^4 + 4a^4)^{-1}$	$F(t) = \sin at \cosh at - \cos at \sinh at$
$f(s) = (s^3 + a^2 s)^{-1}$	$F(t) = a^{-2} - a^{-2} \cos at$
$f(s) = (s^4 + a^2 s^2)^{-1}$	$F(t) = a^{-2} t - a^{-3} \sin at$
$f(s) = (s^2 + a^2)^{-2}$	$F(t) = 0.5\, a^{-3} \sin at$ $- 0.5\, a^{-2} t \cos at$
$f(s) = (s^2 - a^2)^{-2}$	$F(t) = 0.5\, a^{-2} t \cosh at$ $- 0.5\, a^{-3} \sinh at$
$f(s) = s(s^2 + a^2)^{-2}$	$F(t) = 0.5\, a^{-1} t \sin at$
$f(s) = s(s^2 - a^2)^{-2}$	$F(t) = 0.5\, a^{-1} t \sinh at$
$f(s) = s^2 (s^2 + a^2)^{-2}$	$F(t) = 0.5\, a^{-1} \sin at + 0.5\, t \cos at$
$f(s) = s^2 (s^2 - a^2)^{-2}$	$F(t) = 0.5\, a^{-1} \sinh at + 0.5\, t \cosh at$
$f(s) = s^3 (s^2 + a^2)^{-2}$	$F(t) = \cos at - 0.5\, at \sin at$
$f(s) = s^3 (s^2 - a^2)^{-2}$	$F(t) = \cosh at + 0.5\, at \sinh at$
$f(s) = (s^2 - a^2)(s^2 + a^2)^{-2}$	$F(t) = t \cos at$
$f(s) = (s^2 + a^2)(s^2 - a^2)^{-2}$	$F(t) = t \cosh at$
$f(s) = (s^2 + a^2)^{-3}$	$F(t) = (3/8)\, a^{-5} \sin at$ $- (1/8)\, a^{-3} t^2 \sin at$ $- 3/8\, a^{-4} t \cos at$
$f(s) = (s^2 - a^2)^{-3}$	$F(t) = (3/8)\, a^{-5} \sinh at$ $+ (1/8)\, a^{-3} t^2 \sinh at$ $- 3/8\, a^{-4} t \cosh at$
$f(s) = s\,(s^2 + a^2)^{-3}$	$F(t) = (1/8)\, a^{-3} t \sin at$ $- (1/8)\, a^{-2} t^2 \cos at$
$f(s) = s\,(s^2 - a^2)^{-3}$	$F(t) = (1/8)\, a^{-2} t^2 \cosh at$ $- (1/8)\, a^{-3} t \sinh at$
$f(s) = s^2 (s^2 + a^2)^{-3}$	$F(t) = (1/8)\, a^{-3} \sin at$ $+ (1/8)\, a^{-1} t^2 \sin at$ $- 1/8\, a^{-2} t \cos at$
$f(s) = s^2 (s^2 - a^2)^{-3}$	$F(t) = (1/8)\, a^{-1} t^2 \sinh at$ $- (1/8)\, a^{-3} \sinh at$ $+ 1/8\, a^{-2} t \cosh at$
$f(s) = s^3 (s^2 + a^2)^{-3}$	$F(t) = (3/8)\, a^{-1} t \sin at$ $+ (1/8)\, t^2 \cos at$
$f(s) = s^3 (s^2 - a^2)^{-3}$	$F(t) = (3/8)\, a^{-1} t \sinh at$ $+ (1/8)\, t^2 \cosh at$
$f(s) = s^4 (s^2 + a^2)^{-3}$	$F(t) = (3/8)\, a^{-1} \sin at$ $- (1/8)\, at^2 \sin at + 5t \cos at$
$f(s) = s^4 (s^2 - a^2)^{-3}$	$F(t) = (3/8)\, a^{-1} \sinh at$ $+ (1/8)\, at^2 \sinh at + 5t \cosh at$
$f(s) = s^5 (s^2 + a^2)^{-3}$	$F(t) = \cos at - (1/8)\, a^2 t^2 \cos at$ $- (7/8)\, at \sin at$
$f(s) = s^5 (s^2 - a^2)^{-3}$	$F(t) = \cosh at + (1/8)\, a^2 t^2 \cosh at$ $+ (7/8)\, at \sinh at$

TABLE 6.13 One-dimensional Laplace Transforms* (*Continued*)

Image function	Original function
$f(s) = (3s^2 - a^2)(s^2 + a^2)^{-3}$	$F(t) = 0.5\, a^{-1}t^2 \sin at$
$f(s) = (3s^2 + a^2)(s^2 - a^2)^{-3}$	$F(t) = 0.5\, a^{-1}t^2 \sinh at$
$f(s) = (s^2 - 3a^2s)(s^2 + a^2)^{-3}$	$F(t) = 0.5\, t^2 \cos at$
$f(s) = (s^2 + 3a^2s)(s^2 - a^2)^{-3}$	$F(t) = 0.5\, t^2 \cosh at$
$f(s) = 8a^3s^2 (s^2 - a^2)^{-3}$	$F(t) = \sin at + a^2t^2 \sin at - at \cos at$
$f(s) = (s^4 - 6a^2s^2 + a^4)(s^2 + a^2)^{-4}$	$F(t) = (1/6)\, t^3 \cos at$
$f(s) = (s^4 + 6a^2s^2 + a^4)(s^2 - a^2)^{-4}$	$F(t) = (1/6)\, t^3 \cosh at$
$f(s) = (s^3 - a^2s)(s^2 + a^2)^{-4}$	$F(t) = (1/24)\, a^{-1}t^3 \sin at$
$f(s) = (s^3 + a^2s)(s^2 - a^2)^{-4}$	$F(t) = (1/24)\, a^{-1}t^3 \sinh at$
$f(s) = (s^2 + 2sa + a^2 + b^2)^{-1}$	$F(t) = b^{-1} e^{-at} \sin bt$
$f(s) = (s^2 - 2sa + a^2 + b^2)^{-1}$	$F(t) = b^{-1} e^{at} \sin bt$
$f(s) = (s^2 - 2sa + a^2 - b^2)^{-1}$	$F(t) = b^{-1} e^{at} \sinh bt$
$f(s) = (s + a)(s^2 + 2sa + a^2 + b^2)^{-1}$	$F(t) = e^{-at} \cos bt$
$f(s) = (s - a)(s^2 - 2sa + a^2 + b^2)^{-1}$	$F(t) = e^{at} \cos bt$
$f(s) = (s - a)(s^2 - 2sa + a^2 - b^2)^{-1}$	$F(t) = e^{at} \cosh bt$
$f(s) = (s + a)^{-1}(s + b)^{-1}$ when $a \neq b$	$F(t) = (b - a)^{-1} e^{-at} - (b - a)^{-1} e^{-bt}$
$f(s) = s(s + a)^{-1}(s + b)^{-1}$ when $a \neq b$	$F(t) = (a - b)^{-1}a\, e^{-at} - (a - b)^{-1}b\, e^{-bt}$
$f(s) = (s + a)^{1/2} - (s + b)^{1/2}$	$F(t) = 0.5\, \pi^{-1/2}\, t^{-3/2} (e^{-bt} - e^{-at})$
$f(s) = s(s^2 + a^2)^{-1}(s^2 + b^2)^{-1}$ when $a^2 \neq b^2$	$F(t) = (b^2 - a^2)^{-1} \cos at - (b^2 - a^2)^{-1} \cos bt$
$f(s) = ((s + a)^{1/2} + (s + b)^{1/2})^{-1}$	$F(t) = (2b - 2a)^{-1/2}\, \pi^{-1/2}\, t^{-3/2}\, e^{-bt} - (2b - 2a)^{-1/2}\, \pi^{-1/2}\, t^{-3/2}\, e^{-at}$
$f(s) = s^{-3/2}\, e^{-a/s}$	$F(t) = \pi^{-1/2}\, a^{-1/2} \sin(2a^{1/2}\, t^{1/2})$
$f(s) = \ln((s + a)^{-1}(s + b))$	$F(t) = t^{-1} e^{-at} - t^{-1} e^{-bt}$
$f(s) = \ln((s - a)^{-1}(s - b))$	$F(t) = t^{-1} e^{at} - t^{-1} e^{bt}$
$f(s) = \ln((s^2 + a^2)^{-1}(s^2 + b^2))$	$F(t) = 2\, t^{-1} \cos bt - 2\, t^{-1} \cos at$
$f(s) = \ln(1 + s^{-2}a^2)$	$F(t) = 2\, t^{-1} - 2\, t^{-1} \cos at$
$f(s) = \ln(1 - s^{-2}a^2)$	$F(t) = 2\, t^{-1} - 2\, t^{-1} \cosh at$
$f(s) = s^{-1} \ln s$	$F(t) = -\ln t - \gamma$
$f(s) = -s^{-1}(\gamma + \ln s)$	$F(t) = \ln t$
$f(s) = s^{-1} \ln^2 s$	$F(t) = \ln^2 t + 2\gamma \ln t + \gamma^2 - (1/6)\, \pi^2$
$f(s) = (1/6)\, s^{-1}\, \pi^2 + s^{-1} \ln^2 s + 2\, s^{-1}\, \gamma \ln s + s^{-1}\, \gamma^2$	$F(t) = \ln^2 t$
$f(s) = \tan^{-1}(s^{-1}\, a)$	$F(t) = t^{-1} \sin at$

*The character F represents the original function; f represents the image function. The character t represents a real variable; s represents a complex variable; a and b represent constants; n represents a positive integer. The character e represents the natural logarithm base, approximately equal to 2.71828. The symbol γ represents Euler's constant, approximately equal to 0.577216. The symbol π represents the circumference-to-diameter ratio of a circle in a plane, approximately equal to 3.14159.

General one-dimensional Laplace transform

$$f(s) = \int_0^\infty e^{-st}\, F(t)\, dt$$

TABLE 6.14 Lowercase Greek Alphabet

Symbol	Character name	Common representations
α	alpha	current gain of bipolar transistor in common-base configuration; alpha particle; angular acceleration; angle; direction angle; transcendental number; scalar coefficient
β	beta	current gain of bipolar transistor in common-emitter configuration; magnetic flux density; beta particle; angle; direction angle; transcendental number; scalar coefficient
γ	gamma	gamma radiation; electrical conductivity; Euler's constant; gravity; direction angle; scalar coefficient; permutation; cycle
δ	delta	derivative; variation of a quantity; point evaluation; support function; metric function; distance function; variation of an integral; Laplacian
ε	epsilon	electric permittivity; natural logarithm base (approximately 2.71828); eccentricity; signature
ζ	zeta	impedance, coefficient; coordinate variable in a transformation
η	eta	electric susceptibility; hysteresis coefficient; efficiency; coordinate variable in a transformation
θ	theta	angle; phase angle; angle in polar coordinates; angle in cylindrical coordinates; angle in spherical coordinates; parameter; homomorphism
ι	iota	definite description (in predicate logic)
κ	kappa	dielectric constant; coefficient of coupling; curvature
λ	lambda	wavelength; Wien Displacement Law constant; ratio; Lebesgue measure; eigenvalue
μ	mu	micro-; magnetic permeability; amplification factor; charge carrier mobility; mean; statistical parameter
ν	nu	frequency; reluctivity; statistical parameter; natural epimorphism
ξ	xi	coordinate variable in a transformation
o	omicron	order
π	pi	ratio of circle circumference to diameter (approximately 3.14159); radian; permutation
ρ	rho	electrical resistivity; variable representing an angle; curvature; correlation; metric; density
σ	sigma	electrical conductivity; Stefan-Boltzmann constant; standard deviation; variance; mathematical partition; permutation; topology

TABLE 6.14 Lowercase Greek Alphabet (*Continued*)

Symbol	Character name	Common representations
τ	tau	time-phase displacement; torsion; mathematical partition; topology
υ	upsilon	—
ϕ or φ	phi	angle; phase angle; dielectric flux; angle in spherical coordinates; Euler phi function; mapping; predicate
φ	chi	magnetic susceptibility; characteristic function; chromatic number; configuration of a body
ψ	psi	angle; mapping; predicate; chart
ω	omega	angular velocity; period; modulus of continuity

TABLE 6.15 Uppercase Greek Alphabet

Symbol	Character name	Common representations
Α	alpha	—
Β	beta	magnetic flux density
Γ	gamma	gamma match; general index set; curve; contour
Δ	delta	delta match; three-phase AC circuit with no common ground; increment; difference quotient; difference sequence; Laplacian
Ε	epsilon	voltage; energy
Ζ	zeta	impedance
Η	eta	efficiency
Θ	theta	order
Ι	iota	current
Κ	kappa	magnetic susceptibility; degrees Kelvin
Λ	lambda	general index set
Μ	mu	mutual inductance
Ν	nu	Avogadro's number (6.022169×10^{23})
Ξ	xi	—
Ο	omicron	order
Π	pi	product; infinite product; homotopy
Ρ	rho	Power
Σ	sigma	summation; series; infinite series
Τ	tau	time constant; temperature
Υ	upsilon	—
Φ	phi	magnetic flux; Frattini subgroup
Χ	chi	reactance
Ψ	psi	dielectric flux
Ω	omega	ohms; volume of a body

TABLE 6.16 General Mmathematical Symbols and Their Common Meanings*

Symbol	Character name	Common representations
.	decimal or radix point	separates integral part of number from fractional part
∀	universal qualifier	read "for all"
#	pound sign	number; pounds
∃	existential qualifier	read "there exists" or "for some"
%	per cent sign	read "parts per hundred" or "percent"
‰	per mil sign	read "parts per thousand" or "permil"
&	ampersand	logical AND operation
@	at sign	read "at the rate of" or "at the cost of"
( )	parentheses	encloses elements defining coordinates of a point; encloses elements of a set of ordered numbers; encloses bounds of an open interval
[ ]	brackets	encloses a group of terms that includes one or more groups in parentheses; encloses elements of an equivalence class; encloses bounds of a closed interval
{ }	braces	encloses a group of terms that includes one or more groups in brackets; encloses elements comprising a set
[ ) or ( ]	half-brackets	encloses bounds of a half-open interval
] [	inside-out brackets	encloses bounds of an open interval
( ) or [ ]	parenthesis or brackets (enlarged)	encloses elements of a matrix
*	asterisk	multiplication; logical AND operation
×	cross	multiplication; logical AND operation; vector (cross) product of two vectors
Π	uppercase Greek letter pi (enlarged)	product of many values

TABLE 6.16 General Mmathematical Symbols and Their Common Meanings* (*Continued*)

Symbol	Character name	Common representations
•	dot	multiplication; logical AND operation; scalar (dot) product of two vectors
+	plus sign	addition; logical OR operation
Σ	uppercase Greek letter sigma (enlarged)	summation of many values
,	comma	separates large numbers by thousands; separates elements defining coordinates of a point; separates elements of a set of ordered numbers; separates bounds of an interval
−	minus sign	subtraction; logical NOT symbol
±	plus/minus sign	read "plus or minus" and defines the extent to which a value can deviate from the nominal value
/	slash or slant	division; ratio; proportion; separates parts of a Web site uniform resource locator (URL)
÷	—	division
:	colon	ratio; separates hours from minutes; separates minutes from seconds
::	double colon	mean
!	exclamation mark	factorial
≤	inequality sign	read "is less than or equal to"
<	inequality sign	read "is less than"
<<	inequality sign	read "is much less than"
=	equal sign	read "is equal to"; logical equivalence
≥	inequality sign	read "is greater than or equal to"
>	inequality sign	read "is greater than"
>>	inequality sign	read "is much greater than"
≅	congruence sign	read "is congruent with"

TABLE 6.16 General Mmathematical Symbols and Their Common Meanings* (*Continued*)

Symbol	Character name	Common representations
$\neq$	unequal sign	read "is not equal to"
$\equiv$	equivalence sign	read "is logically equivalent to"
$\approx$	approximation sign	read "is approximately equal to"
$\propto$	—	read "is proportional to"
$\sim$	squiggle	read "is similar to"
. . .	triple dot	read "and so on" or "and beyond"
$\vert$	vertical line	read "is exactly divisible by"
$\vert\ \vert$	vertical lines	absolute value of quantity between lines; length of vector quantity denoted between lines; distance between two points; cardinality of number; modulus
$\Bigg\vert$	vertical line (elongated)	denotes limits of evaluation for a function
$\Bigg\vert\ \Bigg\vert$	vertical lines (elongated)	determinant of matrix whose elements are enumerated between lines
$\aleph$	uppercase Hebrew letter aleph	transfinite cardinal number; Continuum Hypothesis
$\cap$	intersection sign	set-intersection operation
$\cup$	union sign	set-union operation
$\varnothing$	null sign	set containing no elements (empty set or null set)
$\in$	—	read "is an element of"
$\notin$	—	read "is not an element of"
$\subset$	—	read "is a proper subset of"
$\supset$	implication sign	read "logically implies"

TABLE 6.16 General Mmathematical Symbols and Their Common Meanings* (*Continued*)

Symbol	Character name	Common representations
$\subseteq$	—	read “is a subset of”
$\not\subset$	—	read “is not a proper subset of”
$\angle$	angle sign	angle; angle measure
$\perp$	—	read “is perpendicular to”
∇	del or nabla	vector differential operator
$\surd$	radical or surd	root; square root
$\Leftrightarrow$ or $\leftrightarrow$	double arrow	read “if and only if” or “is logically equivalent to”
$\Rightarrow$	right arrow	logical implication
$\therefore$	three dots	read “therefore”
$\rightarrow$	right arrow	logical implication; convergence; mapping
$\uparrow$	upward arrow	read “above” or “increasing”
$\downarrow$	downward arrow	read “below” or “decreasing”
∂	—	partial derivative; Jacobian; surface of a body
$\int$	—	integral
$\iint$	—	double integral
$\int_E$	—	Riemann integral

TABLE 6.16 General Mmathematical Symbols and Their Common Meanings* (*Continued*)

Symbol	Character name	Common representations
$\int_{\Gamma}$	—	contour integral
$\iint_{S}$	—	surface integral
$\iiint$	—	triple integral
$^{\circ}$	degree sign (superscript)	degree of angle; degree of temperature
∞	infinity sign	infinity; an arbitrarily large number; an arbitrarily great distance away

*For meanings of Greek letters, refer to Tables 6.14 and 6.15.

TABLE 6.17 Comparison of Values in Decimal, Binary, Octal, and Hexadecimal Numbering Systems for Decimal 0 through 256

Decimal	Binary	Octal	Hexadecimal
0	0	0	0
1	1	1	1
2	10	2	2
3	11	3	3
4	100	4	4
5	101	5	5
6	110	6	6
7	111	7	7
8	1000	10	8
9	1001	11	9
10	1010	12	A
11	1011	13	B
12	1100	14	C
13	1101	15	D
14	1110	16	E
15	1111	17	F
16	10000	20	10
17	10001	21	11
18	10010	22	12
19	10011	23	13
20	10100	24	14
21	10101	25	15
22	10110	26	16
23	10111	27	17
24	11000	30	18
25	11001	31	19
26	11010	32	1A
27	11011	33	1B
28	11100	34	1C
29	11101	35	1D
30	11110	36	1E
31	11111	37	1F
32	100000	40	20
33	100001	41	21
34	100010	42	22
35	100011	43	23
36	100100	44	24
37	100101	45	25
38	100110	46	26
39	100111	47	27
40	101000	50	28
41	101001	51	29
42	101010	52	2A
43	101011	53	2B
44	101100	54	2C
45	101101	55	2D

TABLE 6.17 Comparison of Values in Decimal, Binary, Octal, and Hexadecimal Numbering Systems for Decimal 0 through 256 (*Continued*)

Decimal	Binary	Octal	Hexadecimal
46	101110	56	2E
47	101111	57	2F
48	110000	60	30
49	110001	61	31
50	110010	62	32
51	110011	63	33
52	110100	64	34
53	110101	65	35
54	110110	66	36
55	110111	67	37
56	111000	70	38
57	111001	71	39
58	111010	72	3A
59	111011	73	3B
60	111100	74	3C
61	111101	75	3D
62	111110	76	3E
63	111111	77	3F
64	1000000	100	40
65	1000001	101	41
66	1000010	102	42
67	1000011	103	43
68	1000100	104	44
69	1000101	105	45
70	1000110	106	46
71	1000111	107	47
72	1001000	110	48
73	1001001	111	49
74	1001010	112	4A
75	1001011	113	4B
76	1001100	114	4C
77	1001101	115	4D
78	1001110	116	4E
79	1001111	117	4F
80	1010000	120	50
81	1010001	121	51
82	1010010	122	52
83	1010011	123	53
84	1010100	124	54
85	1010101	125	55
86	1010110	126	56
87	1010111	127	57
88	1011000	130	58
89	1011001	131	59
90	1011010	132	5A
91	1011011	133	5B

TABLE 6.17 Comparison of Values in Decimal, Binary, Octal, and Hexadecimal Numbering Systems for Decimal 0 through 256 (*Continued*)

Decimal	Binary	Octal	Hexadecimal
92	1011100	134	5C
93	1011101	135	5D
94	1011110	136	5E
95	1011111	137	5F
96	1100000	140	60
97	1100001	141	61
98	1100010	142	62
99	1100011	143	63
100	1100100	144	64
101	1100101	145	65
102	1100110	146	66
103	1100111	147	67
104	1101000	150	68
105	1101001	151	69
106	1101010	152	6A
107	1101011	153	6B
108	1101100	154	6C
109	1101101	155	6D
110	1101110	156	6E
111	1101111	157	6F
112	1110000	160	70
113	1110001	161	71
114	1110010	162	72
115	1110011	163	73
116	1110100	164	74
117	1110101	165	75
118	1110110	166	76
119	1110111	167	77
120	1111000	170	78
121	1111001	171	79
122	1111010	172	7A
123	1111011	173	7B
124	1111100	174	7C
125	1111101	175	7D
126	1111110	176	7E
127	1111111	177	7F
128	10000000	200	80
129	10000001	201	81
130	10000010	202	82
131	10000011	203	83
132	10000100	204	84
133	10000101	205	85
134	10000110	206	86
135	10000111	207	87
136	10001000	210	88
137	10001001	211	89

TABLE 6.17 Comparison of Values in Decimal, Binary, Octal, and Hexadecimal Numbering Systems for Decimal 0 through 256 (*Continued*)

Decimal	Binary	Octal	Hexadecimal
138	10001010	212	8A
139	10001011	213	8B
140	10001100	214	8C
141	10001101	215	8D
142	10001110	216	8E
143	10001111	217	8F
144	10010000	220	90
145	10010001	221	91
146	10010010	222	92
147	10010011	223	93
148	10010100	224	94
149	10010101	225	95
150	10010110	226	96
151	10010111	227	97
152	10011000	230	98
153	10011001	231	99
154	10011010	232	9A
155	10011011	233	9B
156	10011100	234	9C
157	10011101	235	9D
158	10011110	236	9E
159	10011111	237	9F
160	10100000	240	A0
161	10100001	241	A1
162	10100010	242	A2
163	10100011	243	A3
164	10100100	244	A4
165	10100101	245	A5
166	10100110	246	A6
167	10100111	247	A7
168	10101000	250	A8
169	10101001	251	A9
170	10101010	252	AA
171	10101011	253	AB
172	10101100	254	AC
173	10101101	255	AD
174	10101110	256	AE
175	10101111	257	AF
176	10110000	260	B0
177	10110001	261	B1
178	10110010	262	B2
179	10110011	263	B3
180	10110100	264	B4
181	10110101	265	B5
182	10110110	266	B6
183	10110111	267	B7

TABLE 6.17 Comparison of Values in Decimal, Binary, Octal, and Hexadecimal Numbering Systems for Decimal 0 through 256 (*Continued*)

Decimal	Binary	Octal	Hexadecimal
184	10111000	270	B8
185	10111001	271	B9
186	10111010	272	BA
187	10111011	273	BB
188	10111100	274	BC
189	10111101	275	BD
190	10111110	276	BE
191	10111111	277	BF
192	11000000	300	C0
193	11000001	301	C1
194	11000010	302	C2
195	11000011	303	C3
196	11000100	304	C4
197	11000101	305	C5
198	11000110	306	C6
199	11000111	307	C7
200	11001000	310	C8
201	11001001	311	C9
202	11001010	312	CA
203	11001011	313	CB
204	11001100	314	CC
205	11001101	315	CD
206	11001110	316	CE
207	11001111	317	CF
208	11010000	320	D0
209	11010001	321	D1
210	11010010	322	D2
211	11010011	323	D3
212	11010100	324	D4
213	11010101	325	D5
214	11010110	326	D6
215	11010111	327	D7
216	11011000	330	D8
217	11011001	331	D9
218	11011010	332	DA
219	11011011	333	DB
220	11011100	334	DC
221	11011101	335	DD
222	11011110	336	DE
223	11011111	337	DF
224	11100000	340	E0
225	11100001	341	E1
226	11100010	342	E2
227	11100011	343	E3
228	11100100	344	E4
229	11100101	345	E5

TABLE 6.17 Comparison of Values in Decimal, Binary, Octal, and Hexadecimal Numbering Systems for Decimal 0 through 256 (*Continued*)

Decimal	Binary	Octal	Hexadecimal
230	11100110	346	E6
231	11100111	347	E7
232	11101000	350	E8
233	11101001	351	E9
234	11101010	352	EA
235	11101011	353	EB
236	11101100	354	EC
237	11101101	355	ED
238	11101110	356	EE
239	11101111	357	EF
240	11110000	360	F0
241	11110001	361	F1
242	11110010	362	F2
243	11110011	363	F3
244	11110100	364	F4
245	11110101	365	F5
246	11110110	366	F6
247	11110111	367	F7
248	11111000	370	F8
249	11111001	371	F9
250	11111010	372	FA
251	11111011	373	FB
252	11111100	374	FC
253	11111101	375	FD
254	11111110	376	FE
255	11111111	377	FF
256	100000000	400	100

TABLE 6.18 Flip-flop States

A: R-S flip-flop			
R	**S**	**Q**	**−Q**
0	0	Q	−Q
0	1	1	0
1	0	0	1
1	1	?	?
B: J-K flip-flop			
J	**K**	**Q**	**−Q**
0	0	Q	−Q
0	1	1	0
1	0	0	1
1	1	−Q	Q

TABLE 6.19 Logic Gates and Their Characteristics

Gate type	Number of inputs	Remarks
NOT	1	Changes state of input.
OR	2 or more	Output high if any inputs are high. Output low if all inputs are low.
AND	2 or more	Output low if any inputs are low. Output high if all inputs are high.
NOR	2 or more	Output low if any inputs are high. Output high if all inputs are low.
NAND	2 or more	Output high if any inputs are low. Output low if all inputs are high.
XOR	2	Output high if inputs differ. Output low if inputs are the same.
XNOR	2	Output low if inputs differ. Output high if inputs are the same.

TABLE 6.20A American Wire Gauge (AWG) Diameters

AWG	Millimeters	Inches
1	7.35	0.289
2	6.54	0.257
3	5.83	0.230
4	5.19	0.204
5	4.62	0.182
6	4.12	0.163
7	3.67	0.144
8	3.26	0.128
9	2.91	0.115
10	2.59	0.102
11	2.31	0.0909
12	2.05	0.0807
13	1.83	0.0720
14	1.63	0.0642
15	1.45	0.0571
16	1.29	0.0508
17	1.15	0.0453
18	1.02	0.0402
19	0.912	0.0359
20	0.812	0.0320
21	0.723	0.0285
22	0.644	0.0254
23	0.573	0.0226
24	0.511	0.0201
25	0.455	0.0179
26	0.405	0.0159
27	0.361	0.0142
28	0.321	0.0126
29	0.286	0.0113
30	0.255	0.0100
31	0.227	0.00894
32	0.202	0.00795
33	0.180	0.00709
34	0.160	0.00630
35	0.143	0.00563
36	0.127	0.00500
37	0.113	0.00445
38	0.101	0.00398
39	0.090	0.00354
40	0.080	0.00315

TABLE 6.20B British Standard Wire Gauge (NBS SWG) Diameters

NBS SWG	Millimeters	Inches
1	7.62	0.300
2	7.01	0.276
3	6.40	0.252
4	5.89	0.232
5	5.38	0.212
6	4.88	0.192
7	4.47	0.176
8	4.06	0.160
9	3.66	0.144
10	3.25	0.128
11	2.95	0.116
12	2.64	0.104
13	2.34	0.092
14	2.03	0.080
15	1.83	0.072
16	1.63	0.064
17	1.42	0.056
18	1.22	0.048
19	1.02	0.040
20	0.91	0.036
21	0.81	0.032
22	0.71	0.028
23	0.61	0.024
24	0.56	0.022
25	0.51	0.020
26	0.46	0.018
27	0.42	0.0164
28	0.38	0.0148
29	0.345	0.0136
30	0.315	0.0124
31	0.295	0.0116
32	0.274	0.0108
33	0.254	0.0100
34	0.234	0.0092
35	0.213	0.0084
36	0.193	0.0076
37	0.173	0.0068
38	0.152	0.0060
39	0.132	0.0052
40	0.122	0.0048

TABLE 6.20C Birmingham Wire Gauge (BWG) Diameters

BWG	Millimeters	Inches
1	7.62	0.300
2	7.21	0.284
3	6.58	0.259
4	6.05	0.238
5	5.59	0.220
6	5.16	0.203
7	4.57	0.180
8	4.19	0.165
9	3.76	0.148
10	3.40	0.134
11	3.05	0.120
12	2.77	0.109
13	2.41	0.095
14	2.11	0.083
15	1.83	0.072
16	1.65	0.064
17	1.47	0.058
18	1.25	0.049
19	1.07	0.042
20	0.889	0.035

TABLE 6.21 Maximum Safe Continuous DC Carrying Capacity, in Amperes, for Various Wire Sizes (AWG) in Open Air

Wire size, AWG	Current, A
8	73
9	63
10	55
11	47
12	41
13	36
14	31
15	26
16	22
17	18
18	15
19	13
20	11

TABLE 6.22 Resistivity of Various Gauges of Solid Copper Wire, in Micro-ohms per Meter, to Three Significant Figures

Wire size, AWG	Resistivity, $\mu\Omega/m$
2	523
4	831
6	1320
8	2100
10	3340
12	5320
14	8450
16	13,400
18	21,400
20	34,000
22	54,000
24	85,900
26	137,000
28	217,000
30	345,000

TABLE 6.23 Permeability Figures for Some Common Materials

Substance	Permeability (approx.)
Aluminum	Slightly more than 1
Bismuth	Slightly less than 1
Cobalt	60–70
Ferrite	100–3000
Free space	1
Iron	60–100
Iron, refined	3000–8000
Nickel	50–60
Permalloy	3000–30,000
Silver	Slightly less than 1
Steel	300–600
Super permalloys	100,000 to 1,000,000
Wax	Slightly less than 1
Wood, dry	Slightly less than 1

TABLE 6.24 Common Types of Solder Used in Industry

Solder type	Melting point	Common uses
Tin/lead 50/50 rosin core	430 F 220 C	Electronics
Tin/lead 60/40 rosin core, low heat	370 F 190 C	Electronics
Tin/lead 63/37 rosin core, low heat	360 F 180 C	Electronics
Silver high heat, high current	600 F 320 C	Electronics
Tin/lead 50/50 acid core	430 F 220 C	Sheet-metal bonding

TABLE 6.25 Bands in the Radio Spectrum

Standard designation	Frequency range	Wavelength range
Very Low (VLF)	3 kHz–30 kHz	100 km–10 km
Low (LF)	30 kHz–300 kHz	10 km–1 km
Medium (MF)	300 kHz–3 MHz	1 km–100 m
High (HF)	3 MHz–30 MHz	100 m–10 m
Very High (VHF)	30 MHz–300 MHz	10 m–1 m
Ultra High (UHF)	300 MHz–3 GHz	1 m–100 mm
Super High (SHF)	3 GHz–30 GHz	100 mm–10 mm
Extremely High (EHF)	30 GHz–300 GHz	10 mm–1 mm

TABLE 6.26 Schematic Symbols Used in Electronics

Component	Symbol
Ammeter	 A
Amplifier general	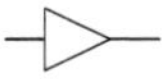
Amplifier, inverting	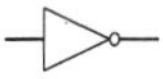
Amplifier, operational	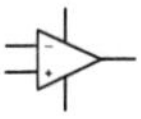
AND gate	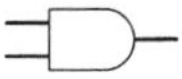
Antenna, balanced	
Antenna, general	
Antenna, loop	
Antenna, loop, multiturn	
Battery	 − +
Capacitor, feedthrough	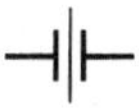
Capacitor, fixed	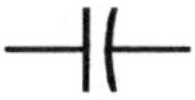
Capacitor, variable	
Capacitor, variable, split-rotor	
Capacitor, variable, split-stator	

TABLE 6.26 Schematic Symbols Used in Electronics (*Continued*)

Component	Symbol
Cathode, electron-tube, cold	
Cathode, electron-tube, directly heated	
Cathode, electron-tube indirectly heated	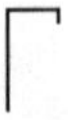
Cavity resonator	
Cell, electrochemical	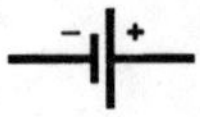 – +
Circuit breaker	
Coaxial cable	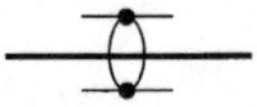
Crystal, piezoelectric	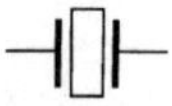
Delay line	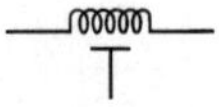
Diac	
Diode, field-effect	
Diode, general	
Diode, Gunn	
Diode, light-emitting	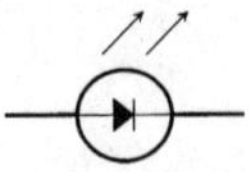
Diode, photosensitive	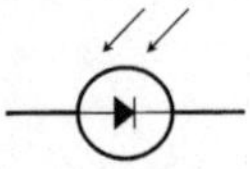

TABLE 6.26 Schematic Symbols Used in Electronics (*Continued*)

Component	Symbol
Diode, PIN	
Diode, Schottky	
Diode, tunnel	
Diode, varactor	
Diode, zener	
Directional coupler	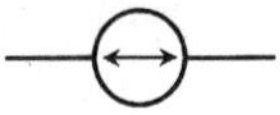
Directional wattmeter	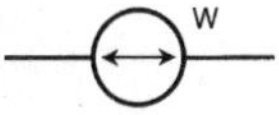 W
Exclusive-OR gate	
Female contact, general	
Ferrite bead	
Filament, electron-tube	
Fuse	
Galvanometer	 G
Grid, electron-tube	
Ground, chassis	

TABLE 6.26 Schematic Symbols Used in Electronics (*Continued*)

Component	Symbol
Ground, earth	
Handset	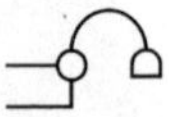
Headset, double	
Headset, single	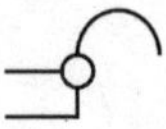
Headset, stereo	 L R
Inductor, air core	
Inductor, air core, bifilar	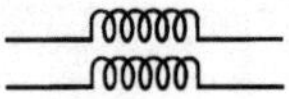
Inductor, air core, tapped	
Inductor, air core, variable	
Inductor, iron core	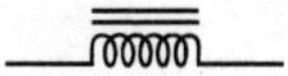
Inductor, iron core, bifilar	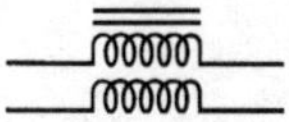
Inductor, iron core, tapped	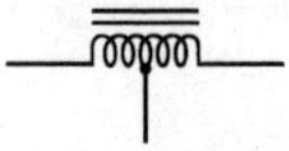
Inductor iron core, variable	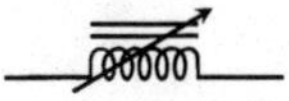
Inductor, powdered-iron core	
Inductor, powdered-iron core, bifilar	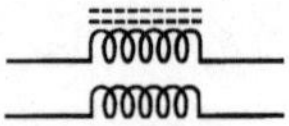

TABLE 6.26 Schematic Symbols Used in Electronics (*Continued*)

Component	Symbol
Inductor, powdered-iron core, tapped	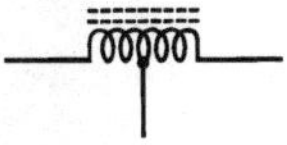
Inductor, powdered-iron core, variable	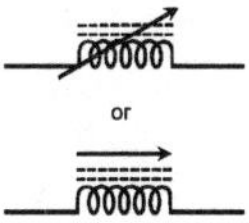 or
Integrated circuit, general	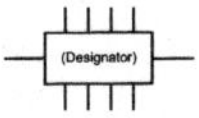 (Designator)
Jack, coaxial or phono	
Jack, phone, two-conductor	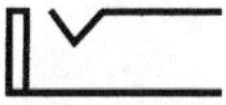
Jack, phone, three-conductor	
Key, telegraph	
Lamp, incandescent	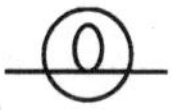
Lamp, neon	
Male contact, general	
Meter, general	
Microammeter	 μA
Microphone	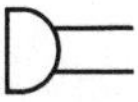

TABLE 6.26 Schematic Symbols Used in Electronics (*Continued*)

Component	Symbol
Microphone, directional	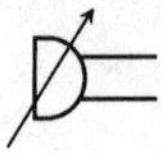
Milliammeter	 mA
NAND gate	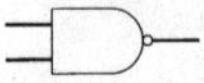
Negative voltage connection	–
NOR gate	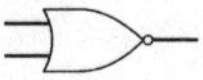
NOT gate	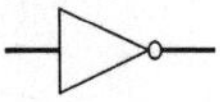
Optoisolator	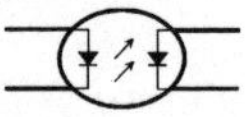
OR gate	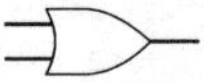
Outlet, two-wire, nonpolarized	
Outlet, two-wire, polarized	
Outlet, three-wire	
Outlet, 234-V	
Plate, electron-tube	
Plug, two-wire, nonpolarized	

TABLE 6.26 Schematic Symbols Used in Electronics (*Continued*)

Component	Symbol
Plug, two-wire, polarized	
Plug, three-wire	
Plug, 234-V	
Plug, coaxial or phono	
Plug, phone, two-conductor	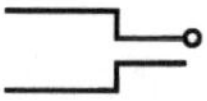
Plug, phone, three-conductor	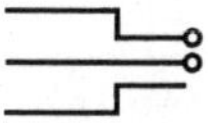
Positive voltage connection	 +
Potentiometer	
Probe, radio-frequency	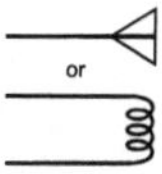 or
Rectifier, gas-filled	
Rectifier, high-vacuum	
Rectifier, semiconductor	
Rectifier, silicon-controlled	

TABLE 6.26 Schematic Symbols Used in Electronics (*Continued*)

Component	Symbol
Relay, double-pole, double-throw	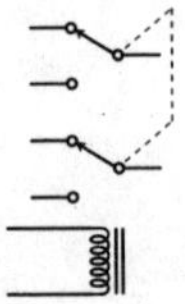
Relay, double-pole, single-throw	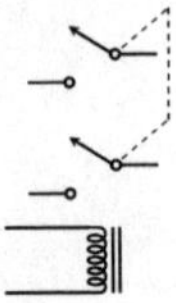
Relay, single-pole, double-throw	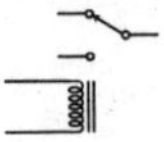
Relay, single-pole, single-throw	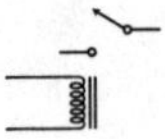
Resistor, fixed	
Resistor, preset	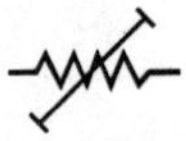
Resistor, tapped	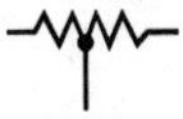
Resonator	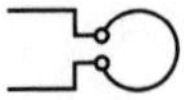
Rheostat	
Saturable reactor	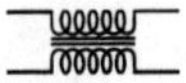
Signal generator	
Solar battery	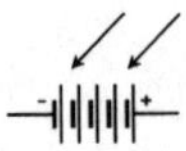

TABLE 6.26 Schematic Symbols Used in Electronics (*Continued*)

Component	Symbol
Solar cell	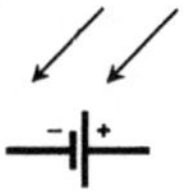
Source, constant-current	
Source, constant-voltage	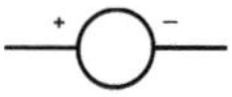
Speaker	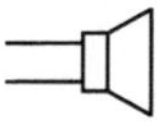
Switch, double-pole, double-throw	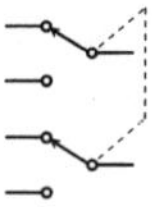
Switch, double-pole, rotary	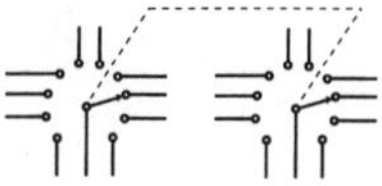
Switch, double-pole, single-throw	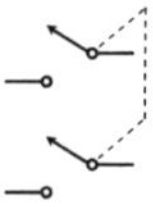
Switch, momentary-contact	
Switch, silicon-controlled	
Switch, single-pole, rotary	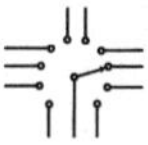
Switch, single-pole, double-throw	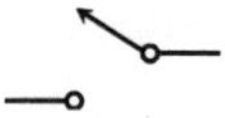
Switch, single-pole, single-throw	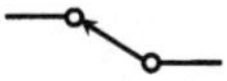

TABLE 6.26 Schematic Symbols Used in Electronics (*Continued*)

Component	Symbol
Terminals, general, balanced	
Terminals, general, unbalanced	
Test point	TP
Thermocouple	or
Transformer, air core	
Transformer, air core, step-down	
Transformer, air core, step-up	
Transformer, air core, tapped primary	
Transformer, air core, tapped secondary	
Transformer, iron core	
Transformer, iron core, step-down	
Transformer, iron core, step-up	
Transformer, iron core, tapped primary	
Transformer, iron core, tapped secondary	

TABLE 6.26 Schematic Symbols Used in Electronics (*Continued*)

Component	Symbol
Transformer, powdered-iron core	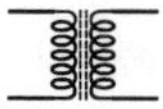
Transformer, powdered-iron core, step-down	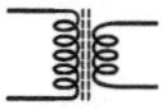
Transformer, powdered-iron core, step-up	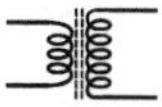
Transformer, powdered-iron core, tapped primary	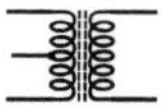
Transformer, powdered-iron core, tapped secondary	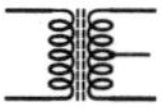
Transistor, bipolar, *NPN*	
Transistor, bipolar, *PNP*	
Transistor, field-effect, *N*-channel	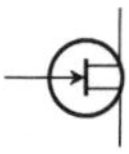
Transistor, field-effect, *P*-channel	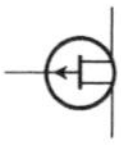
Transistor, MOS field-effect, *N*-channel	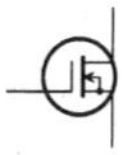
Transistor, MOS field-effect, *P*-channel	
Transistor, photosensitive, *NPN*	
Transistor, photosensitive, *PNP*	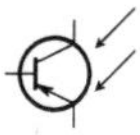

TABLE 6.26 Schematic Symbols Used in Electronics (*Continued*)

Component	Symbol
Transistor, photosensitive, field-effect, *N*-channel	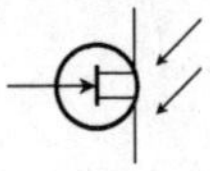
Transistor, photosensitive, field-effect, *P*-channel	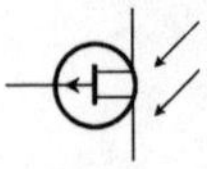
Transistor, unijunction	
Triac	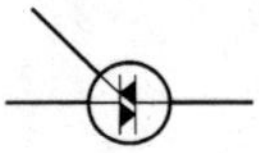
Tube, diode	
Tube, heptode	
Tube, hexode	
Tube, pentode	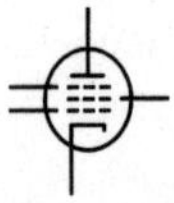
Tube, photosensitive	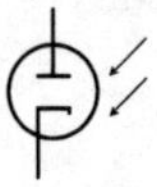
Tube, tetrode	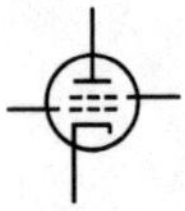

TABLE 6.26 Schematic Symbols Used in Electronics (*Continued*)

Component	Symbol
Tube, triode	
Voltmeter	 V
Wattmeter	 W
Waveguide, circular	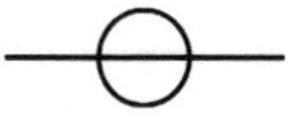
Waveguide, flexible	
Waveguide, rectangular	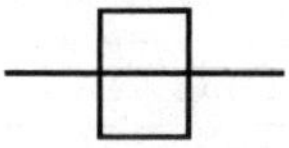
Waveguide, twisted	
Wires, crossing, connected	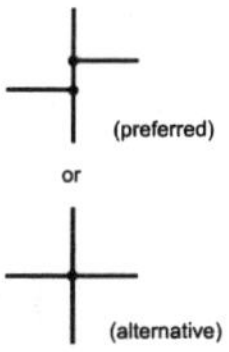 (preferred) or (alternative)
Wires, crossing, not connected	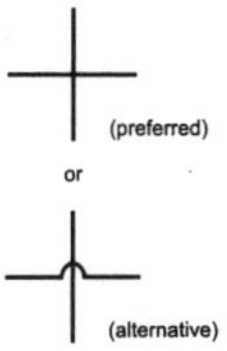 (preferred) or (alternative)

TABLE 6.27A VHF Television Broadcast Channels

Channel	Frequency, MHz
2	54–60
3	60–66
4	66–72
5	76–82
6	82–88
7	174–180
8	180–186
9	186–192
10	192–198
11	198–204
12	204–210
13	210–216

TABLE 6.27B UHF Television Broadcast Channels

Channel	Frequency, MHz
14	470–476
15	476–482
16	482–488
17	488–494
18	494–500
19	500–506
20	506–512
21	512–518
22	518–524
23	524–530
24	530–536
25	536–542
26	542–548
27	548–554
28	554–560
29	560–566
30	566–572
31	572–578
32	578–584
33	584–590
34	590–596
35	596–602
36	602–608
37	608–614
38	614–620
39	620–626
40	626–632

TABLE 6.27B UHF Television Broadcast Channels (*Continued*)

Channel	Frequency, MHz
41	632–638
42	638–644
43	644–650
44	650–656
45	656–662
46	662–668
47	668–674
48	674–680
49	680–686
50	686–692
51	692–698
52	698–704
53	704–710
54	710–716
55	716–722
56	722–728
57	728–734
58	734–740
59	740–746
60	746–752
61	752–758
62	758–764
63	764–770
64	770–776
65	776–782
66	782–788
67	788–794
68	794–800
69	800–806

TABLE 6.28 Common Q Signals and Their Meanings

Signal	Query and response
QRA	What is the name of your station? The name of my station is —.
QRB	How far from my station are you? I am — miles or — kilometers from your station.
QRD	From where are you coming, and where are you going? I am coming from —, and am going to —.
QRG	What is my frequency, or that of —? Your frequency, or that of —, is— (kHz, MHz, GHz).
QRH	Is my frequency unstable? Your frequency is unstable.
QRI	How is the tone of my signal? Your signal tone is: 1 (good), 2 (fair), 3 (poor).
QRK	How readable is my signal? Your signal is: 1 (unreadable), 2 (barely readable), 3 (readable with difficulty), 4 (readable with almost no difficulty), 5 (perfectly readable).
QRL	Are you busy? Or, Is this frequency in use? I am busy. Or, This frequency is in use.
QRM	Are you experiencing interference from other stations? I am experiencing interference from other stations.
QRN	Is your reception degraded by sferics or electrical noise? My reception is degraded by sferics or electrical noise.
QRO	Should I increase my transmitter output power? Increase your transmitter output power.
QRP	Should I reduce my transmitter output power? Reduce your transmitter output power.
QRQ	Should I send (Morse code) faster? Send (Morse code) faster.
QRS	Should I send (Morse code) more slowly? Send (Morse code) more slowly.
QRT	Shall I stop transmitting? Or, Are you going to stop transmitting? Stop transmitting. Or, I am going to stop transmitting.
QRU	Do you have information for me? I have no information for you.
QRV	Are you ready for —? I am ready for —.
QRW	Should I tell — that you are calling him/her/them? Tell — that I am calling him/her/them.
QRX	When will you call me again? I will call you again at —.

TABLE 6.28 Common Q Signals and Their Meanings (*Continued*)

Signal	Query and response
QRY	What is my turn in order? Your turn is number — in order.
QRZ	Who is calling me? You are being called by —.
QSA	How strong are my signals? Your signals are: 1 (almost inaudible), 2 (weak), 3 (fairly strong), 4 (strong), 5 (very strong).
QSB	Are my signals varying in strength? Your signals are varying in strength.
QSD	Are my signals mutilated? Or, is my keying bad? Your signals are mutilated. Or, your keying is bad.
QSG	Should I send more than one message? Send — messages.
QSJ	What is your charge per word? My charge per word is —.
QSK	Can you hear me between your signals? Or, Do you have full break-in capability? I can hear you between my signals. Or, I have full break-in capability.
QSL	Do you acknowledge receipt of my message? I acknowledge receipt of your message.
QSM	Should I repeat my message? Repeat your message.
QSN	Did you hear me on — (frequency, channel, or wavelength)? I heard you on — (frequency, channel, or wavelength).
QSO	Can you communicate with —? I can communicate with —.
QSP	Will you send a message to —? I will send a message to —.
QSQ	Is there a doctor there? Or, Is — there? There is a doctor here. Or, — is here.
QSU	On what frequency, channel, or wavelength should I reply? Reply on — (frequency, channel, or wavelength).
QSV	Shall I transmit a series of V's for test purposes? Transmit a series of V's for test purposes.
QSW	On which frequency, channel, or wavelength will you transmit? I will transmit on — (frequency, channel, or wavelength).
QSX	Will you listen for me? Or, Will you listen for —? I will listen for you. Or, I will listen for —.
QSY	Should I change frequency, channel, or wavelength? Change frequency, channel, or wavelength to —.

TABLE 6.28 Common Q Signals and Their Meanings (*Continued*)

Signal	Query and response
QSZ	Should I send each word or word group more than once? Send each word or word group more than once.
QTA	Should I cancel message number —? Cancel message number —.
QTB	Does your word count agree with mine? My word count disagrees with yours.
QTC	How many messages do you have to send? I have — messages to send.
QTE	What is my bearing relative to you? Your bearing relative to me is — (azimuth degrees).
QTH	What is your location? My location is —.
QTJ	What is the speed at which your vehicle is traveling? My vehicle is traveling at — (miles or kilometers per hour).
QTL	In what direction are you headed? I am headed toward —. Or, My heading is — (azimuth degrees).
QTN	When did you leave —? I left — at —.
QTO	Are you airborne? I am airborne.
QTP	Do you intend to land? I intend to land.
QTR	What is the correct time? The correct time is — Coordinated Universal Time (UTC).
QTX	Will you stand by for me? I will stand by for you until —.
QUA	Do you have information concerning —? I have information concerning —.
QUD	Have you received my urgent signal, or that of —? I have received your urgent signal, or that of —.
QUF	Have you received my distress signal, or that of —? I have received your distress signal, or that of —.

TABLE 6.29 Ten-Code Signals and Their Meanings

Ten-code signals used in the Citizens Radio Service.

Signal	Query and response
10-1	Are you having trouble receiving my signals? I am having trouble receiving your signals.
10-2	Are my signals good? Your signals are good.
10-3	Shall I stop transmitting? Stop transmitting.
10-4	Have you received my message completely? I have received your message completely.
10-5	Shall I relay a message to —? Relay a message to —.
10-6	Are you busy? I am busy; stand by until —.
10-7	Is your station out of service? My station is out of service.
10-8	Is your station in service? My station is in service.
10-9	Shall I repeat my message? Or, Is reception poor? Repeat your message. Or, Reception is poor.
10-10	Are you finished transmitting? I am finished transmitting.
10-11	Am I talking too fast? You are talking too fast.
10-12	Do you have visitors? I have visitors.
10-13	How are your weather and road conditions? My weather and road conditions are —.
10-14	What is the local time, or the time at —? The local time, or the time at —, is —.
10-15	Shall I pick up — at —? Pick up — at —.
10-16	Have you picked up —? I have picked up —.
10-17	Do you have urgent business? I have urgent business.
10-18	Have you any information for me? I have some information for you; it is —.
10-19	Have you no information for me? I have no information for you.

TABLE 6.29 Ten-Code Signals and Their Meanings (*Continued*)

Ten-code signals used in the Citizens Radio Service.	
Signal	Query and response
10-20	Where are you located? I am located at —.
10-21	Shall I call you on the telephone? Call me on the telephone.
10-22	Shall I report in person to —? Report in person to —.
10-23	Shall I stand by? Stand by until —.
10-24	Are you finished with your last assignment? I am finished with my last assignment.
10-25	Are you in contact with —? I am in contact with —.
10-26	Shall I disregard the information you just sent? Disregard the information I just sent.
10-27	Shall I move to channel —? Move to channel —.
10-30	Is this action legal or is it illegal? This action is illegal.
10-33	Do you have an emergency message? I have an emergency message.
10-34	Do you have trouble? I have trouble.
10-35	Do you have confidential information? I have confidential information.
10-36	Is there an accident? There is an accident at —.
10-37	Is a tow truck needed? A tow truck is needed at —.
10-38	Is an ambulance needed? An ambulance is needed at —.
10-39	Is there a convoy at —? There is a convoy at —.
10-41	Shall we change channels? Change channels.
10-60	Please give me your message number. My message number is —.
10-63	Is this net directed? This net is directed.

TABLE 6.29 Ten-Code Signals and Their Meanings (*Continued*)

Signal	Query and response
	Ten-code signals used in the Citizens Radio Service.
10-64	Do you intend to stop transmitting? I intend to stop transmitting.
10-65	Do you have a net message for —? I have a net message for —.
10-66	Do you wish to cancel your messages number ——through —? I wish to cancel my messages number — through —.
10-67	Shall I stop transmitting to receive a message? Stop transmitting to receive a message.
10-68	Shall I repeat my messages number — through —? Repeat your messages number — through —.
10-70	Have you a message? I have a message.
10-71	Shall I send messages by number? Send messages by number.
10-79	Shall I inform — regarding a fire at —? Inform — regarding a fire at —.
10-84	What is your telephone number? My telephone number is —.
10-91	Are my signals weak? Your signals are weak.
10-92	Are my signals distorted? Your signals are distorted.
10-94	Shall I make a test transmission? Make a test transmission.
10-95	Shall I key my microphone without speaking? Key your microphone without speaking.
	Ten-code signals used in law enforcement
10-1	Are you having trouble receiving my signals? I am having trouble receiving your signals.
10-2	Are my signals good? Your signals are good.
10-3	Shall I stop transmitting? Stop transmitting.
10-4	Have you received my message in full? I have received your message in full.
10-5	Shall I relay a message to —? Relay a message to —.
10-6	Are you busy? I am busy; stand by until —.

TABLE 6.29 Ten-Code Signals and Their Meanings (*Continued*)

Ten-code signals used in law enforcement	
Signal	Query and response
10-7	Is your station out of service? My station is out of service.
10-8	Is your station in service? My station is in service.
10-9	Shall I repeat my message? Repeat your message.
10-10	Is there a fight in progress at your location? There is a fight in progress at my location.
10-11	Do you have a case involving a dog? I have a case involving a dog.
10-12	Shall I stand by? Or, Shall I stand by until —? Stand by. Or, Stand by until —.
10-13	How are your weather and road conditions? My weather and road conditions are —.
10-14	Have you received a report of a prowler? I have received a report of a prowler.
10-15	Is there a civil disturbance at your location? There is a civil disturbance at my location.
10-16	Is there domestic trouble at your location? There is domestic trouble at my location.
10-17	Shall I meet the person who issued the complaint? Meet the person who issued the complaint.
10-18	Shall I hurry to finish this assignment? Hurry to finish this assignment.
10-19	Shall I return to —? Return to —.
10-20	What is your location? My location is —.
10-21	Shall I call — by telephone? Call — by telephone.
10-22	Shall I ignore the previous information? Ignore the previous information.
10-23	Has — arrived at —? —has arrived at —.
10-24	Have you finished your assignment? I have finished my assignment.
10-25	Shall I report in person to —? Report in person to —.

TABLE 6.29 Ten-Code Signals and Their Meanings (*Continued*)

Ten-code signals used in law enforcement	
Signal	Query and response
10-26	Are you detaining a subject? I am detaining a subject.
10-27	Do you have data on driver license number —? Here is data on driver license number —.
10-28	Do you have data on vehicle registration number —? Here is data on vehicle registration number —.
10-29	Shall I check records to see if — is a wanted person? Check records to see if — is a wanted person.
10-30	Is — using a radio illegally? — is using a radio illegally.
10-31	Is there a crime in progress at your location (or at —)? There is a crime in progress at my location (or at —).
10-32	Is there a person with a gun at your location (or at —)? There is a person with a gun at my location (or at —).
10-33	Is there an emergency at your location (or at —)? There is an emergency at my location (or at —.)
10-34	Is there a riot at your location (or at —)? There is a riot at my location (or at —).
10-35	Do you have an alert concerning a major crime? I have an alert concerning a major crime.
10-36	What is the correct time? The correct time is — local (or — UTC).
10-37	Shall I investigate a suspicious vehicle? Investigate a suspicious vehicle.
10-38	Are you stopping a suspicious vehicle? I am stopping a suspicious vehicle (of type —).
10-39	Is your (or this) situation urgent? My (or this) situation is urgent, use lights and/or siren.
10-40	Shall I refrain from using my light or siren? Refrain from using your light or siren.
10-41	Are you just starting duty? I am just starting duty.
10-42	Are you finishing duty? I am finishing duty.
10-43	Do you need, or are you sending, data about —? I need, or am sending, data about —.
10-44	Do you want to leave patrol? I want to leave patrol and go to —.

TABLE 6.29 Ten-Code Signals and Their Meanings (*Continued*)

Ten-code signals used in law enforcement	
Signal	Query and response
10-45	Is there a dead animal at your location (or at —)? There is a dead animal at my location (or at —).
10-46	Shall I assist a motorist at my location (or at —)? Or, are you assisting a motorist at your location (or at —)? Assist a motorist at your location (or at —). Or, I am assisting a motorist at my location (or at —).
10-47	Are road repairs needed now at your location (or at —)? Road repairs are needed now at my location (or at —).
10-48	Does a traffic standard need to be fixed at your location (or at —)? A traffic standard needs to be fixed at my location (or at —).
10-49	Is a traffic light out at your location (or at ——? A traffic light is out at my location (or at —).
10-50	Is there an accident at your location (or at —)? There is an accident at my location (or at —).
10-51	Is a tow truck needed at your location (or at —)? A tow truck is needed at my location (or at —).
10-52	Is an ambulance needed at your location (or at —)? An ambulance is needed at my location (or at —).
10-53	Is the road blocked at your location (or at —)? The road is blocked at my location (or at —).
10-54	Are there animals on the road at your location (or at —)? There are animals on the road at my location (or at —).
10-55	Is there a drunk driver at your location (or at —)? There is a drunk driver at my location (or at —).
10-56	Is there a drunk pedestrian at your location (or at —)? There is a drunk pedestrian at my location (or at —).
10-57	Has there been a hit-and-run accident at your location (or at —)? There has been a hit-and-run accident at my location (or at —).
10-58	Shall I direct traffic at my location (or at —)? Direct traffic at your location (or at —).
10-59	Is there a convoy at your location (or at —)? Or, does — need an escort? There is a convoy at my location (or at —). Or, —needs an escort.
10-60	Is there a squad at your location (or at —)? There is a squad at my location (or at —).
10-61	Are there personnel in your vicinity (or in the vicinity of —)? There are personnel in my vicinity (or in the vicinity of —).
10-62	Shall I reply to the message of —? Reply to the message of —.

TABLE 6.29 Ten-Code Signals and Their Meanings (*Continued*)

Ten-code signals used in law enforcement	
Signal	Query and response
10-63	Shall I make a written record of —? Make a written record of —.
10-64	Is this message to be delivered locally? This message is to be delivered locally.
10-65	Do you have a net message assignment? I have a net message assignment.
10-66	Do you want to cancel message number —? I want to cancel message number —.
10-67	Shall I clear for a net message? Clear for a net message.
10-68	Shall I disseminate data concerning —? Disseminate data concerning —.
10-69	Have you received my messages numbered — through —? I have received your messages numbered — through —.
10-70	Is there a fire at your location (or at —)? There is a fire at my location (or at —).
10-71	Shall I advise of details concerning the fire at my location (or at —)? Advise of details concerning the fire at your location (or at —).
10-72	Shall I report on the progress of the fire at my location (or at —)? Report on the progress of the fire at your location (or at —).
10-73	Is there a report of smoke at your location (or at —)? There is a report of smoke at my location (or at —).
10-74	(No query) Negative.
10-75	Are you in contact with —? I am in contact with —.
10-76	Are you going to —? I am going to —.
10-77	When do you estimate arrival at —? I estimate arrival at — at — local time (or — UTC).
10-78	Do you need help? I need help at this location (or at —).
10-79	Shall I notify a coroner of —? Notify a coroner of —.
10-82	Shall I reserve a hotel or motel room at —? Reserve a hotel or motel room at —.
10-85	Will you (or —) be late? I (or —) will be late.

TABLE 6.29 Ten-Code Signals and Their Meanings (*Continued*)

Ten-code signals used in law enforcement	
Signal	Query and response
10-87	Shall I pick up checks for distribution? Pick up checks for distribution. Or, I am picking up checks for distribution.
10-88	What is the telephone number of —? The telephone number of — is —.
10-90	Is there a bank alarm at your location (or at —)? There is a bank alarm at my location (or at —).
10-91	Am I using a radio without cause? Or, Is — using a radio without cause? You are using a radio without cause. Or, — is using a radio without cause.
10-93	Is there a blockade at your location (or at —)? There is a blockade at my location (or at —).
10-94	Is there an illegal drag race at your location (or at —)? There is an illegal drag race at my location (or at —).
10-96	Is there a person acting mentally ill at your location (or at —)? There is a person acting mentally ill at my location (or at —).
10-98	Has someone escaped from jail at your location (or at —)? Someone has escaped from jail at my location (or at —).
10-99	Is — wanted or stolen? — is wanted or stolen. Or, there is a wanted person or stolen article at —.

TABLE 6.30 The International Morse Code

Character	Symbol
A	·–
B	–···
C	–·–·
D	–··
E	·
F	··–·
G	––·
H	····
I	··
J	·–––
K	–·–
L	·–··
M	––
N	–·
O	–––
P	·––·
Q	––·–
R	·–·
S	···
T	–
U	··–
V	···–
W	·––
X	–··–
Y	–·––
Z	––··
0	–––––
1	·––––
2	··–––
3	···––
4	····–
5	·····
6	–····
7	––···
8	–––··
9	––––·
Period	·–·–·–
Comma	––··––
Query	··––··
Slash	–··–·
Dash	–····–
Break (pause)	–···–
Semicolon	–·–·–·
Colon	–––···

TABLE 6.31 Phonetic Alphabet as Recommended by the International Telecommunication Union (ITU). Uppercase Indicates Syllable Emphasis

Letter	Phonetic
A	AL-fa
B	BRAH-vo
C	CHAR-lie
D	DEL-ta
E	ECK-o
F	FOX-trot
G	GOLF
H	ho-TEL
I	IN-dia
J	Ju-li-ETTE
K	KEE-low
L	LEE-ma
M	MIKE
N	No-VEM-ber
O	OS-car
P	pa-PA
Q	Que-BECK
R	ROW-me-oh
S	see-AIR-ah
T	TANG-go
U	YOU-ni-form
V	VIC-tor
W	WHIS-key
X	X-ray
Y	YANK-key
Z	ZOO-loo

TABLE 6.32 Coordinated Universal Time (UTC) with Conversion for the Various Time Zones within the United States*

UTC	EDT	EST/CDT	CST/MDT	MST/PDT	PST
0000	2000*	1900*	1800*	1700*	1600*
0100	2100*	2000*	1900*	1800*	1700*
0200	2200*	2100*	2000*	1900*	1800*
0300	2300*	2200*	2100*	2000*	1900*
0400	0000	2300*	2200*	2100*	2000*
0500	0100	0000	2300*	2200*	2100*
0600	0200	0100	0000	2300*	2200*
0700	0300	0200	0100	0000	2300*
0800	0400	0300	0200	0100	0000
0900	0500	0400	0300	0200	0100
1000	0600	0500	0400	0300	0200
1100	0700	0600	0500	0400	0300
1200	0800	0700	0600	0500	0400
1300	0900	0800	0700	0600	0500
1400	1000	0900	0800	0700	0600
1500	1100	1000	0900	0800	0700
1600	1200	1100	1000	0900	0800
1700	1300	1200	1100	1000	0900
1800	1400	1300	1200	1100	1000
1900	1500	1400	1300	1200	1100
2000	1600	1500	1400	1300	1200
2100	1700	1600	1500	1400	1300
2200	1800	1700	1600	1500	1400
2300	1900	1800	1700	1600	1500
2400	2000	1900	1800	1700	1600

*Asterisks indicate previous day from UTC.

Suggested Additional References

Books and CD-ROMs

Basu, D., *Dictionary of Pure and Applied Physics* (Boca Raton, FL: CRC Press, 2000)

Clark, D., *Dictionary of Analysis, Calculus, and Differential Equations* (Boca Raton, FL: CRC Press, 2000)

Crowhurst, N. and Gibilisco, S., *Mastering Technical Mathematics—2nd Edition* (New York, NY: McGraw-Hill, 1999)

Dorf, R., *Electrical Engineering Handbook—2nd Edition* (Boca Raton, FL: CRC Press, 1997)

Lide, D., *Handbook of Chemistry and Physics—81st Edition* (Boca Raton, FL: CRC Press, 2000)

Shackelford, J. and Alexander, W., *Materials Science and Engineering Handbook—3rd Edition* (Boca Raton, FL: CRC Press, 2000)

Van Valkenburg, M., *Reference Data for Engineers: Radio, Electronics, Computer and Communications* (Indianapolis, IN: Howard W. Sams & Co., 1998)

Veley, V., *The Benchtop Electronics Reference Manual* (New York, NY: McGraw-Hill, 1994)

Weisstein, E., *Concise Encyclopedia of Mathematics CD-ROM* (Boca Raton, FL: CRC Press, 1999)

Web sites

TechTarget.com, Inc., *whatis.com* (www.whatis.com)

Weisstein, E., *Eric Weisstein's Treasure Troves of Science* (www.treasure-troves.com)

Index

ABOUT THE AUTHOR

Stan Gibilisco is one of McGraw-Hill's most versatile, prolific, and best-selling authors. His clear, friendly, easy-to-read writing style makes his electronics titles accessible to a wide audience, and his background in mathematics and research makes him an ideal handbook editor. He is the author of *The TAB Encyclopedia of Electronics for Technicians and Hobbyists, Teach Yourself Electricity and Electronics,* and *The Illustrated Dictionary of Electronics. Booklist* named his *McGraw-Hill Encyclopedia of Personal Computing* one of the "Best References of 1996."